Silver Nanoparticles - Fundamentals, Properties, Synthesis, and Applications

Edited by Phuong V. Pham

Published in London, United Kingdom

Silver Nanoparticles - Fundamentals, Properties, Synthesis, and Applications
http://dx.doi.org/10.5772/intechopen.1011189
Edited by Phuong V. Pham

Contributors
Abrar Khalid Aloufi, Abubaker M. Hamad, Adepero Olubukola Awolesi, Ally Kafesa, Ana García-Villaraco, Beatriz Ramos-Solano, Bernard Opatimidi Patani, Davi L. Rios, Fares Fenniche, Farhan Baehaki, Farhan Tariq, Francisco Javier Gutiérrez-Mañero, Hafsi Zoulikha, José Antonio Lucas, Laraib Javed, Luiz P. da Costa, Maribel Maldonado-Muñiz, Martha Guadalupe Nieto Lopez, Mireya Tapia Salazar, Mohammed Hashim Albashir, Muhammad Pervaiz, Olutayo Ademola Adeleye, Sara E. Ibrahim, Sofiane Khane, Sonia A. Soto-Rodriguez, Svitlana Plokhovska, Vitória O. M. Duval, Yasmina Khane, Yi Feng, Zhe Li

First published in London, United Kingdom, 2026 by IntechOpen
IntechOpen is the global imprint of INTECHOPEN LIMITED, registered in England and Wales, registration number: 11086078, 167-169 Great Portland Street, London, W1W 5PF, United Kingdom

For EU product safety concerns: IN TECH d.o.o., Prolaz Marije Krucifikse Kozulić 3, 51000 Rijeka, Croatia, info@intechopen.com or visit our website at intechopen.com.

British Library Cataloguing-in-Publication Data
A catalogue record for this book is available from the British Library

Silver Nanoparticles - Fundamentals, Properties, Synthesis, and Applications
Edited by Phuong V. Pham
p. cm.

This title is part of the Nanotechnology and Nanomaterials Book Series, Volume 15
Topic: Nanomaterials and Nanostructures
Series Editor: Jung Huang
Topic Editors: Jose De Jesus Perez-Bueno and Maria Luisa Mendoza López

Print ISBN 978-1-80631-454-6
Online ISBN 978-1-80631-453-9
eBook (PDF) ISBN 978-1-80631-455-3
ISSN 3029-0538

If disposing of this product, please recycle the paper responsibly.

IntechOpen Book Series

Nanotechnology and Nanomaterials

Volume 15

Aims and Scope of the Series

Humans face growing challenges in environmental sustainability. We need better materials, more effective devices, and higher computing power to conquer these challenges. Researchers have long explored the realm of science at the nanometer scale for developing novel problem-solving technologies. In the past decades, breakthroughs in multiscale computer simulation, novel hierarchical material and device structures, nanometrology, and new functionalities from interfacial and quantum size effects have enhanced our ability to utilize nanotechnology to solve real-world problems.

This book series addresses important advancements in nanotechnology and nanomaterials. It showcases how nanotechnology is being continually developed and implemented in a variety of domains of science and technology. The two main topics this series covers are Nanotechnology and Nanodevices, and Nanomaterials and Nanostructures.

Meet the Series Editor

Jung Y. Huang, a university educator and researcher, has been working to unravel the structures and functional properties of materials and cellular events in living cells with varying optical methodologies. He has co-authored hundreds of journal papers and five book chapters. He also holds tens of patents in laser techniques and single-molecule/hyperspectral imaging and has developed architectural photonics based on hierarchically structured materials. Currently, his research focuses on the integration of artificial intelligence methodology with optics to automatically discover meaningful information from optical sensing/imaging data cubes. He has been an editorial board member and reviewer for several scientific journals. As a member of the global scientific community, he sincerely supports and endeavors to promote the spread of scientific knowledge.

Meet the Volume Editor

Phuong V. Pham is a Professor at the Department of Physics, National Sun Yat-sen University (NSYSU), Taiwan. He earned his Ph.D. from the SKKU Advanced Institute of Nanotechnology (SAINT) at Sungkyunkwan University (SKKU), South Korea. Currently, he leads the Materials Physics and Electronic Surfaces Group at Physics-NSYSU. Professor Pham has collaborated on industrial projects with major organizations, including SAMSUNG, POSCO, and government agencies such as the Korean Government, the Ministry of Education, the Ministry of Science and Technology of China, and the NSTC Grant Program of Taiwan. He has held various research positions, including Postdoctoral Researcher, Research Fellow, and Distinguished Research Fellow, at prestigious institutions like the School of Advanced Material Science and Engineering, SKKU, the Center for Multidimensional Carbon Materials (Director: Prof. Rodney Ruoff) at the Institute for Basic Science, South Korea, and Hangzhou Global Scientific and Technological Innovation Center, Zhejiang University, China. Professor Pham is a recipient of several prestigious awards, including the NSF Career Award, the National Postdoctoral Award for Outstanding Young Scientists in China (2019), and NSTC Grant Awards in Taiwan (2023, 2025). Additionally, he serves as an Academic Editor for IntechOpen and a Reviewer for over 80 Journals. He has authored approximately 100 peer-reviewed articles/books/chapters such as *Chemical Reviews, Chemical Society Reviews, Advanced Materials, Materials Today, ACS Nano, Nano Energy, Small, ACS Materials Letters, Coordination Chemistry Reviews, Chemical Engineering Journal, 2D Materials, Carbon, Nanoscale Horizons, Nanoscale, Journal of Materials Chemistry C, Chinese Journal of Physics, Journal of Catalysis, Scientific Reports, Advanced Materials Interfaces, Sensors & Actuators A: Physical, Materials Chemistry & Physics, Optical Materials, and Optics Communications*, among others. His research interests span a wide range of fields, including nanotechnology, materials physics, surface physics, chemical engineering, atomic film synthesis, 2D heterostructures, nanofilm transfer, donor/acceptor systems, nanocomposites, plasma engineering for advanced electronics/optoelectronics, energy storage, environment, and extending to predictive simulations associated with next-generation nanomaterials processes (kinetics, growths).

Contents

Preface

Nanotechnology has become one of the most influential scientific and technological fields of the modern era, enabling the manipulation of matter at dimensions where unique physicochemical and biological properties emerge. Among the many engineered nanomaterials investigated over recent decades, silver nanoparticles (AgNPs) have attracted exceptional scientific and industrial attention due to their remarkable optical, electrical, catalytic, antimicrobial, and therapeutic properties. Their multifunctional nature has opened new opportunities across interdisciplinary fields, including materials science, chemistry, physics, medicine, environmental engineering, agriculture, biotechnology, and industrial manufacturing.

The book *Silver Nanoparticles - Fundamentals, Properties, Synthesis, and Applications* presents a comprehensive overview of current advances in the rapidly evolving field of silver nanotechnology. This edited volume gathers contributions from researchers working in diverse scientific disciplines and highlights recent developments in nanoparticle synthesis, physicochemical characterization, functional engineering, biomedical applications, environmental technologies, and safety assessment. The objective of this book is to provide readers with both fundamental knowledge and emerging perspectives regarding the design, application, and responsible development of AgNP-based technologies.

The volume is organized into four thematic sections that guide readers from fundamental concepts toward advanced applications and future challenges.

The first section, *Fundamentals, Synthesis, and Physicochemical Properties*, establishes the scientific foundation of silver nanoparticle research. Chapter 1 introduces the synthesis methods, characterization techniques, physicochemical properties, and cutting-edge applications of AgNPs, providing readers with a broad overview of nanosilver technologies. Chapters 2 and 3 focus on green and biological synthesis approaches, emphasizing environmentally friendly fabrication methods using plants, microorganisms, algae, and biomolecules as reducing and stabilizing agents. These sustainable strategies have attracted increasing interest because they minimize the use of toxic chemicals while improving biocompatibility and environmental safety. Chapter 4 further discusses the optical, antimicrobial, and structural properties of silver nanoparticles and highlights how nanoparticle size and morphology can influence performance in medical, industrial, and optical applications.

The second section, *Advanced Functional and Catalytic Properties*, explores the engineering of silver nanoparticles to enhance their catalytic and electronic performance. Chapter 5 examines the preparation of silver nanoparticles with crystalline defects and their applications in electrocatalysis. By discussing the influence of crystal defects, coordination environments, and lattice strain on electronic structure and catalytic activity, this chapter demonstrates how nanoscale engineering can significantly improve the functional performance of AgNPs in energy-related applications.

The third section, *Biomedical and Therapeutic Applications*, highlights the growing importance of silver nanoparticles in healthcare and nanomedicine. Chapter 6 discusses the application of green-synthesized AgNPs in topical drug delivery systems, including wound healing, antimicrobial therapy, anti-inflammatory treatment, and controlled drug release. The chapter illustrates how nanoscale materials can improve localized therapeutic efficacy while reducing systemic side effects. Chapter 7 addresses the critical issue of neurotoxicity associated with silver nanoparticle exposure. Through histopathological, genetic, and microbiological perspectives, the chapter explores oxidative stress, inflammatory responses, neural damage, and gut–brain interactions induced by AgNPs. Together, these chapters provide a balanced discussion of the therapeutic potential and biosafety concerns of nanomedicine applications.

The final section, *Agricultural, Environmental, and Aquaculture Applications*, focuses on emerging applications of silver nanoparticles in sustainable agriculture and food production systems. Chapter 8 presents innovative bio-functionalized silver nanoparticles coated with bacterial elicitors to enhance plant adaptation to drought stress, demonstrating the potential of nanotechnology to improve crop resilience and agricultural productivity. Chapter 9 discusses the application of green silver nanoparticles in shrimp aquaculture, highlighting their antimicrobial effectiveness, immune-enhancing properties, biosafety considerations, and ecological impacts. These studies illustrate the important role silver nanotechnology may play in addressing global challenges in food security, environmental sustainability, and responsible agricultural innovation.

One of the major strengths of this volume is its interdisciplinary perspective, integrating nanomaterial synthesis, physicochemical characterization, biological interactions, toxicity assessment, environmental considerations, and technological implementation into a unified scientific framework. Such integration is increasingly important as nanotechnology advances from laboratory research toward industrial-scale production and clinical translation.

As the Editor, I sincerely hope that this book will serve as a valuable reference for researchers, graduate students, engineers, clinicians, and industry professionals interested in the expanding field of silver nanotechnology. I want to express my deepest appreciation to all contributing authors for their outstanding scientific contributions and collaborative efforts. I also extend sincere gratitude to the editorial and publishing team at IntechOpen for their continuous professional support throughout the preparation of this volume.

It is hoped that this book will contribute meaningfully to future advances in nanoscience and inspire continued interdisciplinary collaboration toward the safe, sustainable, and innovative development of silver nanoparticle technologies for the benefit of science and society.

Phuong V. Pham
Department of Physics,
Center for Quantum Computing and Quantum Materials,
National Sun Yat-Sen University,
Kaohsiung, Taiwan

Section 1

Fundamentals, Synthesis, and Physicochemical Properties

Chapter 1

Silver Nanoparticles: From Synthesis to Cutting-Edge Applications

Muhammad Pervaiz, Farhan Tariq and Laraib Javed

Abstract

A single atom can behave nothing like its larger form; that tiny scale is where nanotechnology steps in. At that stage, silver nanoparticles stand out, mainly because they interact with light in unusual ways, carry electricity well, and fight many kinds of microbes. Their story begins long before use, rooted deeply in how they are made. Some methods smash bigger chunks into small pieces using lasers – this one's called top-down. Others build up particles from molecules through chemical reactions, which grow them step by step. Lately, though, something different has caught attention: turning plants or tiny living organisms into natural factories for making these particles. It skips harsh chemicals, leaning on biology instead. What happens after creation matters just as much; the full journey counts. Looking closely at how we identify key features, certain tools stand out: UV-visible spectroscopy gives clues about light absorption, XRD reveals crystal structure, while TEM offers clear images of tiny shapes and sizes. Moving into real-world uses, silver nanoparticles appear in medicine as fighters against bacteria and cancer cells. They also show up in materials like flexible circuits, filters that clean water, and even wraps that keep food fresh longer. A closer look brings concern: these particles may harm cells by releasing ions, triggering stress reactions, and building up where they shouldn't. Because of this, planning safety early in design feels necessary. Even with wide potential across fields, moving forward depends less on excitement and more on weighing usefulness against careful study of risks.

Keywords: silver nanoparticles, green synthesis, antimicrobial efficacy, safe by design, nanotechnology commercialization

1. Introduction

1.1 Overview of nanotechnology and silver nanoparticles

Looking into tiny things changes how we see materials. When stuff gets shrunk to sizes between one and a hundred billionths of a meter, its behavior shifts in surprising ways. Properties like heat flow, chemical reactions, or light interaction become stronger or act differently than larger versions. Silver particles built at this level stand out among metals made so far. These ultra-small silver bits hold their shape well, carry electricity efficiently, yet also stop microbes from growing across many types. Because of these traits, more research focuses on them compared to others, and they already appear in products today.

What makes silver nanoparticles (AgNPs) act differently comes mostly from how much surface they have compared to their size. As particles get smaller, a larger share of atoms ends up on the outside, boosting both chemical activity and contact with living tissues [1]. Light hitting the metal surface sets off collective electron waves known as Surface Plasmon Resonance which leads to intense color effects through absorption and scatter; tweaking shape or dimensions shifts these optical traits. Biomedical uses stand out most clearly. Because of that, AgNPs now play key roles across areas such as light-based tech, disease detection tools, and clean power solutions (**Figure 1**).

1.2 Historical context

Long before we named tiny science, people already trusted silver. Back then, Greeks filled vessels with liquid, slipped silver inside, and kept everything clean – it worked every time. Romans did much the same, learning through years of trial what stayed pure. Even Phoenician travelers carried silverware just to be safe. Empires like Macedonia served meals on metal dishes not for show; they knew it blocked sickness. Persians followed close behind with similar habits. Wounds? Ulcers? Doctors such as Hippocrates turned to silver without hesitation. Proof hides in old texts, page after page, showing how deeply this practice ran.

Back then, scientists started looking into colloidal silver toward the end of the 1800s. By 1889, M.C. Lea managed to make stable silver mixtures using citrate. This later turned into a medicine called Collargol by 1897 [2]. Early in the next century, solutions such as Argyrol became common tools against infections, including syphilis and gonorrhea. After Fleming discovered penicillin in 1928, antibiotic development took off, pushing silver aside in hospitals [3].

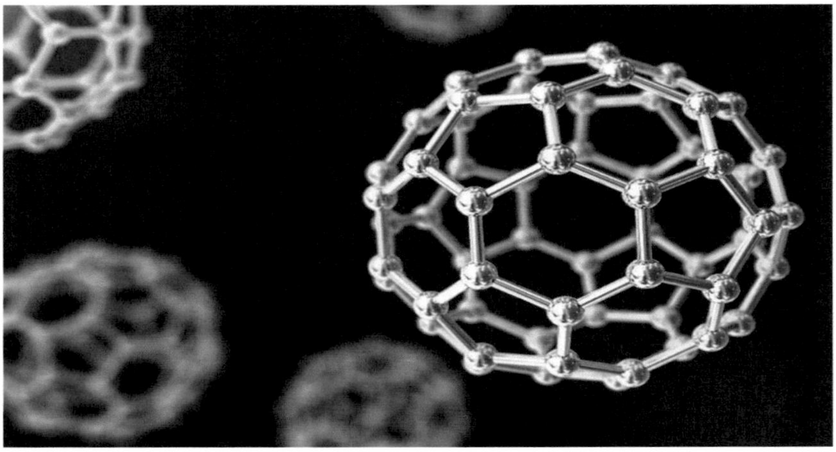

Figure 1.
Silver Nanoparticles.

By the late 1900s, attention turned again to silver. As bacteria began shrugging off antibiotics, scientists looked for alternatives. The rise of the so called post antibiotic age brought old options into fresh view. AgNPs reemerged as a subject worth studying. Most antibiotics go after one specific process inside microbes. These tiny particles act differently hitting multiple targets at once. They weaken cell membranes. Oxidative stress follows. Genetic material gets tangled during copying. Because they strike in several ways simultaneously, bugs struggle to adapt. Resistance doesn't take hold easily. Evidence supports this shift in strategy [4]. Between 2015 and 2025, AgNP research has increasingly emphasized sustainable synthesis, application specific design, mechanistic toxicity studies, and regulatory considerations. This evolution reflects a transition from exploratory nanomaterial research toward application driven and risk informed scientific inquiry.

1.3 Scope and industrial significance

Right now, tiny bits of silver lead the way in everyday tech you can buy. About half actually 55.4% of all nano powered items on shelves contain these particles, turning up in clothing that resists germs, wraparound films for food, even special ink used in lightweight electronic circuits [5]. Worth nearly 16.5 billion dollars last year alone, the worldwide trade in such advanced materials keeps climbing fast; much of that push comes from countries across Asia and the Pacific, where factories need more of it for power systems and gadgets [6].

Even though AgNPs sell well, their rapid spread raises concerns about producing them at scale and their impact on nature. Toxic chemicals and high energy consumption characterize traditional production methods, prompting science to explore cleaner alternatives using plants or microorganisms instead [7]. As more of these particles enter water systems, understanding their behavior over time in natural environments now motivates much laboratory research.

A fresh look at AgNPs begins here, exploring how they are made through top-down methods, bottom-up approaches, or ecofriendly techniques. Their physical and chemical traits come under close inspection next. Uses span across medical treatments and industrial processes, showing wide reach. Yet, concerns about health risks pop up just as fast. Rules guiding their safe use remain uncertain, shifting with new findings. What comes after depends on balancing innovation with caution.

2. Synthesis methods: Mechanisms, kinetics, and optimization

A single reaction can shift outcomes when making silver particles; tiny changes alter everything. Methods split two ways: one breaks big pieces into smaller ones, while another builds from atoms up. Getting every particle identical demands closely watching the early formation moments. How they grow depends on barriers that are hard to predict. Staying stable later means controlling surfaces with care. Each phase links tightly to the next, without pause. Abbas and his team showed how fine adjustments matter the most.

2.1 Top-down approaches

Breaking big chunks of silver into tiny pieces happens when outside forces supply energy. Though shapes can turn out uneven, making some frown, they still matter a lot when mass output is needed. Purity stands above all in those cases.

2.1.1 Laser ablation in liquid

Using lasers in liquid creates AgNPs cleanly, skipping harmful chemicals. This method leaves particle surfaces untouched by additives. Pure results come from physics, not reactions. No soapy residues stick around after processing.

A sudden burst happens as intense laser light hits silver sitting in liquid. Energy piles up right where solid meets fluid, sparking a tiny fireball. This hot spot can soar past 1,000 degrees Kelvin. Pressure inside climbs beyond a billion pascals. Heat and force unfold fast, trapped near the metal's surface.

A cloud of hot gas spreads out, losing heat fast. Tiny specks of silver form when the vapor gets too dense to stay airborne. Cooling speed shapes what happens next no chemicals involved. When liquid pulls heat quickly, newborn particles end up finer. Size depends on how fast things chill, not on added substances.

Picture how liquid affects reactions; solvents do more than just sit there. Take silver, blasted into bits while swimming in pure water – it rounds out, thanks to that tight skin on the droplet's surface. Swap in acetone instead, and things stretch oddly, going lopsided when the fluid itself breaks apart and leaves behind carbon traces. Ethanol does similar tricks too. That twist in form? It comes from the liquid falling apart during the hit [8].

2.1.2 Mechanical milling, ball milling

Fragments of silver break apart when hit hard inside a fast spinning mill. Yet those crashes twist the metal's inner structure, leaving behind tiny flaws. Over time, pieces get smaller and this goes on until they start sticking together just as often as they crack, as shown in **Figure 2**. The push to merge matches the force that splits them, holding sizes steady in the end.

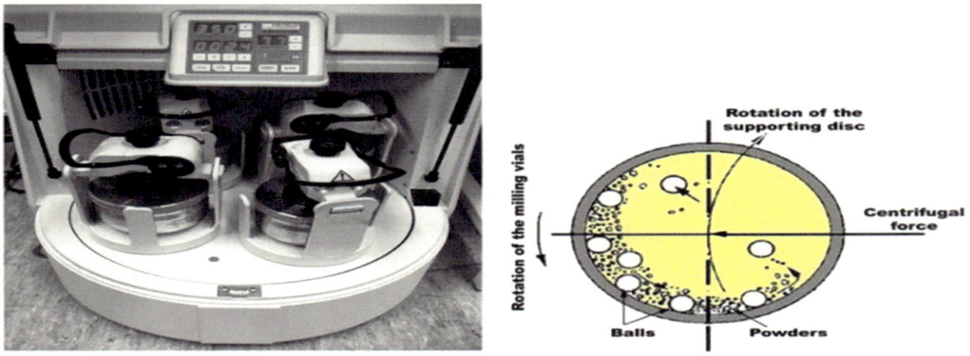

Figure 2.
Mechanical milling, ball milling.

Balls made of zirconia or steel wear down during grinding. This wearing leaves tiny bits behind, which mix into the AgNPs being made. Impurities like these shift how the particles behave electrically. One big problem arises from this kind of pollution inside the mixture.

2.2 Chemical reduction using bottom-up methods

Starting small builds today's tiny tech, thanks to atoms arranging themselves without help. One way this happens? Silver particles form when charged silver bits lose their charge through chemistry. The schematic representation is shown in **Figure 3**.

2.2.1 The LaMer model explains how particles begin to form

A sudden spark kicks off particle formation, this idea comes from LaMer's take on how chemicals suddenly cluster. His model explains that moment when nothing becomes something, fast.

A sudden jump in silver atom levels occurs once the reducer meets the silver salt mix. This buildup begins right before particles start forming. Things shift when enough atoms gather. The stage is set quietly, then changes.

Beads of matter start clumping when levels cross a key tipping point. As soon as that line is passed, tiny clusters lock into place fast. Pressure from excess density begins fading once these seeds take hold. Stability emerges out of chaos, piece by small piece.

When levels fall under Ccrit yet stay past equilibrium, fresh clusters do not appear. Old ones take up leftover silver pieces instead. Splitting these stages helps match particle dimensions closely [1].

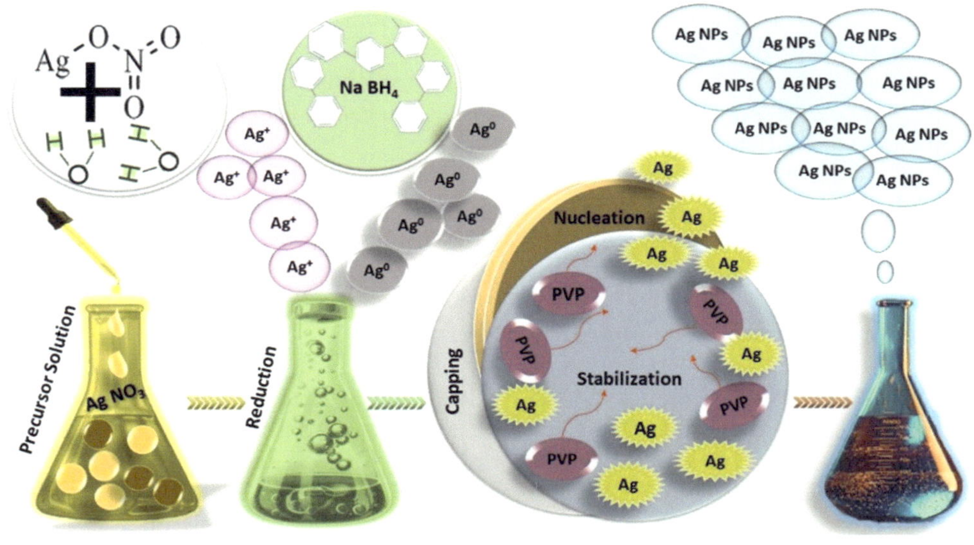

Figure 3.
Schematic diagram for chemical synthesis.

2.2.2 Chemical reducing agents and reaction speeds

A different reducer changes how fast the reaction goes; this speed shapes the final particle dimensions. What you pick at the start sets the pace that defines the outcome.

2.2.2.1 Strong reductants (sodium borohydride)

Right away, sodium borohydride (NaBH4) gives off electrons fast. Since every bit of silver ion gets used up at the start, almost nothing stays behind to help particles grow as shown in **Figure 3**. That's why they stay tiny usually between 2 and 10 nanometers across.

2.2.2.2 Weak reductants

Starting off, citrate serves dual roles: cutting down metals while wrapping around particle edges. It moves at a crawl, needing heat – sometimes boiling levels – to get going. Because seeds form gradually, crystals have time to stretch out. The result? Particles on the bigger side, ranging from 20 to 100 nanometers wide.

How acidity affects reactions matters more than it seems. When the environment becomes less acidic, turning alkaline, citrate breaks down faster. This shift speeds things up, and smaller round particles tend to form when that happens.

2.2.3 Stabilizers and caps

Smaller AgNPs tend to vanish when there's no stabilizer around. That happens because their surface holds more energy than bigger ones. Material shifts from tiny particles to larger ones over time. This movement follows natural physical laws. The system ends up with fewer, but bulkier, particles as a result.

A cloud of charged ions wraps each particle, forming a shield. When that electric push gets stronger than 30 millivolts positive or negative the particles keep apart. Their own force field keeps clumping at bay.

A layer forms when polymers such as PVP or PEG stick to surfaces; this blocks particles from touching one another. Shape control gets easier with PVP because it latches onto certain crystal faces. That preference guides development toward rods or cubes instead of round shapes [7].

2.3 Green synthesis meets biological interfaces

A different kind of chemistry shapes these tiny particles nature does the work. What sets them apart isn't only being gentle on Earth, yet lies more in the living layer they carry around each speck. This wrapping comes alive through natural reactions, built molecule by molecule without force or harshness. The result? A soft shell formed not by machines, but microbes and plants doing their quiet tasks. **Figure 4** illustrate the green synthesis approach using biological materials.

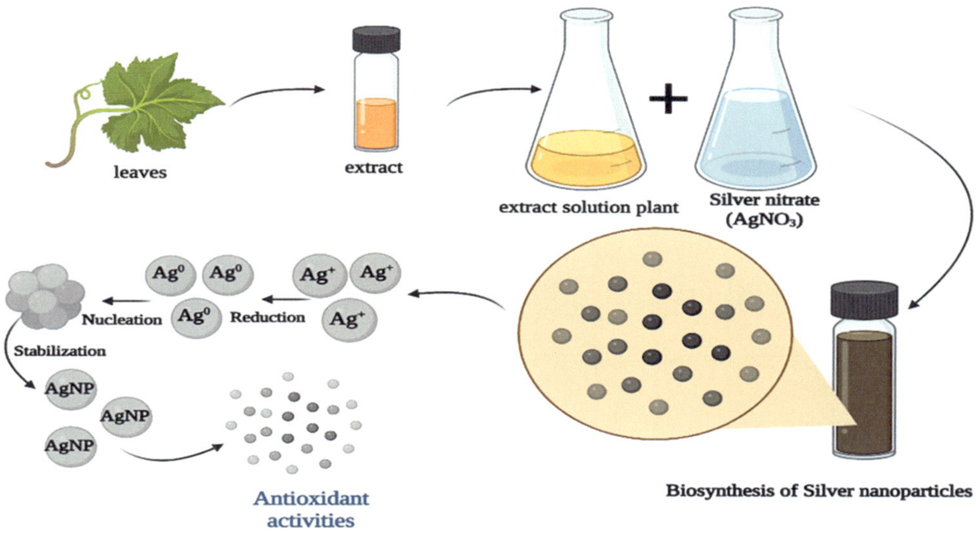

Figure 4.
Green synthesis of silver nanoparticles.

2.3.1 Plant mediated reduction mechanisms

A single leaf holds countless natural chemicals, working together quietly. Tiny green cells cook up complex molecules without any fuss.

Something interesting happens when plant chemicals meet oxidation. Flavonoids, terpenoids, and polyphenols take part most actively. What makes them effective is their ability to give up electrons without breaking down completely. Groups like OH and carbonyls sit right in the molecular structure, ready to react. Their readiness to change marks the start of electron transfer processes. Some flavonoids shift between forms one with oxygen bonded to hydrogen, another where that shifts to a double bond this movement sets free hydrogens able to turn silver ions into pure metal. A change like this opens the door for reduction without extra agents stepping in.

Examples in Literature: Bursting with Nimbin and Quercetin, Azadirachta indica often called neem delivers swift creation of AgNPs from its leaves. Minutes pass before the extract completes the task. Speed comes naturally here, driven by potent plant compounds doing unseen work. From aloe vera comes an extract full of acemannan which helps form round particles that hold their shape well because the sugar chains wrap tightly around them [9].

2.3.2 Microbial synthesis and enzymatic pathways

Bacteria build tiny particles when stressed by toxic metals. These small structures help them survive harsh conditions. Sometimes, survival means turning poison into something less harmful. The process happens naturally, without human help. Tiny life forms adapt in surprising ways. Protection comes in microscopic sizes.

Moving inside microbes, silver ions slip through tiny gates in the membrane. Once inside, proteins fueled by NADH hand off electrons one at a time to those ions.

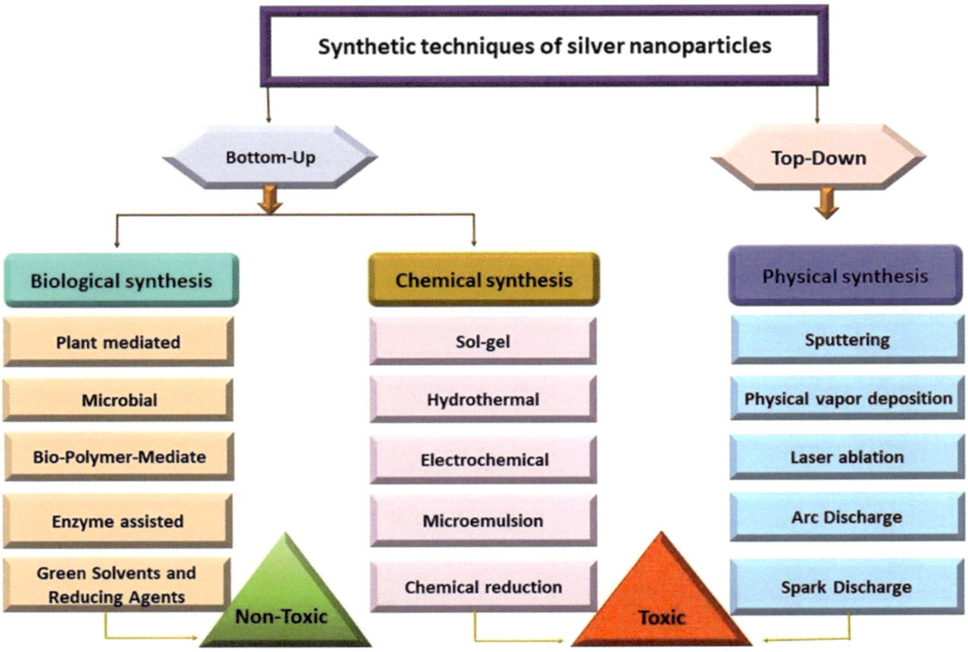

Figure 5.
Synthetic techniques of silver nanoparticles.

What forms next hides between layers of the inner wall, stuck until the cell bursts open. Only then can the tiny clusters be taken out.

Fungi like Fusarium oxysporum release enzymes outside their cells. Because of this, industries favor the method as it makes cleaning up easier after production. Filtering out the leftover fungus becomes straightforward when the proteins are already in the liquid around them [10]. A comparative overview of all synthetic techniques is presented in **Figure 5**.

2.4 What influences how well synthesis works

Getting the most out of AgNPs means fine tuning key factors. One thing that matters a lot is temperature control during synthesis. Another factor with strong influence is pH level it shapes particle size. Reaction time plays a role too; longer isn't always better. The source of silver ions affects results just as much. Even mixing speed changes how particles form. Each step needs careful adjustment for consistent output. Faster reactions usually come with warmer conditions that's what the Arrhenius pattern shows. When molecules move more, they bump into each other harder and more often. This speeds up the start of particle formation. Yet in plant based methods, too much warmth weakens natural coating substances. Once those break down, particles tend to clump together. Too much $AgNO_3$ might cause uneven particle sizes. When levels are high, particles start forming while others keep growing at the same time. Over time, slower reactions might cause particles to clump or shift shape. Triangular silver prisms could gradually become spherical, driven by a drop in surface energy [6].

Technique	Information obtained	Key limitations
UV–Vis	SPR, aggregation trends	Indirect, non specific
XRD	Crystal structure, crystallite size	Cannot determine particle size alone
SEM	Surface morphology	Limited resolution for small AgNPs
TEM	Size, shape, crystallinity	Limited sampling, beam damage
FTIR	Surface functional groups	Qualitative interpretation
DLS	Hydrodynamic size	Sensitive to aggregation
Zeta potential	Colloidal stability	Condition dependent

Table 1.
Comparison of characterization techniques.

3. Characterization techniques

Creating AgNPs marks just the start. What follows matters more; checking every detail counts. Size, shapes, and behavior – so does form. Surface traits matter too. One tool never shows enough. Clarity comes by matching results across methods. Light readings help, crystal patterns add clues, and electron images reveal structure. Together, they answer key questions: Is it really silver? Does the size fit the goal? Is the outer layer steady? Only then can confidence grow [11]. **Table 1** provides a comprehensive comparison of the key characterization techniques used for AgNPs.

3.1 UV-visible spectroscopy optical analysis

Looking at AgNPs often begins with UV-visible spectroscopy, thanks to a special light reaction called Surface Plasmon Resonance. Bulk silver just reflects light like a mirror, but tiny silver particles respond very differently. Their outer electrons move when hit by certain light waves, especially if those match their own vibration rhythm. This sync happens because incoming light pushes electrons while the core pulls them back. With round AgNPs, that interaction creates a clear spike in absorption usually somewhere from 400 to 450 nanometers.

Start anywhere. Picture light hitting tiny things – how they scatter matters. That scattering ties back to Mie Theory; know that first. Peaks appear when plasmons resonate; their look reveals what the particles are doing. One clean spike? Likely uniform in size. If smeared out or doubled up, things got messy during growth. Bigger spheres drag the color toward red; slower phase changes spread the response. When shapes twist – not just round – the peak splits without warning. Two peaks show up if rods or triangles formed instead. Each shift tells whether the method worked or failed. Duman and team saw this pattern repeat across trials.

3.2 Structural analysis using X-ray diffraction

Not every method shows everything. What you see under UV Vis only hints at silver being there. To check how neatly the atoms are packed, XRD steps in. It runs on an old rule Bragg's: $n\lambda = d\theta$. When X rays hit orderly layers inside crystals, they line up just right at certain angles. Silver particles leave their mark through sharp spikes around set 2θ spots. Those marks match up with faces labeled (111), then (200), later (220), also (311). Their positions tell us the metal settled into a face centered cube form. That kind of arrangement is what silver naturally prefers when it forms solid tiny bits.

Not just for ID work, XRD helps figure out how big crystals are inside materials. Using the Debye Scherrer formula on the widest part of the strongest signal often from the (111) layer gives a number for typical grain span. Still, what matters is this: XRD sees only sections where atoms line up neatly, not whole particles you'd see under a lens. When one tiny particle holds many stuck together micro crystals, the method shows smaller sizes than images caught by microscope views [7].

3.3 Morphological analysis with electron microscopy TEM and SEM

Seeing tiny things far below what regular light can reveal requires special tools. A method using beams of electrons instead of light helps overcome that limit. One version moves a thin stream of these charged particles over an object's outer layer. The result produces something like a height map shaped by how the surface bends and curves. It works well when you want to see where small pieces sit in relation to each other. Clumping together becomes obvious under this view. Still, if those specks dip beneath 10 billionths of a meter, clear edges start to blur. Exact borders between atoms remain out of reach even then.

Tiny details show up best under Transmission Electron Microscopy (TEM), the top choice for precise size analysis. A powerful beam of electrons usually 100 to 300 thousand volts moves through an extremely thin specimen. Because of this, scientists see inside nanoparticles clearly, spotting exact width down to less than one nanometer. With high resolution TEM, it goes further: atomic layers become visible as fine lines, making space between crystal planes measurable by eye. Looking at green made particles, a blurry ring often appears around the solid metal center in these images. That fuzzy outline marks the natural coating left behind by plant or microbe material used during creation [11].

3.4 Surface chemistry stability Fourier Transform Infrared Spectroscopy (FTIR), dynamic light scattering

What coats the outside of AgNPs shapes how they act in living systems. To spot those coating chemicals, scientists often turn to a tool called FTIR. This technique shows exactly which parts of molecules stick to the particle surface. When plants are used to make these particles, checking with FTIR becomes essential. The proof lies in the differences seen when you line up two readings: one from raw plant juice, another from cleaned nanoparticles. Certain peaks move position, like those tied to carbon-oxygen double bonds or oxygen-hydrogen links, and that shift reveals who did the work forming and holding the tiny bits.

Looking at how tiny bits move in liquid? That method is called Dynamic Light Scattering (DLS). While TEM sees just the solid metal part, DLS catches the full reach particle included, surrounded by clinging liquid bits and charged fragments. So what you get from DLS tends to be bigger numbers compared to TEM results. The machine doing the DLS work checks another thing too, the push or pull coming off the outside, known as Zeta Potential. When particles carry a strong negative or positive charge, they push apart hard, which often means the mix stays stable over time. On the flip side, if the charge is close to nothing, clumping and sinking become likely proof the process did not work right [4].

4. Physicochemical properties

What makes AgNPs useful isn't just silver as an element, but how its behavior changes wildly at the tiny scale of atoms and molecules. When particles shrink near the size where electrons move freely, strange quantum forces kick in, along with much more surface area compared to volume. This reshapes how they handle light, heat, and electricity. Their traits aren't fixed; adjusting how they're made can shift their features on purpose, opening paths for custom uses in real-world tech.

4.1 Light and metal interactions explained through plasmon behavior

What grabs attention about AgNPs is how they play with light thanks to something called localized surface plasmon resonance. Bulk silver just bounces back visible light, looking metallic and bright. Tiny silver particles behave differently, soaking up or deflecting certain colors instead. That happens due to loose electrons moving around on the particle's outer layer. If incoming light pulses at the same rate as those jiggling electrons, everything lines up just right. Then, absorption spikes sharply, making the particles block much more light than expected.

A single shift in shape alters how electrons move inside tiny particles. Around 400 nanometers marks the usual peak for round ones between 10 and 20 nanometers wide. That range gives off a vivid yellow shade. Light interaction shifts when the form deviates from a sphere to a rod. Silver rods show two separate responses instead of one. One comes from the side-to-side motion of charge. The longer version stretches along the axis. Stretching the rod pushes that response further into infrared zones. Penetrating deeper into body tissues becomes possible under such wavelengths. Medical strategies rely on this reach, especially those using heat triggered by light. References tie this effect to practical medical designs [12].

4.2 Lower melting points in small particles

Tiny silver particles melt differently than regular silver because of something called the Gibbs Thomson Effect. Regular silver melts at exactly 961.8°C. When these bits get smaller, more atoms sit on the outside instead of inside. Atoms on the edge aren't as stable they bond with less nearby atoms and carry extra energy. That changes how heat affects them.

A smaller amount of heat breaks surface atom bonds more easily, which means tiny particles melt at lower temps. Particles of silver just 2.5 nanometers wide may begin melting near 600°C over 350 below normal. That shift matters most when building bendable circuits. Lower fusion points let ink made from silver join into working traces on delicate bases such as paper or polymer films. Heat stays mild enough so the base doesn't warp or burn [13].

4.3 Electrical behavior and flow of charge

Electrons usually travel far without interruption inside solid silver. Yet, when scaled down to tiny particles, something shifts. Boundaries within the material begin to bump into moving electrons. That disruption tends to slow them down. The smaller the piece, the more these edges get in the way. Resistance creeps up not because of the substance, but because of how confined it becomes. What flows easily in large chunks stumbles in narrow spaces.

Even so, AgNPs show up a lot in conductive inks since they link together easily. Once printed and heated, these tiny bits merge into solid pathways that carry current. How well they transmit electricity hinges mostly on how tightly packed they are and whether the coating around them gets cleared away those coatings tend to block conduction. In early 2024, researchers zeroed in on methods to sinter at room temperature, applying chemicals to dissolve the outer layer, bringing back strong conductivity without heat. This shift opens doors for flexible tech applications where high temps won't work [14].

4.4 Surface chemistry and reactivity

A burst of surface energy doesn't just drop the melting point; it drives sharp chemical responses too. Right when AgNPs hit bodily fluids, such as blood or lab solutions, they grab nearby proteins fast. That gathering forms what scientists call a protein corona around each tiny grain. Instead of reacting to bare metal, living cells respond to this outer cloak. The true face seen by tissue is shaped entirely by those clinging biomolecules.

What happens on the outside of AgNPs shapes how fast they release silver ions, the particles responsible for killing microbes. Oxygen triggers a shift from neutral silver to its charged form, especially when conditions become more acidic. That slow release of ions turns these tiny particles into lasting germ fighters, yet keeping them stable over time isn't easy. Without careful coating, they break down too soon, losing their effectiveness before they can work properly.

5. Applications of AgNPs

What makes AgNPs stand out is how they fit into so many different industries, from everyday products to large scale manufacturing. Still, it's in medicine where most progress happens, thanks to the way tiny silver particles behave inside living tissues.

5.1 Biomedical applications

5.1.1 How antimicrobials work and how well they work

Silver is coming back into medical use because bacteria are increasingly resisting regular drugs. Traditional medicines usually target one specific process inside microbes, such as building proteins or constructing walls around cells. AgNPs, however, work in multiple ways simultaneously, making it harder for germs to adapt and survive. Their method involves several layers of action occurring together without pause. Upon contact, these tiny particles stick directly to the outer layer of bacteria. When AgNPs encounter gram-negative bacteria, they latch onto sulfur-rich proteins in the membrane. These particles disrupt the movement of materials across the cell. The phospholipid layer is severely affected, suffering permanent damage. Holes begin forming in the outer wall, making the barrier leaky. The contents inside start spilling out, and cells cannot survive that kind of breakage [4].

Inside the cell, silver ions move quickly toward vital enzyme sites marked by sulfur atoms. These +1 charged particles stick tightly to those spots, shutting down key energy processes like those involving NADH dehydrogenase. When that happens, the microbe cannot breathe at a cellular level. Its power supply drops without oxygen flow. Beyond enzymes, the metal also reaches into genetic material, locking onto nitrogen-rich parts of DNA strands. That interference blocks copying steps needed for reproduction. A separate path shows how silver sparks harmful molecules such as superoxide or hydroxyl fragments. Such unstable forms tear through internal structures, adding damage on top of ion effects. When AgNPs meet cells, they spark reactions that flood bacteria with harmful molecules. These overwhelm the microbe's natural shields. Proteins falter. Fats break down. Genetic material gets harmed. New research shows this three-part attack works on tough germs, even ones resisting many drugs. MRSA falls short. So does Pseudomonas aeruginosa. Evidence grows [7].

5.1.2 Anticancer properties

Not just limited to microbes, AgNPs can also act against tumors. What drives their effect on cancer? They trigger programmed cell death. Tumors burn energy faster, grab more materials that means these particles gather inside them more easily. After slipping into cells through membrane wrapping, they interfere with power producing parts of the cell. Something shifts when the mitochondria lose their inner balance, letting Cytochrome C spill out into the fluid part of the cell. Once that happens, certain proteins wake up Caspase 9 first, then Caspase 3 setting off a chain reaction inside. These molecules take apart the cell piece by piece, like a slow breakdown from the core.

Scientists studied AgNPs made with plant materials during 2024. These tiny particles demonstrated strong effects on certain types of cancer cells, such as those found in breast and cervical cancers. Since they are coated with natural substances, they tend to be gentler on normal body cells while remaining harmful to malignant ones. When researchers used yew tree extract to create them, higher amounts resulted in fewer living cancer cells. This occurred because the particles disrupted the cancer's growth cycle right before division could take place [6].

5.1.3 Healing skin and growing new tissue

Now silver plays a quiet role in treating burns, thanks to tiny tech advances. Inside gels, cloths, and fine fiber mats, little silver bits do steady work. They block harmful germs the kind that often harm burned skin while helping tissue grow back. When these particles move through damaged areas, swelling goes down faster than before. Signals inside cells shift, turning off harsh reactions that slow recovery. Wound healing gets a boost when these particles push fibroblasts to shift into myofibroblasts this tightens the injury site faster, reshaping damaged areas more efficiently. Lately, many medical supplies mix silver nanoparticles with frameworks like collagen or chitosan; such bases give fresh skin something solid to grow on, all while blocking germs from settling in [13].

5.1.4 Drug delivery systems

Tiny silver particles work well for carrying medicine because their surfaces can be adjusted easily. Chemotherapy the old way hits both sick and healthy cells without distinction. Attaching special molecules to these particles lets scientists aim straight at tumor cells. Hidden like a weapon inside a gift, the toxic substance rides on the particle's outside. A tiny carrier moves the medicine straight to the tumor. Once there, the sour surroundings around cancer cells cause it to let go of the treatment inside. Because it goes exactly where needed, the drug works better. At the same time, damage to healthy parts of the body drops sharply compared to regular chemo methods [11].

5.2 Environmental applications

5.2.1 Water treatment and disinfection

Water that is safe to drink is becoming harder to find around the world. Still, tiny silver particles are now helping clean water right where people use it. Chlorine and similar cleaners sometimes leave dangerous residues after they work. On the flip side, using silver avoids those harmful traces. These small silver bits are placed inside materials like clay filters, foam pads, or paper-like sheets. Their job? To stop bacteria from building up and clogging things. That gunk buildup has a name: biofouling, and silver fights it well.

Water cleaning works because tiny silver particles slowly release charged atoms. As liquid moves through a filter full of these particles, those atoms steadily slip out. They destroy harmful microbes like *E. coli*, *Vibrio cholerae*, and *Salmonella typhi*. New versions mix silver materials with titanium dioxide films. Light energy activates the titanium part, triggering reactions. Meanwhile, silver captures loose electrons, preventing them from disappearing too quickly. Together, they create powerful oxygen fragments that eliminate germs and break down pollutants simultaneously [15].

5.2.2 Catalytic degradation of organic pollutants

Out in factories, dirty water carries tough chemical dyes such as Methylene Blue, Congo Red, and even Methyl Orange that resist breakdown, harm life, and may cause cancer. Tiny silver particles step in quietly, breaking those stubborn compounds apart; what remains is just water, CO_2, and nothing sharp left behind.

Electrons move more easily when AgNPs step into the process. A big gap in energy levels usually slows down the reaction – on one side, BH4; on the other, the target molecule. These tiny particles act like bridges, grabbing electrons from the donor first. Then, they pass them forward, almost like handing off a package. With that help, less push is needed to get things going. That means the reaction speeds up without requiring extra force. Tiny bits of AgNPs cut dye breakdown time from hours to minutes. Because they pack so much surface area relative to their size, reactions find more spots to happen. Molecules stick and transform faster where space meets substance. This boost comes from how much room exists on such small particles. Evidence points to quicker cleanup when nanosilver enters the mix. Activity spreads across countless tiny zones during degradation. Speed rises thanks to open real estate at the microscopic level. Particles offer ground zero for chemical change in polluted water. More contact means swifter results in removing color toxins. Efficiency climbs without requiring extra material input. Smallness creates opportunities for rapid environmental repair. Reaction pace jumps once these specks are part of the process. Space availability drives performance far beyond expectations. Minutes replace hours when surfaces work this hard. Degradation gains momentum through the sheer density of interaction points.

5.3 Industrial and consumer uses

5.3.1 Textiles and apparel

Right now, the biggest buyer of tiny silver particles happens to be clothing makers. These small bits get mixed into cloth so it can clean itself and stay fresh longer. Smell from people does not come from wetness alone. It shows up when tiny living things break down moisture on skin. Two types often involved are Micrococcus and Corynebacterium. When silver specks go inside man made threads like polyester while they're melted and stretched, protection sticks around. Another way uses cotton dipped then dried in a mix holding these metals. That also stops microbes without washing away easily.

Most chemical treatments fade fast. Not so with AgNPs locked into fibers they keep working past 50 washes. Yet here lies a problem: tiny bits of silver slip away each time, winding up in wastewater. Because of this, scientists are testing natural binders from plants to hold the particles tighter on cloth [7].

5.3.2 Food packaging and preservation

Freshness lasts longer when tiny silver particles are incorporated into wraps that hold groceries. Some wrappers break down over time, while others stick around, yet both kinds can carry these particles. Instead of sitting quietly, the wrapping works by altering what happens inside the package.

Freshness slips away quickly when mold and yeast take hold on food surfaces. Yet silver particles step in, halting those unwelcome guests. These tiny agents also tackle ethylene, a gas produced by plants that accelerates ripening. Instead of allowing it to build up, the material transforms the gas through oxidation. That shift extends the ripening timeline far beyond normal. During shipping or storage, this delay makes a visible difference. One look at strawberries tells the story: wrapped in special film laced with chitosan and silver, they retained their shape and goodness much longer. Fifteen full days passed before changes set in, while others faded after only five.

5.3.3 Electronics with conductive inks

Fine threads of silver help carry current in bendable gadgets. Tiny particles mix into fluid pastes that flow through printers. Plastic sheets take these patterns without warping under heat. Machines stamp out ID labels this way, fast and thin. Screens that roll up start with such prints too. Sun power collectors gain better contact layers by similar means.

Heat damages flexible materials when making conductive prints. Removing protective layers around tiny silver particles allows them to connect. Usually, that requires 200°C, which is too hot for most plastics. However, these nanosized bits melt more easily, enabling bonding at just 150°C. Some methods now use special liquids instead of heat, breaking down coatings without warmth. This creates opportunities to write electronics on delicate surfaces like leaves or skin. Paper can hold circuit patterns when treated this way.

5.4 From lab to market: Evidence of pilot-scale translation and commercial bottlenecks

While laboratory-scale research on AgNPs is extensive, evidence for their translation to pilot-scale and real-world deployment is concentrated on specific sectors, most notably textiles, food packaging, and water treatment.

5.4.1 Evidence for real-world deployment

(1) Antimicrobial textiles: This is the most mature commercial sector. Pilot-scale production often utilizes "pad-dry-cure" methods or ultrasonic-assisted spray coating to integrate AgNPs into synthetic and natural fibers. Companies like *SilverSocks* and various sportswear brands have successfully deployed AgNP-enhanced fabrics that maintain antimicrobial efficacy for over 50 wash cycles.

(2) Food packaging: Real-world deployment is evidenced by the development of AgNP-incorporated polyethylene and starch-based films. Pilot studies have demonstrated that these films can extend the shelf life of perishable goods (e.g., strawberries, meat) by 3–5 days by inhibiting common spoilage organisms like *Escherichia coli* and *Staphylococcus aureus*.

(3) Water purification: Pilot-scale wastewater treatment plants (WWTPs) have begun deploying immobilized AgNPs on ceramic or polymeric membranes. These systems demonstrate high efficacy in "point-of-use" (POU) disinfection, though full-scale municipal adoption remains limited by concerns over silver leaching.

The "Valley of Death" in AgNP Translation

The transition from a 100 mL laboratory flask to a 100 L pilot reactor often reveals non-linear scaling effects. For instance, maintaining the precise temperature and mixing required for the LaMer nucleation model becomes exponentially difficult at scale, leading to polydisperse particles that fail to meet strict industrial quality standards. Addressing these bottlenecks requires a shift toward "Quality-by-Design" (QbD) manufacturing frameworks and the development of robust, high-throughput characterization tools. The major bottlenecks in commercialization are summarized in **Table 2**.

6. Toxicology and safety evaluation

A shift toward the widespread use of tiny silver particles brings big hopes alongside serious health concerns. What makes these bits work so well against microbes – their minuscule shape, large outer zone, and steady shedding of

Bottleneck category	Specific challenges	Impact on translation
Manufacturing Consistency	High batch-to-batch variability in size and shape during scale-up.	Complicates regulatory approval, which requires predictable efficacy and safety profiles.
Economic Viability	High costs of physical synthesis (e.g., laser ablation) and low yields of green synthesis.	Limits AgNPs to "premium" products rather than mass-market commodity items.
Regulatory Barriers	Inconsistent classification (e.g., as "pesticides" in some regions or "medical devices" in others).	Creates a long and expensive path to market, often exceeding 5–10 years for new applications.
Long-term Stability	Propensity for AgNPs to aggregate or oxidize in complex real-world environments.	Reduces the functional lifespan of products and increases the risk of uncontrolled ion release.
Public Perception & Safety	Concerns over environmental bioaccumulation and "nanophobia."	Can lead to restrictive labeling requirements or consumer boycotts.

Table 2.
Major bottlenecks in the commercialization of AgNP-based products.

ions – also lets them interfere with human tissues and wildlife alike. With more being made every year, mapping how they harm living systems, plus defining safe contact levels, now drives urgent research efforts [16].

6.1 Cytotoxicity and how cells take up substances

Beyond size alone, what stands out is how AgNPs slip through body defenses meant to block bigger particles. When people breathe them in, swallow them, or absorb them through the skin, these tiny bits move into blood pathways. From there, they settle where it matters most: liver, kidneys, even deep inside the brain. The route changes, but the result holds steady across exposures.

Inside cells, harm begins when nanoparticles sneak in like hidden invaders. The cell pulls them inside via a process called endocytosis. When they reach the lysosome, a space filled with acid, the coating on the particles breaks down. That breakdown sets free silver ions in a sudden burst. Those released ions then travel through different routes, interfering with the cell's balance.

Now here comes trouble silver ions stir up reactive oxygen species, like superoxide and hydrogen peroxide. Once those build up past what the cell can handle, membranes start breaking down through oxidation. Damage piles up, hitting fats in the wall like layers and messing with genetic material too.

Mitochondria struggle when AgNPs build up inside them. As these particles interfere with electron flow, energy production falters. The inner membrane's charge weakens without proper chain function. Cells run low on power because ATP levels drop sharply. Energy supply dwindles as if fuel lines were cut.

Tiny particles might slip into the nucleus through its pores, then reach the genetic material inside. This contact can snap DNA strands or twist chromosomes out of shape [17]. One way they cause harm is by bumping right up against active cell structures. Damage appears as broken pieces or odd forms in the hereditary code. These changes occur because their size allows them to go places larger bits cannot.

6.2 Environmental ecotoxicity

Out in rivers and lakes, tiny silver particles are turning up more often. Wastewater outflows carry them there, especially near cities. Clothes and skincare products release these bits when people use them daily. They travel through drains toward water-cleaning stations. Some settle at the bottom during processing, trapped in thick mud. Others slip past, flowing onward into natural water bodies.

Tiny silver particles harm small water life such as algae, tiny shrimp, and young fish. Gill damage happens when these particles clump together on the tissue, making breathing harder. Instead of moving smoothly through water systems, silver builds up in creatures higher in the food web from floating plants to little animals to bigger fish. People who eat contaminated fish might face health concerns. Surprisingly, land studies show similar issues; one recent look into dirt found that microbes helping plants grow struggle when exposed. These germs fix nitrogen, a key nutrient, yet their activity drops near polluted areas [18].

6.3 Regulation and challenges in safety assessment

Even though dangers exist, controlling tiny materials puzzles groups such as the FDA in America and ECHA in Europe. What makes it hard is how we define exposure levels. Normally, harmful effects depend on weight, but when dealing with ultra-small bits, total outer surface area and count matter much more than bulk. Five-nanometer specks at low weights pack a punch because they expose way more area compared to 50-nanometer ones, even if both weigh the same.

Rules keep shifting to catch up. In Europe, REACH now demands detailed info on how tiny materials behave. Still missing are agreed ways to run tests the same everywhere. Different lab methods, coating choices, and clumping levels mess up results when comparing harm levels across studies. One idea gaining ground involves building protective layers that stop metal ions from escaping too soon only releasing them where needed. Experts see this strategy, shaping safety right into the material, as the clearest route ahead [19].

6.4 Implementing safe by design for risk mitigation

Safe by Design (SbD) shifts the safety focus from post-production testing to the initial design phase. Three core strategies enable this without reducing the efficacy of AgNPs:

- Surface functionalization: By coating AgNPs with biocompatible molecules like Polyethylene Glycol (PEG) or antioxidants, the "corona" of the nanoparticle is modified. This masks the silver core from immune recognition and suppresses nanoparticle induced oxidative stress, reducing cytotoxicity while maintaining the particle's ability to reach its target.

- Controlled release: Designing AgNPs that only release silver ions (Ag^+) in response to specific environmental triggers (such as the acidic pH of a bacterial infection site or tumor) prevents the "burst release" of toxic ions in healthy tissues. This sustains therapeutic efficacy over longer periods while minimizing systemic exposure [20].

- Encapsulation: Enclosing AgNPs in a protective shell, such as silica (SiO_2), physically isolates the toxic metal core from direct contact with cells. This reduces bioavailability in the environment and the human body, significantly lowering toxicity without interfering with the particles' optical or conductive properties.

6.5 Challenges in data comparability and the need for standardization

The reliability of comparing toxicological data across AgNP studies is currently limited by significant heterogeneity in experimental design. Discrepancies often arise from three primary factors:

(1) **Biological Model Variability:** Responses in *in vitro* immortalized cell lines frequently fail to replicate the complex physiological interactions seen in in vivo mammalian models or primary cell cultures.

(2) **Exposure Conditions:** Variations in media composition (e.g., presence of fetal bovine serum), pH, and temperature drastically alter the "biological corona" and aggregation state of AgNPs, shifting their effective dose and bioavailability.

(3) **Measurement Endpoints**: Different assays (e.g., MTT vs. LDH for cytotoxicity) can yield conflicting results due to AgNPs interfering with optical signals or chemical reagents in the assays themselves [21].

To overcome these challenges, a Standardized Evaluation Framework (SEF) is proposed in **Table 3**, focusing on Physicochemical Biological Integrated Characterization.

6.5.1 Proposed standardized comparison metric

Authors propose the use of the "Relative Toxicity Index (RTI)", which scales the observed effect against a known ionic silver ($AgNO_3$) control and a standard

Framework pillar	Standardized requirement	Rationale for reliability
Pristine Characterization	DLS, TEM, and XPS in dry and stock states.	Establishes a baseline for size, shape, and surface chemistry.
***In Situ* Characterization**	Characterization in the actual exposure medium (UV Vis, Zeta Potential).	Accounts for aggregation and protein corona formation during the study.
Dosimetry Standardization	Reporting dose in mass/cm^2 (surface area) rather than just mg/L.	Provides more accurate cellular level exposure metrics for sedimentation prone NPs.
Interference Controls	Mandatory "nanoparticle-only" and "assay-only" controls.	Prevents false positives/negatives caused by AgNP optical interference.
Biological Benchmark	Use of standard reference materials (e.g., NIST AgNPs).	Allows for cross laboratory validation and relative toxicity scaling.

Table 3.
Proposed standardized evaluation framework for AgNP toxicity assessment.

reference AgNP. This index allows researchers to distinguish between particle-specific toxicity and toxicity derived from dissolved silver ions, regardless of the biological model used.

7. Looking ahead final thoughts

7.1 Next generation AgNPs

AgNPs exhibit excellent properties that enable extraordinary applications in a variety of fields. However, further improvements are needed, such as avoiding the use of toxic chemicals, employing simple and cost-effective techniques, and ensuring high quality. Now that studies on AgNPs are advancing, attention moves beyond basic creation and analysis toward intricate systems with several roles. What shapes nanotech in the years ahead may hinge on blending AgNPs into combined materials and responsive gadgets.

What happens when silver meets cutting edge materials? Blending it with things like Graphene Oxide or Carbon Nanotubes brings out strong teamwork one side holds things together, spreads things out, the other fights microbes or speeds up reactions. Jump to another mix: slip AgNPs into Metal Organic Frameworks. Suddenly, storing gases or speeding chemical changes gets way more efficient. Clumping used to mess everything up now these frameworks lock particles in place. Exact spots for reactions form right where needed [22].

A sudden shift has taken place; old ways of tweaking lab conditions through guesswork are fading. Instead, computers trained to learn patterns now guide how nanoparticles form. One recent run showed predictions hitting close to spot-on when guessing silver particle dimensions. Temperature, acidity, and ingredient amounts steer outcomes more precisely than before. With such smart models running simulations inside machines, fewer physical tests become necessary. Less wasted effort means quicker paths from small experiments to large production lines. Patel's team pointed it out clearly back in 2025 – numbers speak louder now.

To remain competitive against emerging 2D materials and carbon-based nanotechnologies, AgNP research must address several unresolved scientific and technological hurdles over the next decade:

- **Standardization of "Real World" Testing:** Most safety data comes from simplified lab models. There is an urgent need for standardized protocols that simulate complex biological and environmental "media" (e.g., soil, seawater, or human blood) to accurately predict long term behavior and clumping [23].

- **Precision Scaling of Green Synthesis:** While plant mediated synthesis is eco friendly, achieving the same "batch to batch" consistency as chemical methods remains difficult. Moving from small lab samples to industrial scale production without losing control over size and shape is a primary engineering challenge.

- **Predictive AI Integration:** Shifting from trial and error to AI-driven design is essential. Developing machine learning models that can accurately predict how a specific coating will affect both the antimicrobial performance and the environmental degradation of an AgNP will drastically speed up commercialization.

- **Life Cycle Circularity:** As AgNPs become ubiquitous in textiles and electronics, developing efficient "nano recycling" methods to recover silver from wastewater and end-of-life products is critical for both economic competitiveness and environmental sustainability [24].

8. Conclusion

What began long ago with old world containers meant to keep microbes at bay now shapes high end gadgets and healing tools today. Tiny silver pieces stand out as a major win in tiny tech progress over recent decades. Their actions depend less on size alone, but more on hidden forces like atomic arrangement, energy shifts at the outer layer, and particle squeeze effects deep inside. These details guide how they work across fields far apart in purpose yet linked by design.

From laser ablation's accuracy to using plants for greener nanoparticle creation methods, adaptations fit exact tasks. Not limited to one role, these particles fight drug-resistant microbes, work inside bendable screen circuits, and break down harmful chemicals in nature. AgNPs respond where modern problems demand answers.

Still, using this technology means taking accountability. Even though AgNPs help in many ways, they can cause harm too. Silver ions escaping into water systems create dangerous conditions; reactive molecules form, threatening both people and wildlife when levels go unchecked. Success down the road won't come only from finding fresh uses; it'll hinge on creating strict rules everyone follows to prevent danger. Building safety right into early development stages makes sense – one small step now might protect entire environments later.

Nomenclature

AgNPs	Silver Nanoparticles
SPR	Surface Plasmon Resonance
λ	X ray wavelength
θ	Bragg angle
d	Interplanar spacing of the crystal
n	Integer representing the order of reflection
Ccrit	Critical concentration for nucleation
PVP	Polyvinylpyrrolidone (stabilizer)
$NaBH_4$	Sodium Borohydride (strong reductant)
TEM	Transmission Electron Microscopy
XRD	X Ray Diffraction

Acknowledgment

The author acknowledges the use of Manus AI for language polishing and for improving the structural coherence of the manuscript.

Author details

Muhammad Pervaiz*, Farhan Tariq and Laraib Javed
Department of Basic & Applied Chemistry, Faculty of Science & Technology, University of Central Punjab, Lahore, Pakistan

*Address all correspondence to: muhammad.perviaz@ucp.edu.pk

References

[1] Duman H, Eker F, Akdaşçi E, Witkowska AM, Bechelany M, Karav S. Silver nanoparticles: A comprehensive review of synthesis methods and chemical and physical properties. Nanomaterials. 2024;**14**(18):1527. DOI: 10.3390/nano14181527

[2] Sim W, Barnard RT, Blaskovich MAT, Ziora ZM. Antimicrobial silver in medicinal and consumer applications: A patent review of the past decade (2007–2017). Antibiotics. 2018;**7**(4):93. DOI: 10.3390/antibiotics7040093

[3] Ebrahiminezhad A, Raee MJ, Manafi Z, Sotoodeh Jahromi A, Ghasemi Y. Ancient and novel forms of silver in medicine and biomedicine. Journal of Advanced Medical Sciences and Applied Technologies. 2016;**2**(1):122. DOI: 10.18869/nrip.jamsat.2.1.122

[4] Rodrigues AS, Batista JGS, Rodrigues MÁV, Thipe VC, Minarini LAR, Lopes PS, Lugão AB. Advances in silver nanoparticles: A comprehensive review on their potential as antimicrobial agents and their mechanisms of action elucidated by proteomics. Frontiers in Microbiology. 2024;**15**:1440065. DOI: 10.3389/fmicb.2024.1440065

[5] Hossain MM, Robinson Junior NA, Mok YS, Wu S. Investigation of silver nanoparticle synthesis with various nonthermal plasma reactor configurations. Arabian Journal of Chemistry. 2023;**16**(10):105174. DOI: 10.1016/j.arabjc.2023.105174

[6] Duman H, Eker F, Akdaşçi E, Witkowska AM, Bechelany M, Karav S. Silver Nanoparticles: A comprehensive review of synthesis methods and chemical and physical properties. Nanomaterials. 2024;**14**(18):1527. DOI: 10.3390/nano14181527

[7] Sati A, Ranade TN, Mali SN, Ahmad Yasin HK, Pratap A. Silver nanoparticles (AgNPs): Comprehensive insights into bio/synthesis, key influencing factors, multifaceted applications, and toxicity—A 2024 update. ACS Omega. 2025;**10**(3):2500–2515. DOI: 10.1021/acsomega.4c11045

[8] Rafique M, Rafique M, Kalsoom U, Afzal A, Butt SH, Usman A. Laser ablation synthesis of silver nanoparticles in water and dependence on laser nature. Optical and Quantum Electronics. 2019;**51**(6). DOI: 10.1007/s11082-019-1902-0

[9] Abdelghany TM, Al-Rajhi AMH, Abboud MAA, Alawlaqi MM, Magdah AG, Helmy EAM, Mabrouk AS. Recent advances in green synthesis of silver nanoparticles and their applications: About future directions. A review. BioNanoScience. 2017;**8**(1):5–16. DOI: 10.1007/s12668-017-0413-3

[10] Panwar MS, Pal P, Joshi D. Advances in Green Synthesis of silver nanoparticles: Sustainable approaches and applications. Journal of Drug Delivery and Therapeutics. 2024;**14**(11):177–184. DOI: 10.22270/jddt.v14i11.6854

[11] Abbas R, Luo J, Qi X, Naz A, Khan IA, Liu H, Yu S, Wei J. Silver nanoparticles: Synthesis, structure, properties and applications. Nanomaterials (Basel, Switzerland).

2024;**14**(17):1425. DOI: 10.3390/nano14171425

[12] Danai L, Rolband LA, Perdomo VA, Skelly E, Kim T, Afonin KA. Optical, structural and antibacterial properties of silver nanoparticles and DNA-Templated silver nanoclusters. Nanomedicine. 2023;**18**(9):769–782. DOI: 10.2217/nnm-2023-0082

[13] Kaushal A, Khurana I, Yadav P, Allawadhi P, Banothu AK, Neeradi D, Thalugula S, Barani PJ, Naik RR, Navik U, Bharani KK, Khurana A. Advances in therapeutic applications of silver nanoparticles. Chemico-Biological Interactions. 2023;**382**:110590. DOI: 10.1016/j.cbi.2023.110590

[14] Ibrahim N, Akindoyo JO, Mariatti M. Recent development in silver-based ink for flexible electronics. Journal of Science: Advanced Materials and Devices. 2021;7(1):100395. DOI: 10.1016/j.jsamd.2021.09.002

[15] Bhardwaj AK, Sundaram S, Yadav KK, Srivastav AL. An overview of silver nano-particles as promising materials for water disinfection. Environmental Technology & Innovation. 2021;**23**:101721. DOI: 10.10 16/j.eti.2021.101721

[16] Melo RM, Albuquerque GM, Monte JP, Pereira GAL, Pereira G. Recent advances in the application of silver nanoparticles for enhancing phototherapy outcomes. Pharmaceuticals. 2025;**18**(7):970. DOI: 10.3390/ph18070970

[17] AshaRani PV, Mun GLK, Hande MP, Valiyaveettil S. Cytotoxicity and genotoxicity of silver nanoparticles in human cells. ACS Nano. 2008;**3** (2):279–290. DOI: 10.1021/nn800596w

[18] Kumari M, Pandey S, Mishra SK, Nautiyal CS, Mishra A. Effect of biosynthesized silver nanoparticles on native soil microflora via plant transport during plant–pathogen–nanoparticles interaction. 3 Biotech. 2017;**7**(5):345. DOI: 10.1007/s13205-017-0988-y

[19] Ghobashy MM, Elkodous MA, Shabaka SH, Younis SA, Alshangiti DM, Madani M, Al-Gahtany SA, Elkhatib WF, Noreddin AM, Nady N, El-Sayyad GS. An overview of methods for production and detection of silver nanoparticles, with emphasis on their fate and toxicological effects on human, soil, and aquatic environment. Nanotechnology Reviews. 2021;**10**(1):954–977. DOI: 10.1515/ntrev-2021-0066

[20] Yan L, Zhao F, Wang J, Zu Y, Gu Z, Zhao Y. A safe-by-design strategy towards safer nanomaterials in nanomedicines. Advanced Materials. 2019;**31**(45):1805391. DOI: 10.1002/adma.201805391

[21] Hosny S, Gaber GA, Ragab MS, Ragheb MA, Anter M, Mohamed LZ. A comprehensive review of silver nanoparticles (AGNPs): Synthesis strategies, toxicity concerns, biomedical applications, AI-Driven advancements, challenges, and future perspectives. Arabian Journal for Science and Engineering. 2025;**50**:3435–3455. DOI: 10.1007/s13369-025-10612-0

[22] Hassan PB, Ameen SSM, Mohammed L, Ameen SMM, Omer KM. Enhanced antibacterial activity of a novel silver-based metal organic framework towards multidrug-resistant Klebsiella pneumonia. Nanoscale Advances. 2024;**6**(15):3801–3808. DOI: 10.1039/d4na00037d

[23] Schmutz M, Borges O, Jesus S, Borchard G, Perale G, Zinn M, Sips ÄAJAM, Soeteman Hernandez LG, Wick P, Som C. A methodological safe by design approach for the development of nanomedicines. Frontiers in Bioengineering and Biotechnology. 2020;**8**:258. DOI: 10.3389/fbioe.2020.00258

[24] Huang G, Guo Y, Chen Y, Nie Z. Application of machine learning in material synthesis and property prediction. Materials. 2023;**16**(17):5977. DOI: 10.3390/ma16175977

Chapter 2

Green Synthesis of Silver Nanoparticles

Ally Kafesa and Farhan Baehaki

Abstract

Green synthesis of silver nanoparticles (AgNPs) has become an important strategy in developing safer, more economical, and sustainable nanomaterials. This technique replaces dangerous chemicals used in traditional procedures with biomolecules from plants, microbes, algae, and polysaccharides as reducing and stabilizing agents. Phytochemicals like flavonoids, polyphenols, terpenoids, and organic acids reduce Ag^+ ions to Ag^0 during the biosynthesis process. This is followed by nucleation, particle development, and stabilization by biomolecule capping. The size, shape, colloidal stability, and biological activity of the final nanoparticles are greatly impacted by changes in synthesis parameters, such as $AgNO_3$ concentration, extract composition, pH, temperature, and reactant ratio. Strong antibacterial, antibiofilm, antioxidant, anti-inflammatory, and anticancer properties are demonstrated by the green manufacture of AgNPs. Several studies have reported improved biocompatibility as a result of the presence of organic compounds on the nanoparticle surface. The AgNPs are also frequently used in chemical sensors, food packaging, environmental catalysis, and wound healing. However, their possible toxicity to environmental organisms and mammalian cells is still a significant problem that needs thorough investigation, especially in relation to the release of Ag^+ ions and the development of oxidative stress. Variations in biological extract composition, production repeatability, and issues with industrial scalability are further hurdles. All things considered, green synthesis presents a viable method for creating more ecologically friendly nanosilver; however, before it can be extensively used in biomedical and environmental technology applications, process standardization, production parameter optimization, and long-term risk assessment are necessary.

Keywords: green synthesis, silver nanoparticles, biological activity, toxicity, biomedical applications

1. Introduction

Over the last 20 years, nanotechnology has advanced quickly and emerged as one of the most fruitful areas of research in the disciplines of biomedicine, pharmaceuticals, food, and the environment. Silver nanoparticles (AgNPs) are the most extensively researched type of metal nanoparticles because of their strong antimicrobial activity, intriguing plasmonic characteristics, and potential for a wide range of uses as environmental catalysts, as well as anti-inflammatory, anticancer, and wound-healing agents [1].

Hazardous organic solvents, high temperatures and energies, and toxic chemical reductants/stabilizers (such as $NaBH_4$, hydrazine, and synthetic surfactants) are typically needed for conventional synthesis methods (physical and chemical). This raises important concerns about chemical waste, toxicity, and process sustainability [2].

A green synthesis strategy, which uses ecologically friendly biological agents such as plant extracts, microbes, polysaccharides, or biomolecules as reducing agents, capping agents, and stabilizers, has been developed in response to these problems. This approach has the potential to be more cost-effective and scalable, minimizes the use of hazardous materials and excessive energy consumption, and is consistent with the principles of green chemistry.

The following topics will be methodically covered in this study:

(1) Fundamental ideas and traits of nanosilver

(2) The nanosilver synthesis reaction principle

(3) Green AgNP synthesis from biological sources

(4) Factors influencing green nanosilver synthesis

(5) Characterization of green synthesized–produced nanosilver

(6) Utilizing green synthesis for nanosilver

(7) Safety and toxicity concerns with nanosilver

(8) Obstacles in the creation of green synthesis nanosilver

2. Fundamental ideas and traits of nanosilver

2.1 Definition and range of sizes

Silver particles with a size range of 1–100 nm are known as AgNPs. When compared to their bulk form, the physical and chemical characteristics of silver metal undergo substantial changes at this scale (see **Figure 1**) [1, 2]. Increased specific surface area, altered optical characteristics (plasmon surface resonance), and improved chemical reactivity and biological activity are some of these changes. Particles with at least one dimension ranging from 1 to 100 nm are referred to as nanoparticles internationally (International Organization for Standardization [ISO], American Society for Testing and Materials [ASTM], Food and Drug Administration [FDA]) [1]. This range is based on changes in material qualities that have been measured experimentally; it is not arbitrary. Below 100 nm, a number of physical property changes take place, such as:

(1) The ratio of surface area to volume rises significantly. Surface atoms become significantly more prominent and exhibit a sharp rise in reactivity as particle

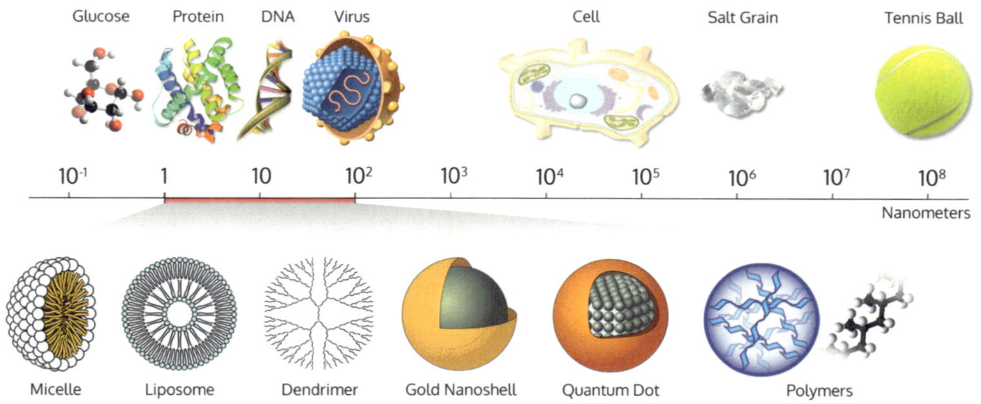

Figure 1.
Particle sizes of nanoparticle.

size decreases. For instance, nanosilver in the 10–50 nm size range [1] and comparatively inert bulk silver particles are both extremely chemically and physiologically active. AgNPs' potency as antimicrobials and anticancer agents stems from this [1, 2].

(2) The appearance of quantum phenomena: Electrons start to experience quantum confinement at particle sizes smaller than ± 100 nm. Color changes (such as a yellow-brown AgNPs solution) are caused by the electron energy becoming discontinuous rather than continuous. Surface plasmon resonance (SPR) is one example of a unique optical property [1]. Microparticles (> 1000 nm) do not experience this impact. One nanometer is the smallest size range, or lower limit, of the nanoscale. A particle is no longer stable as a "particle" if its size is less than 1 nm; instead, it is better referred to as a cluster atom or a molecular atom. The particle is no longer a useful nanomaterial at this size, and thus the rules of molecular chemistry apply. Consequently, the practical limit of nanoparticle stability is 1 nm; 100 nm is the maximum size range, or upper limit, of the nanoscale. The surface area ratio sharply drops, biological and catalytic activities are diminished, and quantum effects are reduced or eliminated if the particle size is larger than 100 nm. For instance, AgNPs have excellent biological effects as antibacterials when their size is between 10 and 20 nm [1]. The antibacterial activity of AgNPs is significantly reduced when their size falls between 200 and 500 nm [1]. Consequently, the boundary between the nano- and micro- (conventional) worlds is 100 nm. A size of 1–100 nm enables nanoparticles to interact with proteins and DNA, pass through cell membranes, and enter cells by endocytosis in the context of particle interactions with biological activity [3]. On the other hand, particles that are overly large (> 100 nm) will find it difficult to enter cells, which will lessen their biological impacts. Particles that are too small (less than 5 nm) may be excessively toxic and are quickly removed by the kidneys.

Because materials display special characteristics that set them apart from their bulk forms at this scale, such as increased surface area, the emergence of quantum effects, and the capacity to interact with biological systems, nanoparticle sizes are therefore restricted to the range of 1–100 nm. While particles larger than 100 nm tend to lose their nanoscale properties and resemble normal materials, particles smaller than 1 nm are unstable as nanoparticles.

2.2 Unique characteristics of nanosilver

When free electrons on the surface of AgNPs interact with the electric field of light, a process known as optical plasmon occurs. The AgNPs have a distinctive absorbance peak in the ultraviolet–visible (UV–vis) spectrum between 380 and 450 nm, which is linked to SPR [4]. Because it happens in individual nanoparticles, it is referred to as localized. Instead of bulk metal, electrons are focused at the nanoscale. When exposed to an electric field on the surface of a nanoparticle, free electrons (conduction) in the upper energy band of silver (Ag), a conducting metal, readily migrate [5]. In comparison to the particle size, the number is extremely high. Optical plasmon in AgNPs is a phenomenon where free electrons on the surface of AgNPs oscillate collectively due to interaction with the electric field of light. The AgNPs show a characteristic absorbance peak in the 380–450 nm region in the UV–vis spectrum, associated with SPR. It is called localized because it occurs in individual nanoparticles. Electrons are concentrated at the nanoscale (not in bulk metal). Silver (Ag) is a conducting metal that has free electrons (conduction) in the upper energy band, and these electrons easily move when exposed to an electric field on the surface of the nanoparticle. The number is very large relative to the particle size [6]. The optical plasmon process consists of several stages, namely: (1) Incident light/visible light hitting the AgNP absorbs the energy of electromagnetic waves, which are the main triggers of plasmon. (2) Surface electron polarization occurs, where the electric field of light pushes free electrons to one side of the nanoparticle, causing the movement of the negative charge side to the positive charge side. A transient electric dipole is created as a result. (3) The Coulomb force causes a collective oscillation of electrons, whereby pushed electrons are "pulled back," leading to a back-and-forth oscillation of electrons on the surface. (4) When maximum resonance takes place, light absorption becomes extremely intense, and light scattering sharply increases. (5) Plasmon energy is released as local heat (photothermal effect), light scattering, energy transfer to nearby molecules, and the production of reactive oxygen species (ROS), which are crucial for antibacterial and anticancer actions [7]. The size, shape, and aggregation of the particles have a significant impact on the location and form of the SPR peak. The AgNPs can interact with microorganisms' DNA and proteins, cause oxidative stress (ROS), and harm cell membranes. The AgNPs can speed up reduction/oxidation processes and the breakdown of organic pigments in wastewater treatment. Due to their electrical characteristics and conductivity, nanosilver is helpful in flexible electronics, sensors, and conductors.

3. The nanosilver synthesis reaction principle

3.1 Fundamentals of conventional method reactions

Using a potent reducing agent, silver ions (Ag^+) are reduced from a precursor (often $AgNO_3$) to zero-valent metallic silver (Ag^0) in order to create AgNPs. After that, the Ag atoms go through nucleation, growth, and stabilization to become nanoparticles [1, 2, 8]. Traditional chemical and physical techniques are as follows:

- Apply potent reducing agents (ethylene glycol, hydrazine, and $NaBH_4$).
- Need synthetic surfactants and organic solvents.
- Use a lot of energy (pressure and temperature).
- Leave behind hazardous chemical residues

Common reactions:

$$Ag^+ + e^- \rightarrow Ag^0$$

A very quick reduction reaction is produced by strong reducing agents that supply electrons quickly, enabling very tiny particle sizes.

(1) AgNP synthesis with sodium borohydride ($NaBH_4$)

$NaBH_4$ is a very potent reducing agent that can form very small AgNPs (2–10 nm) in a quick, exothermic process. The following is the detailed procedure for creating AgNPs with $NaBH_4$ [9]:

(a) Making the precursor solution, $AgNO_3$ is dissolved in water (low concentration: 0.1–1 mM). Low temperatures (0–5 °C) are frequently employed.

(b) Making the $NaBH_4$ solution, dissolve $NaBH_4$ in freezing water. The solution needs to be fresh because it breaks down quickly followed:

$$BH_4^- + 2\boldsymbol{H_2O} \rightarrow \boldsymbol{BO_2^-} + 4\boldsymbol{H_2}\uparrow$$

(c) Quick reduction response

- The $NaBH_4$ solution is gradually added to $AgNO_3$ while being stirred aggressively.
- A response takes place:

$$Ag^+ + BH_4^- + 3\boldsymbol{H_2O} \rightarrow Ag^0 + \boldsymbol{BO_3^{3-}} + 3.5\boldsymbol{H_2}$$

The solution's hue shifts from clear to yellow to brown.

(d) Nucleation explosion

- Extremely quick reduction results in the simultaneous formation of many Ag nuclei.
- Growth is subordinated to nucleation.

(e) Maintaining stability

- Polyvinylpyrrolidone (PVP), citrate, or polyethylene glycol (PEG) is examples of capping agents that are necessary.

- In the absence of a stabilizer, agglomeration occurs quickly.

(2) AgNP synthesis with hydrazine (N_2H_4)

In addition to being extremely poisonous and carcinogenic, hydrazine is a potent reducing agent that produces a quick but more regulated reaction than $NaBH_4$ [9]. A sign of the reaction is the basic reaction mechanism, which releases N_2 gas.

$$4Ag^+ + N_2H_4 \rightarrow 4Ag^0 + N_2 + 4H^+$$

The following is the detailed procedure for creating AgNPs with N_2H_4:

(a) $AgNO_3$ solution preparation: It uses a medium concentration (0.5–2 mM). Alcohol or water is utilized as a solvent.

(b) Hydrazine addition: Hydrazine is gradually introduced, and maintaining a basic pH is advised (NaOH is supplied).

(c) Growth and reduction: Rapid nucleation is caused by high reduction rates. Continued growth produces particles bigger than those made by $NaBH_4$.

(d) Maintain stability: To maintaining stability, PVP or other surfactants are typically used.

(3) AgNP synthesis with ethylene glycol (polyol method)

High temperatures (120–180°C) are needed for the process, and ethylene glycol serves as both a solvent and a mild reducing agent. The following high temperatures are necessary for the reduction mechanism [10]:

$$HO\text{–}CH_2\text{–}CH_2\text{–}OH \rightarrow HO\text{–}CH_2\text{–}CHO + 2H^+ + 2e^-$$

Ag is reduced by these electrons:

$$Ag^+ + e^- \rightarrow Ag^0$$

The following is the detailed procedure for creating AgNPs with ethylene glycol:

a. $AgNO_3$ dissolving in ethylene glycol:

- Water-free system.
- A comparatively high level of focus.

b. Heating needs 140–180 degrees Celsius. Slow reduction leads to effective growth control.

c. Regulated growth and nucleation. Nucleation that is not explosive, resulting in a larger size due to dominant growth.

d. Including a capping agent:

- PVP is frequently utilized.
- Managing the form (wire, cubic, or spherical).

3.2 Fundamentals of the nanosilver reaction in green synthesis

The green synthesis method typically consists of three steps and uses eco-friendly plant extracts [10, 11]:

(1) Activation (reduction of Ag^+ ions to Ag_2)
Ag^+ nuclei are created when phytochemicals (polyphenols, flavonoids, terpenoids, ascorbic acid, proteins) or microbial metabolites convert Ag^+ ions from precursors ($AgNO_3$, Ag_2SO_4).

(2) Particle growth (growth and nucleation)
Ag_3 nuclei develop into nanoparticles with a particular size and shape when more Ag_4 atoms are added.

(3) Capping or stabilization
Phytochemicals, proteins, and polysaccharides are examples of organic molecules that attach to the surface of AgNPs to produce a capping layer, which enhances colloidal stability and inhibits agglomeration:

- Using microorganisms, biomolecules, or plant extracts as capping and reducing agents.
- makes use of solvents based on water.
- It can be used at moderate or room temperature.
- Generates trash that is comparatively safer and biodegradable. AgNPs with antibacterial and anticancer activity equivalent to or even greater than chemically manufactured nanoparticles can be produced via plant-mediated and microbial-mediated synthesis, according to several recent reviews. This is because bioactive molecules are linked to the surface of the nanoparticles.

3.3 Phytochemicals' function as capping and reducing agents

The natural capacity of phytochemical substances present in plant extracts to take the place of hazardous synthetic chemicals is utilized in the green manufacture of AgNPs. In this system, phytochemicals serve as both stabilizing agents (capping agents) [12], which coat the surface of the nanoparticles and prevent agglomeration, as well as reducing agents that transform silver ions (Ag^+) into metallic silver (Ag^0). The primary benefit of green synthesis over traditional chemical techniques is this dual function [8]. Numerous redox-active phytochemicals are found in plant extracts, including:

1. Alkaloids, glycosides, tannins, terpenoids, saponins, flavonoids (quercetin, rutin, catechin), and phenols and polyphenols (gallic acid, chlorogenic acid) [13]. Aromatic hydroxyl groups (–OH) in phenolic and flavonoid compounds are easily oxidized. Ag^+ can be reduced to Ag_2 under reaction circumstances by these –OH groups, releasing hydrogen atoms or electrons. Electrons can also be donated by other molecules having functional groups such as –COOH, $-NH_2$, and –C=O [14]. In a nutshell, the mechanism can be described as follows:

    a. Plant extract-derived phenols and flavonoids undergo oxidation to produce free electrons, quinones, or other oxidized structures.

    b. Ag^+ ions become Ag when they take on electrons.

Enol–keto tautomerization, which creates reactive hydrogen and promotes electron transfer to silver ions, is frequently involved in this process in flavonoids. The quantity and location of –OH groups in the flavonoid structure have a significant impact on the reduction rate.

2. Vitamin C, ascorbic acid, and other organic acids

One very powerful natural reducing agent is ascorbic acid. This substance easily reduces Ag^+ while oxidizing to dehydroascorbate. The SPR phenomenon causes the color of the solution to shift from clear to yellow or brown, indicating that the presence of ascorbic acid in plant extracts frequently speeds up the nucleation of nanoparticles [15].

Phytochemicals' main function as reducing agents is to supply electrons for the reduction of Ag^+ ions to neutral silver atoms (Ag^0). The first and most important step in the creation of nanoparticles is this process. The thermodynamic system becomes unstable once Ag^+ ions are converted to Ag^0 and form the nanoparticle core. Particles will clump together and lose their nanoscale characteristics if they are not protected [1, 16]. This is where phytochemicals play a critical role as capping agents. Through their functional groups, such as the coordination bond between Ag atoms and –OH or $-COO^-$ groups, the electrostatic interactions between the surface charges of nanoparticles and organic molecules, and the influence of hydrogen bonds and Van der Waals interactions, phytochemical molecules will adsorb or bind in coordination on the surface of AgNPs. This organic layer creates an organic corona, a shield that keeps nanoparticles from coming into direct contact with one another. The phytochemical capping layer offers:

- Steric stabilization: Big organic molecules prevent particles from coming into direct contact with one another.

- Electrostatic stabilization: Particles repel one another when charged groups ($-COO^-$, $-O^-$) raise the zeta potential [5]. Maintaining nanoscale size and narrow particle distribution during storage and biological applications depends on this colloidal stability.

The properties of the resultant AgNPs are mostly determined by the phytochemical content and composition in plant extracts:

- Extracts high in polyphenols can encourage rapid reduction, which leads to dominant nucleation and the formation of tiny particles.
- Extracts high in protein and polysaccharides encourage slower reduction, resulting in controlled growth and the formation of larger particles.
- The equilibrium between nucleation and growth is determined by the ratio of the reducing agent to Ag^+.

This diversity explains why AgNPs of varied sizes and morphologies can be produced by extracts from different plants, even from different plant sections.

3.4 Mechanisms in systems mediated by plants

Using plant extracts as a source of active biomolecules, plant-mediated green synthesis of AgNPs reduces silver ions (Ag^+) to metallic silver (Ag) and stabilizes the resultant nanoparticles. Because it incorporates a variety of phytochemical substances that cooperate in an aqueous reaction medium, this system is complicated. Activation (ion reduction), nucleation, particle development, and stability (capping) are the basic processes that describe the mechanism of AgNP production in plant-mediated systems [17]. The mechanism can be simplified for plant-based systems as follows:

1. Ag^+ ion diffusion into the matrix of plant extract.
2. Interaction with phytochemical substances that have dissolved.
3. Ag_2 nuclei production; this process is distinguished by the creation of an SPR peak, which causes the solution to change color from yellow to dark brown.
4. Controlled agglomeration and development; the nanoparticles' surface is covered by one or more organic molecules (capping). The degree of ionization and electron donor capacity of phytochemicals is influenced by pH and temperature, which, in turn, affects the rate and effectiveness of reduction.

3.5 Microbial system mechanisms

Through enzymatic activity and cellular metabolites, bacteria, fungi, yeast, and algae can convert silver ions (Ag^+) to metallic silver (Ag^0) in a process known as microbial-mediated green synthesis of AgNPs. The oxidoreductase enzymes, coenzymes, proteins, and polysaccharides found in microorganisms' biochemical systems serve as stabilizers and reducing agents for nanoparticles. The AgNPs are often synthesized in microbial systems via two primary channels: intracellular and extracellular mechanisms. Each of these pathways has unique kinetic properties and applications. The reduction process can take place in fungi and bacteria:

1. AgNPs are created in the cytoplasm and on the cell surface when Ag + ions enter the cell and are reduced by oxidoreductase enzymes or NADH/NADPH molecules [18].

2. Extracellular: Ag^+ is directly reduced outside of the cell by enzymes and metabolites released into the culture medium. It has been claimed that enzymes, including dehydrogenase, nitrate reductase, and other oxidoreductases, are important in this process [18]. Then, polysaccharides and microbial proteins serve as capping agents.

Ag ions in the medium will either disperse or be carried into the microbial cell via the cell membrane during the intracellular process. Ag^+ + ions interact with reductase enzymes and electron-donating coenzymes like NADH or NADPH in the cytoplasm [19]. Ag^+ is reduced to Ag_2 as a result of this electron transfer process, which subsequently forms the core of the nanoparticles inside the cell. Depending on the type of microbe, the generated nanoparticles may be found in the cytoplasm, cell wall, or periplasmic region. The AgNPs are then stabilized by intracellular proteins and biomolecules acting as organic capping agents.

The extracellular process, on the other hand, involves microorganisms secreting active metabolites and enzymes into the culture medium. Extracellular enzymes like nitrate reductase, dehydrogenase, and other redox proteins directly decrease silver ions outside the cell [20]. Ag_2 nuclei are created in the medium as a result of this reduction, and AgNPs then gradually develop. The microbes' released proteins, polysaccharides, and biomolecules are adsorbed on the surface of AgNPs, creating a capping layer that improves colloidal stability and inhibits agglomeration. Because nanoparticles are created directly in the medium and are simpler to separate without endangering the cell biomass, the extracellular method is thought to be preferable in terms of application.

The kind of microbe, biomass density, enzyme activity, and environmental factors, including pH, temperature, aeration, and incubation duration, all have a kinetically significant impact on the rate of Ag^+ ion reduction in microbial systems [21]. Due to their capacity to secrete large quantities of proteins and enzymes, fungi typically produce AgNPs with more uniform sizes, whereas bacteria frequently produce nanoparticles with smaller sizes because of their quicker rates of reduction. In addition to serving as a stabilizer, the biomolecular layer that covers the surface of AgNPs also affects the nanoparticles' biological characteristics, such as enhancing their biocompatibility and antibacterial activity. Therefore, the mechanism of AgNPs synthesis in microbial systems is an integrated biological process that includes surface stabilization, enzymatic reduction, nucleation, and nanoparticle development in a single, eco-friendly system with enormous potential for biomedical and environmental applications.

4. Green AgNP synthesis from biological sources

4.1 Plants (synthesis mediated by plants)

Plants are the most popular biological source for the environmentally friendly synthesis of AgNPs because of their abundance, affordability, and high secondary metabolite content. Silver ions (Ag^+) can be reduced to metallic silver (Ag^0) by a variety of redox-active phytochemicals found in plant extracts, including polyphenols, flavonoids, terpenoids, alkaloids, tannins, saponins, and organic acids

[22]. These compounds' aromatic structure and carbon chain serve as capping agents that stabilize the nanoparticles, while functional groups like –OH, –COOH, and –C=O operate as electron donors. The AgNPs can be produced from a variety of plant parts, including leaves, flowers, fruits, seeds, bark, and roots. The phytochemical makeup of each part greatly influences the size and shape of the nanoparticles. The creation of comparatively more biocompatible AgNPs, the ease of the process, and the lack of strict sterile conditions are the primary benefits of the plant-mediated technique. However, standardizing and replicating results is severely hampered by changes in phytochemical composition brought on by seasonal variations, regional factors, and extraction techniques [23]. Extracts have been made from leaves, flowers, bark, fruit, roots, and seeds. Neem (*Azadirachta indica*), green tea (*Camellia sinensis*), betel leaf, *Ocimum gratissimum*, Aloe vera, Magnolia spp., and so forth are frequently mentioned examples. For instance, flower extracts frequently include significant concentrations of flavonoids and anthocyanins, which are excellent reducing and capping agents.

4.2 Microorganisms

Microorganisms, including yeast, fungus, and bacteria, are significant biological sources for the green synthesis of AgNPs because they possess enzymatic systems that can naturally reduce metal ions. Oxidoreductase enzymes, dehydrogenases, and NADH/NADPH coenzymes play a major role as electron donors in microbial-based synthesis, which can take place intracellularly or extracellularly [18]. In external systems, enzymes and metabolites produced by microorganisms directly reduce Ag^+ in the culture media, whereas in intracellular systems, Ag^+ is reduced within the cell and forms nanoparticles localized in the cytoplasm or cell wall. Because they can release enormous amounts of proteins and enzymes that enable the creation of more homogeneous and stable AgNPs, fungi are sometimes regarded as superior to bacteria. Microorganism-derived protein and polysaccharide coatings also serve as natural capping agents, improving the biocompatibility and colloidal stability of nanoparticles. Microbial-based synthesis necessitates more complicated techniques and stringent control of culture conditions than plant-based approaches, despite its great potential for large-scale output through fermentation systems [24].

4.2.1 Bacteria

Through reductase enzymes, certain bacteria, such as Pseudomonas, Bacillus, Escherichia, and Lactobacillus, can facilitate the synthesis of AgNPs. Some benefits are that culture conditions can be used to control the process and are suitable for production on a large scale (bioreactor fermentation). However, it necessitates stringent contamination control and sterile conditions [25].

4.2.2 Yeast

It is well known that fungi like Aspergillus species, Penicillium species, and *Fusarium oxysporum* are very effective at producing AgNP [10] because:

- They generate a lot of extracellular enzymes.
- They yield a lot of biomass and are simple to cultivate.

- Fungi frequently create nanoparticles with a restricted distribution and a comparatively uniform size.

4.3 Cyanobacteria and algae

Algae and cyanobacteria are being studied more and more as alternative biological sources for green AgNP synthesis because of their distinct biomolecule composition, which includes proteins, sulfated polysaccharides, photosynthetic pigments, and phenolic compounds [5, 24]. Through intracellular and extracellular processes involving cellular metabolites and redox enzymes, these species are able to reduce Ag^+. Algae's capacity to be cultivated utilizing sunlight as an energy source, basic nutrients, and even wastewater is one of their main advantages, potentially promoting sustainable production and a circular economy. Algal proteins and polysaccharides are essential capping agents that improve the stability and regulate the size of AgNPs [1]. Nevertheless, there are also drawbacks to using algae, such as the requirement for specialized growth facilities and differences in synthesis yields impacted by pH, light intensity, and nutritional conditions.

Additionally, pigments, polysaccharides, and enzymes that can reduce Ag^+ are abundant in green algae, brown algae, and cyanobacteria. Their benefits include the ability to be cultivated using light as an energy source in photobioreactors. Wastewater can be used as a cultivation medium. The efficiency of synthesis is significantly influenced by variables such as pH, light intensity, and biomass concentration.

4.4 Biomolecules and pure polysaccharides

In the green synthesis of AgNPs, pure natural biomolecules such as polysaccharides (chitosan, starch, derivatized cellulose), proteins, peptides, and amino acids are utilized as biological sources in addition to living organisms and their extracts. Because their compositions are simpler and more defined than those of complex extracts, these biomolecules provide superior control and great reproducibility. For instance, chitosan improves colloidal stability and antibacterial action by acting as a positively charged capping agent and a weak reducing agent [26]. As stabilizing matrices, cellulose and starch prevent the aggregation of nanoparticles. For pharmaceutical and biomedical applications, pure biomolecule-based methods are frequently selected because they are simple to standardize and evaluate for safety. The absence of synergistic effects that are usually obtained with complex phytochemical combinations in plant or microbe extracts is a drawback. A variety of proteins and peptides, as well as polysaccharides like chitosan, starch, and derivatized cellulose, are employed as capping and reducing agents [27]. Chitosan offers extra antibacterial qualities and a positive charge, and to improve stability and regulate the size of AgNPs, encapsulation matrices made of PVA, PVP, and other eco-friendly polymers can be employed.

4.5 The role of microbes, biomolecules, or plant extracts as stabilizing and reducing agents in green synthesis processes

Plant or microorganism extracts offer molecules that adsorb onto the AgNP surface to prevent agglomeration (e.g., proteins, polysaccharides, tannins, saponins) if they are acting as stabilizing (capping) agents, and compounds that transfer electrons to reduce $Ag^+ \rightarrow Ag^0$ (e.g., phenols, flavonoids, terpenoids, sugars, organic acids, proteins/peptides, enzymes). The nucleation rate, size, shape, stability (zeta potential), and surface characteristics of nanoparticles are all determined by this integrated mode of action. The constancy of this process in numerous microorganisms and plants has been confirmed by recent reviews and data. The following molecular mechanisms take place:

1. Reduction (electron donor): Reduction by sugars/alcohols (ribose, glucose) and organic acids is also frequent. Phenolics and flavonoids (phenolic – OH) can contribute electrons by oxidizing phenolic groups to quinones, converting Ag^+ to Ag^0. Reductase enzymes, such as nitrate reductase, can catalytically aid reduction in microorganisms. AgNP production has been linked to a fast shift in the UV–vis spectrum (SPR ~ 400 nm) in experimental studies (e.g., leaf extracts, fruit peels).

2. Stabilization (capping): On the surface of AgNPs, large molecules like proteins, polypeptides, or polysaccharides create an adsorption layer. This layer increases the negative/positive zeta potential and inhibits agglomeration by providing steric hindrance and/or surface charge (electrostatic stability). Additionally, capping affects the release of Ag^+ ions and biological interactions (such as affinity for bacterial membranes).

3. Precursor-to-extract ratio (amount of reductant and capping), temperature (kinetics), pH (modifies phenolic/protein ionization, therefore altering the nucleation rate), and reaction duration are the most important factors. The AgNPs with spherical, rod-shaped, or triangular morphologies, ranging in size from 2 to 100 nm, can be created by adjusting these parameters.

Practical implications are that green procedures frequently produce biocapped AgNPs, which are more stable in biological environments and possibly more biomedically compatible than "naked" AgNPs from chemical reduction, since natural extracts contain a combination of molecules. However, without extract standardization, extract variability creates repeatability issues.

5. Factors influencing green nanosilver synthesis

5.1 $AgNo_3$ concentration of silver precursor

One of the most important factors in the synthesis of AgNPs is the concentration of the silver precursor, which is typically used in the form of a silver nitrate ($AgNO_3$) solution. This is because it directly affects the nucleation mechanism, the reduction rate of silver ions (Ag^+), and the dynamics of particle growth. The number of Ag^+ ions available in the system is restricted at relatively low $AgNO_3$ concentrations, which leads to a slow nucleation rate and a relatively small number of nanoparticle nuclei formed; this condition tends to produce small AgNPs but with a low particle number and a less uniform size distribution. On the other hand, increasing the concentration of the silver precursor will speed up the reduction process and increase the number of Ag_2 nuclei formed quickly, making nucleation more dominant than growth, which under ideal circumstances can result in nanoparticles with a narrow distribution and smaller sizes [12]. However, excess Ag^+ ions will encourage uncontrolled particle growth and increase the likelihood of agglomeration, resulting in large and unstable AgNPs if the concentration of $AgNO_3$ is too high and is not balanced by a sufficient amount of reducing agent and capping agent [19].

This effect becomes more complex in green synthesis systems since the balance between nucleation and particle development is determined by the ratio of $AgNO_3$ concentration to phytochemical or active biomolecule content in biological extracts. A large number of nuclei can develop at the ideal concentration of the silver precursor, and the bioactive molecules available can coat and stabilize the surface of the nanoparticles, producing stable AgNPs with the required size and shape [28]. As a result, regulating the proportion of the silver precursor influences not only the size and distribution of AgNPs but also colloidal stability, optical characteristics including the locations of SPR peaks, and the biological activity of the final nanoparticles. One important factor is the concentration of Ag^+ ions:

- Slow nucleation, caused by an excessively low concentration, produces very small yet few particles.

- Agglomeration and the creation of large particles are risks associated with excessively high concentrations.

$AgNO_3$ concentration and temperature are the two factors that have the biggest effects on AgNP size, both in the chemical and green processes, according to recent modeling studies.

5.2 Biological extract composition and concentration

Biological extract content and composition are important parameters that impact the features and success of AgNP synthesis in the green synthesis technique because biological extracts directly serve as a source of reducing and capping agents. Low extract concentrations frequently lead to a slow reduction process, limited nucleation, and suboptimal nanoparticle formation or the production of AgNPs in small quantities and relatively large sizes because there are not enough bioactive compounds available to reduce all silver ions (Ag^+) [29]. Conversely, increasing the concentration of biological extracts will increase the availability of phytochemicals or redox-active biomolecules, such as proteins, polysaccharides, terpenoids, polyphenols, and flavonoids, which can speed up Ag^+ reduction and increase the number of nanoparticle nuclei formed, tending to produce smaller AgNPs with a more uniform distribution. Nevertheless, employing extracts at excessively high concentrations may also have unfavorable consequences, such as the development of an overly thick organic layer on the surface of the nanoparticles or an increase in the reaction medium's viscosity, which may prevent ion diffusion and cause agglomeration, thereby lowering colloidal stability [30].

The composition of biological extracts is just as significant as concentration because variations in the kinds and amounts of biomolecules present in the extract have a direct impact on the nucleation rate, growth pattern, and reduction process of nanoparticles. Extracts with high protein and polysaccharide content typically result in more regulated nanoparticle growth because of strong capping effects, whereas extracts rich in phenolic and flavonoid chemicals typically promote fast reduction and dominant nucleation [31]. Differences in biological species, the portion of the organism exploited, the extraction technique, and the growing environment can all contribute to these compositional discrepancies. In order to produce AgNPs with consistent size, shape, stability, and biological activity that are suitable for their intended use, it is crucial to optimize the concentration and comprehend the composition of biological extracts. The following are impacted by the plant/biomass extract concentration:

- The quantity of capping and reducing agents that are available.
- The rates of nucleation and reduction of Ag^+

Overly concentrated extracts can produce a large number of nuclei and tiny particles, but if capping is insufficient, they can also result in aggregation. The species, plant part, solvent, and extraction conditions all affect the phytochemical content [30].

UV–vis and Fourier transform infrared (FTIR) spectroscopy studies can be used to experimentally demonstrate how the content and composition of biological extracts affect the creation of AgNPs. The development and intensification of the SPR peak in the UV–vis spectrum, which is typically found in the wavelength range of 380–450 nm for spherical AgNPs, are often indicative of an increase in the concentration of biological extracts. Since there are more reducing agents available in the extract, more nanoparticles are generated when the SPR peak intensity is higher. Additionally, the production of smaller nanoparticles is frequently associated with a blue shift in the SPR peak's position toward shorter wavelengths, which occurs when the concentration of reducing chemicals like polyphenols and flavonoids is sufficiently high to promote dominant nucleation. Conversely, uncontrolled particle development or agglomeration caused by an excess of organic molecules or an imbalance between the concentration of the extract and the silver precursor may be indicated by peak broadening or a shift toward longer wavelengths (red shift) [32].

Additional proof of the biological extract composition's role in the decrease and stabilization of AgNPs is provided by FTIR analysis. The presence of phenolic compounds, flavonoids, proteins, and polysaccharides on the surface of AgNPs is typically indicated by characteristic absorption bands at wavenumbers around 3,200–3,400 cm^{-1} (–OH or –NH stretching), 1,600–1,650 cm^{-1} (C=O stretching or amide bonds), and 1000–1,300 cm^{-1} (C–O stretching) [5, 33]. The direct involvement of these functional groups in the Ag^+ ion reduction and capping process is typically shown by a shift in the band position or a change in absorption intensity when comparing the FTIR spectra of the original biological extract and the produced AgNPs. For instance, the existence of an amide band in the AgNPs spectrum verifies the function of proteins as stabilizing agents, whereas a drop in the strength of the –OH band or a shift in the C=O band following synthesis suggests the oxidation of phenolic compounds during Ag^+ reduction [8]. The evidence that differences in the concentration and composition of biological extracts influence not only the quantity and size of AgNPs formed but also the type of biomolecules adsorbed on the nanoparticle surface and their colloidal stability is thus strengthened by the complementary combination of UV–vis and FTIR data.

5.3 pH

One of the reaction parameters that significantly affects the production of AgNPs is the pH of the solution, particularly in green synthesis techniques that use biological extracts as capping and reducing agents. The ionization level of functional groups of bioactive compounds, such as –OH, –COOH, and $-NH_2$ found in phytochemicals or biomolecules, is directly impacted by pH changes. This determines the compounds' capacity to reduce silver ions (Ag^+) and stabilize the resulting nanoparticles. The majority of functional groups are protonated at acidic pH levels, which limits nucleation and dominant particle growth by reducing the electron donor capacity and slowing down the Ag^+ reduction process [34]. This condition frequently produces larger AgNPs with a wider size distribution and a higher tendency to agglomerate. Deprotonation of hydroxyl and carboxylic groups, on the other hand, increases the negative charge density of biomolecules at neutral to basic pH [19]. This strengthens the reduction ability and electrostatic interactions with Ag^+ ions, which encourages the formation of numerous nanoparticle nuclei and yields smaller and more uniform AgNPs.

Furthermore, by raising the negative zeta potential, which creates repulsive interactions between nanoparticles and prevents agglomeration, alkaline pH further improves colloidal stability. However, an excessively high pH can lead to an abnormally fast reduction process, which could result in the production of aggregates or silver oxide deposits that lower the quality of the nanoparticles. Therefore, pH adjustment is a crucial step in producing nanoparticles with the desired qualities, since it controls the equilibrium between nucleation and particle growth as well as the size, shape, stability, and optical features of AgNPs [21]. The ionization of phytochemical functional groups is influenced by pH:

- Low pH (acidic) causes a lot of protonated groups, a reduced ability to donate electrons, and delayed reduction processes, which frequently lead to larger particles and significant agglomeration.

- Basic pH leads to quick reduction and intense nucleation, which tends to create smaller, more homogeneous particles; higher electron-donating and chelating capability; and increased –OH/–COOH deprotonation.

Changes in the UV–vis spectrum, particularly the shift in the position and intensity of the SPR peak, which is the primary indicator of the formation and optical properties of AgNPs, make it easy to see how solution pH affects the synthesis of AgNPs. The UV–vis spectrum typically exhibits a broadened SPR peak and shifts toward a longer wavelength (red shift) under acidic pH conditions. This indicates the formation of relatively large nanoparticles or the occurrence of agglomeration because biomolecules in protonated conditions have weak reduction and stabilization capabilities [35]. Furthermore, the SPR peak's intensity tends to be lower at low pH, which reflects the comparatively small number of nanoparticles produced as a result of the restricted nucleation rate. On the other hand, increased deprotonation of functional groups like –OH and $-COO^-$ in bioactive compounds improves the reduction and stabilization capabilities at neutral to alkaline pH, which speeds up the formation of nanoparticle nuclei and produces smaller AgNPs with a more uniform distribution. This state is shown in the UV–vis spectrum by an increase in the SPR peak's strength and a blue shift in the peak's direction toward shorter wavelengths, which is typically linked to smaller particle sizes and improved colloidal stability. However, an excessively high pH can cause the SPR peak to widen once again or become unstable because of an excessively quick reduction reaction and the potential for aggregates or other silver species to form, which lowers the nanoparticles' homogeneity. As a result, the examination of the SPR peak shift in the UV–vis spectrum offers compelling experimental proof that solution pH is crucial in regulating the size, distribution, and stability of AgNPs by affecting the particle nucleation and growth mechanisms.

5.4 Reaction time and temperature

In both chemical and green synthesis methods based on biological sources, temperature and reaction time are important kinetic parameters in the synthesis of AgNPs. By raising the kinetic energy of the reactant molecules and the effective collision frequency, increasing the reaction temperature typically speeds up the reduction rate of silver ions (Ag^+) to metallic silver (Ag), which in turn accelerates the creation of nanoparticle nuclei (nucleation) [36]. Larger AgNPs with a wider size distribution are frequently produced at relatively moderate temperatures because the reduction process is uneven and sluggish, producing few nuclei and allowing particle expansion to dominate [6]. In contrast, the reduction rate increases, and nucleation occurs more rapidly and evenly at moderate to high temperatures, which tends to result in smaller and more homogeneous AgNPs. However, due to increased particle mobility and the degradation of biological stabilizing agents, extremely high temperatures can lead to excessive nanoparticle growth, sintering, or agglomeration, which eventually reduces colloidal stability [36]. The equilibrium between nucleation and nanoparticle growth is largely determined by reaction time in addition to temperature. The AgNP core formation predominates in the first stage of the reaction, which is indicated by a shift in the color of the solution and the emergence of a SPR peak in the UV–vis spectrum. The addition of Ag atoms to the core surface or the coalescence of tiny particles, which can result in an increase in particle size if the reaction is allowed to proceed for an extended period of time, are two ways that nanoparticle development occurs as reaction time increases. While an excessively long reaction time may result in agglomeration and reduce colloidal stability, an excessively short reaction time may result in unstable or extremely small AgNPs. To regulate the reduction kinetics, the number of nuclei generated, and the development rate of nanoparticles to produce AgNPs with the proper size, shape, and stability for their intended use, simultaneous tuning of temperature and reaction time is crucial. High temperatures can cause particle growth and sintering (enlargement), but they can also speed up reaction kinetics and nucleation. The equilibrium between nucleation and growth is impacted by reaction time; an extended reaction can lead to an increase in size and agglomeration.

The creation and properties of AgNPs are significantly influenced by changes in temperature and reaction time, according to the interpretation of experimental data. Particularly at short reaction times (5–15 min), the UV–vis spectrum typically displays a SPR peak with low intensity and a relatively large peak width at low temperatures around 30 °C [28, 36]. This suggests that the Ag^+ ion reduction process is slow, and the number of nanoparticles formed is still limited.

The intensity of the SPR peak increases when the reaction time is extended to 60–120 minutes at this temperature. However, this is frequently accompanied by peak broadening or a shift toward longer wavelengths (red shift), which indicates the growth of larger particles because the growth process predominates over nucleation.

AgNP production is more effective at moderate temperatures, such as 50–70 ° C. A sharper and more powerful SPR peak at a wavelength of around 400–420 nm is typically seen in UV–vis spectra taken over a period of 15–60 minutes, indicating the creation of relatively tiny AgNPs with a narrower size distribution. This condition is frequently regarded as the ideal one for synthesis since it shows a good balance between the nucleation rate and nanoparticle growth [37, 38]. The SPR peak may, however, somewhat broaden if the reaction time is increased to 120 minutes at this temperature, signifying the beginning of agglomeration as a result of ongoing particle development.

Ag^+ reduction occurs very quickly at high temperatures (80–90 °C), and the SPR peak develops rapidly (5–10 min) with great intensity, indicating the large-scale and swift synthesis of AgNPs. However, at longer reaction times (≥60 min), the UV–vis spectra frequently exhibit peak broadening or a shift toward longer wavelengths, indicating agglomeration or sintering of nanoparticles[28] due to increased particle mobility and potential breakdown of biological stabilizers. These findings suggest that temperature and reaction time should be adjusted concurrently, with moderate reaction periods and intermediate temperatures producing AgNPs with optimal size, stability, and optical properties.

5.5 Extract: Mixing conditions and precursor ratio

In the production of AgNPs, the ratio of biological extracts to silver precursors is crucial because it establishes the equilibrium between the concentration of silver ions (Ag^+) to be reduced and the quantity of stabilizing and reducing agents that are available. The reduction process is slow and tends to produce larger AgNPs with low colloidal stability and a high tendency to agglomerate at low extract-to-precursor ratios because the amount of bioactive compounds, such as polyphenols, flavonoids, proteins, and polysaccharides, is frequently insufficient to effectively reduce all Ag^+ ions or coat the surface of the formed nanoparticles. Conversely, raising the extract-to-silver precursor ratio makes more reducing and capping agents available, which speeds up the creation of nanoparticle nuclei and enhances surface stability, producing smaller and more homogeneous AgNPs [39]. However, employing an excessively high extract ratio may result in an overabundance of organic molecules in the system, which may form an excessively thick capping layer or increase the viscosity of the reaction medium, preventing the diffusion of Ag^+ ions and resulting in a reduction in reaction efficiency or the formation of organic–metal aggregates. The results of AgNPs synthesis are influenced not only by the reactant ratio but also by mixing parameters such as stirring speed, solution homogeneity, and reactant addition technique. In addition to improving mass transfer and preventing the creation of concentration gradients that could lead to nonuniform particle growth, adequate and even stirring promotes the interaction between Ag^+ ions and active biomolecules [35]. Conversely, insufficient mixing may cause excessive local nucleation or particle agglomeration in specific regions, producing AgNPs with a broad size distribution. In order to control the reduction kinetics, nucleation–growth balance, and colloidal stability of AgNPs and produce nanoparticles with ideal physicochemical properties and biological activity, it is crucial to properly optimize the extract-to-silver precursor ratio in conjunction with efficient mixing conditions. The reducing agent's stoichiometry with respect to the metal ion is determined by the extract's volume/mass ratio to the $AgNO_3$ solution [40]. Stirring well reduces concentration gradients that might lead to changes in particle size and enhances homogeneity and mass transfer.

The changes in the properties of the AgNPs created at ratios of 1:1, 1:5, and 1:10 can be easily observed based on the interpretation of the fluctuation in the ratio of biological extract-to-silver precursor. The number of bioactive compounds that function as capping and reducing agents is comparatively small in relation to the amount of accessible Ag^+ ions at low ratios, such as 1:1, which causes nucleation to occur only partially and the reduction process to proceed more slowly [5, 41]. Due to the dominance of the particle growth process and the propensity for agglomeration, this condition is frequently reflected in the UV–vis spectrum by SPR peaks with low intensity and relatively large peak widths, which indicate the formation of larger AgNPs and a nonuniform size distribution.

The equilibrium between the concentrations of the reducing agent, capping agent, and Ag^+ ions is typically better at intermediate ratios, such as 1:5. The development of smaller AgNPs with a narrower size distribution and improved colloidal stability is typically indicated by a sharper and more intense SPR peak at a wavelength of around 400–420 nm in the UV–vis spectrum at this ratio. A ratio of 1:5 is frequently stated as the ideal condition in many green synthesis methods because this condition reflects the balanced nucleation and development of nanoparticles [41].

On the other hand, at greater extract ratios, like 1:10, there are so many active biomolecules available that Ag^+ reduction can happen very quickly, and nanoparticle nucleation is greatly increased. The UV–vis spectra frequently exhibit SPR peak broadening or a minor shift toward longer wavelengths, indicating interparticle interactions or the production of an abnormally thick organic capping layer, even though these conditions can yield AgNPs with extremely small sizes. Additionally, if mixing conditions are not ideal, the increased viscosity of the medium caused by excess extract can impede ion diffusion and reduce system homogeneity. Therefore, to guarantee a narrow AgNP size distribution, strong colloidal stability, and consistent optical characteristics, the extract-to-silver precursor ratio should be tuned together with the stirring settings.

5.6 Lighting (light-assisted synthesis)

The process and kinetics of AgNP production can be greatly impacted by lighting or light-assisted synthesis, particularly in green synthesis systems that use biological extracts as reducing agents. In biological extracts like polyphenols, flavonoids, and natural pigments, exposure to visible and ultraviolet light can activate photoactive compounds. This increases the formation of reactive species or excited electrons that contribute to the reduction of silver ions (Ag^+) to metallic silver (Ag^0). In the presence of light, the electron transfer process proceeds more quickly than in the absence of light, which raises the nucleation rate and speeds up the production of nanoparticle nuclei. The development of SPR peaks in the UV–vis spectrum in a shorter reaction time, together with an increase in peak strength, suggesting a greater number of AgNPs generated, is frequently used in experiments to illustrate this phenomenon [8, 20].

Lighting can influence the size and distribution of AgNPs in addition to speeding up reduction. Faster nucleation brought on by photochemical activation often results in smaller and more homogeneous nanoparticles. However, light intensity and type must be carefully regulated because excessive or extended light exposure can lead to biological capping compound overgrowth, aggregation, or even photodegradation, which lowers colloidal stability. Therefore, light-assisted synthesis is a useful supplementary method for regulating the reaction kinetics and attributes of AgNPs. Nevertheless, in order to generate nanoparticles with ideal size, stability, and optical properties, lighting conditions must be optimized. Certain methods use visible and UV light to accelerate the rate of reduction (photo-assisted green synthesis). Particle size and shape can be affected by photolysis of phytochemical substances [42], which promotes the production of free radicals that can reduce Ag^+ more quickly.

There are noticeable variations in the reaction rates, particle size, and optical characteristics of the final AgNPs when synthesis is done in the dark versus with light assistance. The rate of AgNP formation is comparatively slower and requires a longer reaction time to achieve significant SPR peak intensity in the UV–vis spectrum because the reduction process of silver ions (Ag^+) in the dark is entirely dependent on the chemical or biochemical activity of reducing compounds in biological extracts. Due to the dominance of the particle development process, the SPR peak generated under these conditions often has a greater peak width and a lower intensity, indicating the creation of AgNPs with a comparatively larger size and a less uniform size distribution. Photochemical activation of bioactive chemicals in biological extracts using visible or ultraviolet light speeds up electron transfer and the reduction of Ag^+ to Ag_4 in light-assisted synthesis. The creation of smaller and more uniform AgNPs is shown experimentally by the SPR peak appearing faster, even at short reaction times, with increased peak intensity and a peak position that is frequently shifted toward shorter wavelengths (blue shift) [36]. The UV–vis spectrum may, however, exhibit an expansion of the SPR peak or a shift toward longer wavelengths (red shift) if the light is delivered excessively or with too much intensity. This could indicate agglomeration or colloidal instability brought on by the biological capping agent's deterioration. As a result, this comparison demonstrates that light may effectively regulate the synthesis of AgNPs as long as the intensity and duration of the light are adjusted to strike a balance between the pace of reduction and the stability of the nanoparticles.

5.7 Comparing the effectiveness of green synthesis and traditional techniques on AgNPs' function

Green-synthesized AgNPs frequently exhibit comparable functional properties and, in certain applications (antimicrobial, biocompatibility, stability in biological matrices), outperform conventionally synthesized AgNPs, according to a number of comparative studies. The size and form of the NP, the type of capping ligand, the release of Ag^+, and the testing procedures (concentration, microbial strain, medium) all affect the outcome. Performance varies, according to recent meta-analytic and review findings; however, there is enough data to suggest superiority or suitability for particular applications. When studying AgNPs as an antibacterial, green AgNPs' inhibition zone and minimum inhibitory concentration (MIC) against Gram-positive and Gram-negative are equivalent to or superior to those of chemical AgNPs of the same size. Mechanisms include the following: (i) The NP physical contact leading to membrane breakdown; (ii) Ag^+ release leading to interactions with proteins and redox systems; and (iii) ROS generation. By slowing down the release of

Ag^{+} or increasing binding to the cell wall, biocapping can modify selective toxicity. Green AgNPs with phytoactive capping occasionally exhibit higher antibacterial activity at lower concentrations, according to experimental comparison examples.

AgNPs function as co-catalysts in pigment degradation or as catalysts for reduction reactions (such as the conversion of 4-nitrophenol to 4-aminophenol) in catalytic/photocatalytic processes. Surface accessibility (thick capping might impede substrates) and surface area (smaller size is better) determine catalytic activity. Green AgNPs with loose or porous capping maintain excellent catalytic activity, according to several studies; however, activity may be somewhat decreased in comparison to chemically "naked" NPs if the capping is excessively thick or polar. Therefore, the capping design affects the outcome for pure catalysis.

Green AgNPs frequently exhibit higher biocompatibility profiles (fibroblast/hepatocyte cells) in biomedical applications (antitumor, antibiofilm, biocompatibility) at particular doses because they possess a biomolecular coating that reduces direct harmful interactions. Green AgNPs have demonstrated promising anticancer/antibiofilm effects in several *in vitro* studies and early *in vivo* trials; nevertheless, compared to AgNPs produced industrially, there remains a lack of comprehensive preclinical testing. Regulations, systemic toxicity, and process requirements continue to pose challenges for clinical applications.

6. Characterization of green synthesis–produced nanosilver

Several characterization methods were employed to guarantee the effective synthesis and functional characteristics of AgNPs as follows.

6.1 UV–vis spectroscopy

The most popular and crucial initial characterization method for verifying the successful synthesis of AgNPs from green synthesis is UV–vis spectroscopy, which can quickly provide information on the formation of nanoparticles, their optical characteristics, and an indication of their size and colloidal stability. Localized surface plasmon resonance (LSPR) [32], or the collective oscillation of free electrons on the surface of AgNPs when they interact with electromagnetic radiation in the ultraviolet to visible light region, is the fundamental idea behind this method. The UV–vis spectra of effectively manufactured AgNPs often display a distinctive absorption peak in the 380–450 nm wavelength region, which clearly confirms that AgNPs have formed. The size and shape of AgNPs have a significant impact on the location of the LSPR peak; smaller nanoparticles typically exhibit an absorption peak at a shorter wavelength, whereas bigger or aggregated particles exhibit a peak shift toward a longer wavelength (red shift) [35]. A rise in LSPR peak intensity with reaction time or increasing reducing agent concentration can be interpreted as an increase in the number of AgNPs in the system since absorption intensity, in addition to peak position, represents the relative amount of nanoparticles generated. Important details regarding the particle size distribution can also be found in the width of the absorption peak; a broad peak denotes size heterogeneity or the beginning of agglomeration, whereas a narrow peak shows a more uniform size distribution. Changes in UV–vis spectral features are also frequently utilized in the context of green synthesis to assess the effects of reaction parameters such as pH, temperature, reaction duration, extract-to-precursor ratio, and lighting. Therefore, UV–vis spectroscopy is not only a crucial indicator for comprehending the synthesis dynamics and refining the reaction conditions to produce nanosilver with the required size, stability, and optical properties [43], but it also functions as a confirmation instrument for the formation of AgNPs.

The features of the LSPR peak were strongly impacted by changes in experimental circumstances, according to an interpretation of the UV–vis spectra of green synthesized–produced nanosilver. For instance, the UV–vis spectra typically showed a relatively broad LSPR peak with low intensity at a wavelength of about 430–440 nm under initial synthesis conditions with acidic pH (pH 4–5) and low temperature (30 °C), suggesting that the reduction process of Ag^+ ions was slow and produced larger nanoparticles with a nonuniform size distribution [44]. The LSPR peak became sharper and more intense when the pH was raised to neutral or slightly alkaline conditions (pH 7–9) at intermediate temperatures (50–70 °C). It also shifted to shorter wavelengths, such as 400–415 nm, indicating the formation of smaller and more homogeneous AgNPs as a result of increased nucleation rates and colloidal stabilization [2].

Reaction time variations also have a noticeable impact; at short reaction times (5–15 minutes), the LSPR peak intensity is still low due to the limited number of nanoparticles formed, whereas at medium reaction times (30–60 minutes), the peak intensity increases significantly without significant broadening, indicating ideal conditions for AgNPs formation. However, the UV–vis spectrum frequently exhibits peak broadening or a faint red shift during very long reaction periods (≥120 minutes), indicating ongoing nanoparticle development or aggregation. The UV–vis spectrum also reflects the comparison of the extract-to-silver precursor ratio, with medium ratios (e.g., 1:5) producing the sharpest and most stable LSPR peaks in contrast to low ratios (1:1), which exhibit low intensity, and high ratios (1:10), which typically show peak broadening due to the excess organic layer.

6.2 Electron microscopy (TEM/SEM)

Silver nanoparticles made by green synthesis can be directly observed for size, shape, and distribution using electron microscopy, particularly scanning electron microscopy (SEM) and transmission electron microscopy (TEM). By creating high-resolution images of the sample surface, SEM is mainly utilized to visualize the surface morphology, aggregate shape, and degree of agglomeration of AgNPs. The SEM pictures of green synthesized–produced AgNPs often reveal particles with a mostly spherical or semi-spherical shape; albeit, depending on the biological extract's content and the synthesis circumstances, different shapes like polygonal or irregular can be detected in some systems. Additionally, SEM offers qualitative data on the degree of agglomeration; low agglomeration shows how well the biomolecules from plant or microbial extracts function as natural capping agents to stop nanoparticle clumping [1]. However, the measured particle sizes are frequently aggregate or surface measurements rather than exact individual particle sizes because of the low resolution of SEM at the very small nanoscale. On the other hand, because TEM can provide resolution down to the atomic scale, it is a better method for figuring out the true size, size distribution, and internal structure of AgNPs. Green-synthesized AgNPs usually show up in TEM images as distinct particles with clearly defined borders and precisely quantifiable sizes, frequently between 5 and 50 nm [1, 45]. The TEM images' consistent particle shape and limited size dispersion show that the biomolecules in the biological extracts effectively stabilize the particles throughout synthesis and that the nucleation and growth of nanoparticles are regulated. Additionally, TEM enables the observation of AgNPs' crystalline structure through high-resolution TEM (HRTEM) images and selected area electron diffraction (SAED) patterns, which usually show rings or diffraction spots typical of silver's face-centered cubic (fcc) structure, thereby confirming the nanoparticles' crystalline nature. A complete image of the physical properties of AgNPs is provided by the combination of SEM and TEM results; SEM emphasizes the morphology and surface conditions, while TEM offers quantitative information on size, shape, and crystallinity. As a result, electron microscopy offers compelling visual proof of the efficacy of green nanosilver synthesis and establishes a connection between the synthesis circumstances and the structural properties of the final nanoparticles.

The successful synthesis of AgNPs by green synthesis is confirmed by the characterization results using electron microscopy (SEM and TEM), which exhibit a significant connection with UV–vis and FTIR spectroscopy data. The formation of small and relatively uniform AgNPs is indicated by a sharp LSPR peak in the UV–vis spectrum in the range of 380–420 nm. This is consistent with the findings of TEM images that show individual spherical particles with dominant sizes, such as in the range of 10–30 nm. While the peak broadening or shift toward longer wavelengths (red shift) [35] in the UV–vis spectrum coincides with the SEM/TEM image indicating bigger particles or agglomeration, the narrow and intense LSPR peak is also compatible with the homogeneous size distribution in the TEM image. Additionally, the SEM image demonstrating a minimal degree of agglomeration confirms the steady UV–vis results with no discernible changes in the LSPR peak position throughout the course of the observation period, indicating good colloidal stability of AgNPs.

The presence of phytochemical compounds or biomolecules on the nanoparticle surface was confirmed by FTIR analysis, where the AgNPs spectrum showed distinctive absorption bands of –OH/–NH groups (about 3200–3400 cm^{-1}), C=O or amide (about 1600–1650 cm^{-1}), and C–O (about 1000–1300 cm^{-1}). Because the biomolecules serve as capping agents that offer steric and electrostatic stability, the existence of this biomolecular layer explains the SEM/TEM data showing highly scattered particles with low agglomeration [45]. The interpretation that the morphological and size stability of AgNPs seen in SEM/TEM results from the adsorption of biomolecules on the nanoparticle surface is thus supported by the FTIR data. Overall, complementary experimental evidence that AgNPs have been successfully formed through green synthesis with controlled size, stability, and biomolecular surface coating is provided by the agreement between the LSPR peaks in UV–vis, the morphological and size images from SEM/TEM, and the identification of functional groups by FTIR. Size, shape (spherical, triangular, rod-shaped, hexagonal), and crystallinity are all detailed by transmission electron microscopy (TEM), which can be used in conjunction with SAED. Surface morphology and aggregation are visible using scanning electron microscopy (SEM).

6.3 X-ray diffraction (XRD)

An essential characterization method for verifying the crystal structure, crystalline phase, and degree of crystallinity of AgNPs produced via green synthesis is X-ray diffraction (XRD). The foundation of XRD is the interaction of X-rays with a material's crystal planes, which, in accordance with Bragg's rule, results in a distinctive diffraction pattern. The XRD pattern of green-synthesized AgNPs typically shows distinctive diffraction peaks of metallic silver with a fcc structure at 2θ angles of roughly 38°, 44°, 64°, and 77°, which correspond to the (111), (200), (220), and (311) crystal planes [5], respectively. The successful reduction of silver ions (Ag^+) to crystalline zero-valent silver (Ag^0) is confirmed by the existence of these peaks. The breadth of the diffraction peaks in the XRD pattern not only identifies the crystalline phase but also provides crucial information on the size of the nanoparticles' crystallites; comparatively broad peaks indicate small crystallite sizes, which are typical of nanoscale materials. The Scherrer equation [46] is frequently used to estimate the crystallite size of AgNPs. The resulting crystallite size is typically in the range of a few to tens of nanometers and is consistent with the findings of electron microscopy investigations [47]. The primary Ag peak in green synthesis systems remains prominent, showing good purity of the silver phase, but XRD patterns may also display other low-intensity peaks that could be related to leftover biomolecules or amorphous phases of biological capping agents. In order to relate the synthesis conditions to the physical characteristics and functional activity of the final nanosilver, XRD analysis is essential for confirming the crystalline structure of AgNPs as well as providing quantitative data on crystallite size and crystal quality.

The Scherrer equation, which is written as follows, can be used to estimate the crystallite size of AgNPs produced via green synthesis from XRD data:

$$\boldsymbol{D} = \frac{\boldsymbol{K\lambda}}{\boldsymbol{\beta \cos \theta}}$$

with the following details:
D = stands for crystallite size (nm).
K = the form factor, often 0.9.
λ = X-ray wavelength (Cu Kα = 0.15406 nm) in radians.
β = the peak width at half maximum (FWHM).
θ = Bragg angle (½ of 2θ) in radians.
For example, the XRD results of AgNPs obtained:

- Main peak at 2θ = 38.1° in plane (111).
- FWHM value (β) = 0.80°.

Steps in calculation

1. FWHM conversion from degrees to radians:

$$\beta = 0,80^{\circ} \times \frac{\pi}{180} = 0,01396 \text{ rad}$$

2. Calculate the angle θ:

$$\theta = \frac{38,1^{\circ}}{2} = 19,05^{\circ}$$

3. Convert to radians.

$$\theta = 19,05^{\circ} \times \frac{\pi}{180} = 0,3325 \text{ rad}$$

4. Enter it into the Scherrer formula:

$$\mathbf{D = 0,9 \times 0,1540 \frac{6}{0,01396 \times \cos(0,3325)}}$$

$$\cos(0,3325) = 0,945$$

$$D = (0,13865)/(0,01319)$$

$$D \approx 10,5 \text{ ''} nm \text{ ''}$$

The green synthesized–produced AgNPs have a nano-sized crystalline structure, with a crystallite size of roughly 10.5 nm determined by XRD analysis using the Scherrer equation. This figure is similar to the particle size shown in TEM pictures, suggesting that one or more tiny crystallites make up each nanoparticle. The existence of a biomolecular capping layer, identified by FTIR analysis, is responsible for the slight discrepancy between the crystallite and particle sizes.

6.4 FTIR spectroscopy

Fourier transform infrared spectroscopy is a crucial characterization technique in the study of green synthesis nanosilver because it can identify functional groups of biomolecules directly involved in the reduction process of silver ions (Ag^+) and stabilization of the nanoparticle surface as a capping agent. Polyphenols, flavonoids, proteins, polysaccharides, and organic acids are among the bioactive substances found in biological extracts from plants, microorganisms, and algae that interact with the surface of AgNPs in green synthesis (see **Table 1**). AgNPs typically exhibit a distinctive absorption band in the 3200–3400 cm^{-1} wavenumber region, which is linked to the stretching of –OH and/or –NH groups. This indicates the presence of phenolic chemicals and proteins that function as stabilizers and reducing agents. The absorption band at around 1600–1650 cm^{-1} is frequently linked to aromatic C=C bonds or C=O stretching (amide I), suggesting that proteins or phytochemical aromatic structures are involved in the synthesis and coating of nanoparticles. Furthermore, the presence of polysaccharides and other organic compounds on the surface of AgNPs is confirmed by the bands in the 1000–1300 cm^{-1} range associated with C–O or C–N stretching. Strong evidence that these functional groups are involved in the Ag^+ reduction process and adsorption on the nanoparticle surface can be found by comparing the FTIR spectra of biological extracts prior to synthesis with the FTIR spectra of AgNPs following synthesis. This comparison frequently reveals a shift in the band position or a change in the absorption intensity. In addition to identifying biomolecules attached to AgNPs, FTIR explains the colloidal stability and biocompatibility of nanosilver made via green synthesis and offers a mechanistic understanding of the function of biological compounds as reducing and capping agents [8, 22].

6.5 Dynamic light scattering (DLS) and zeta potential

Important characterization methods for assessing the hydrodynamic size, size distribution, and colloidal stability of green-synthesized AgNPs in a liquid medium are dynamic light scattering (DLS) and zeta potential analysis. In order to obtain the hydrodynamic diameter value, the effective size of the nanoparticle plus the

No	Wave number (cm^{-1})	Functional groups	Biomolecule sources	Interpretation
1	3200–3400	–OH (hydroxyl), –NH (amine)	Phenols, flavonoids, alcohol, proteins	Serves as a capping agent and an electron donor in the reduction of Ag^+; interaction with the AgNPs' surface is indicated by the decreasing shift/intensity
2	2,920– 2,850	–CH (aliphatic)	Organic compounds, lipids, proteins	The nanoparticles are coated with biomolecular carbon chains
3	1,630–1650	C=O (amide I)/ C=C aromatic	Protein, flavonoid	Proteins or aromatic molecules are involved in the stability of AgNPs
4	1,540–1,560	N–H bending (amide II)	Proteins, enzymes	Illustrates how proteins function as a natural capping agent
5	1,380–1,450	C–N (amine), $–CH_3$/$–CH_2$ bending	Alkaloids, proteins, organic compounds	Verifying the existence of organic biomolecules on AgNP surfaces
6	1,240–1300	C–O (ester/ phenol), C–N	Polyphenols, flavonoids	Shows how phenolic chemicals are involved in stability and reduction
7	1,020–1,120	C–O (alcohol, polysaccharide)	Polysaccharides, sugars	Gives AgNPs steric stability
8	600–700	Ag–O/Ag–N bond (weak)	Ag–biomolecule interactions	Indicates the bond between AgNPs and capping biomolecules

Table 1.
Analysis of nanosilver (AgNPs) FTIR bands from green synthesis.

surrounding biomolecular layer and solvation shell is considered [38]. The DLS measures variations in the intensity of light scattered by nanoparticles undergoing Brownian motion in suspension. Because DLS accounts for the presence of a capping layer of biological agents like polyphenols, proteins, and polysaccharides adsorbed on the nanoparticle surface, the hydrodynamic size obtained from DLS in green-synthesized AgNPs is typically larger than the particle size measured using TEM [45]. A high polydispersity index (PDI) value suggests size heterogeneity or the presence of particle agglomeration, whereas a narrow size distribution with a low PDI value (<0.3) indicates that the synthesis process produces relatively homogeneous and well-dispersed nanoparticles.

Zeta potential testing provides important details about the surface charge and colloidal stability of AgNPs, in addition to their size. The type of biomolecule functional groups covering the surface of the nanoparticle has a significant impact on the zeta potential, which is the electrical potential at the particle's shear plane. The AgNPs in green synthesis typically exhibit either positive or negative zeta potential values, depending on whether the capping biomolecules' $–COO^-$, –OH, or $–NH_3^+$ groups predominate. Because electrostatic repulsion between nanoparticles inhibits agglomeration, a relatively high absolute value of zeta potential, for instance, more than ± 25–30 mV, indicates strong colloidal stability [48]. Conversely, a colloidal system that is less stable and may undergo precipitation or clumping during storage is indicated by a zeta potential value near zero. Therefore, a complete picture of the dispersion conditions, effective size, and stability of AgNPs in liquid media is provided by the combination of DLS and zeta potential results. This reinforces the role of biological biomolecules as capping agents that not only regulate the size but also determine the stability and potential applications of green-synthesized nanosilver.

The green synthesized nanosilver had an average hydrodynamic diameter of 80 nm and a PDI value of 0.22, according to the characterization results using DLS. This suggests that the nanoparticles are distributed rather uniformly in the liquid medium. A limited size distribution and a relatively homogeneous colloidal system with little agglomeration are produced by the synthesis process when the PDI value is less than 0.3. The larger hydrodynamic diameter compared to the particle size obtained from TEM analysis (e.g., 15–30 nm) can be explained by the presence of a biomolecular layer from the biological extract that acts as a capping agent, as well as a water solvation shell that is also measured in the DLS method. This confirms that bioactive compounds such as polyphenols, proteins, and polysaccharides not only stabilize the nanoparticles but also contribute to the effective particle size in suspension.

Furthermore, the findings of the zeta potential measurement revealed a value of - 32 mV, indicating that the colloidal system is stable and the AgNPs' surface is negatively charged. In order to create electrostatic repulsion between particles and prevent agglomeration during storage, zeta potential values with an absolute magnitude greater than 30 mV are typically thought to be adequate [49]. Deprotonated functional groups like $-COO^-$ and $-O^-$ from phenolic compounds or organic acids, as identified by FTIR analysis, are most likely the source of this negative charge. The green-synthesized AgNPs exhibit good dispersion, high colloidal stability, and an efficient biomolecular capping layer, which supports their potential use in biomedical and environmental applications. This is demonstrated by the combination of a hydrodynamic diameter of 80 nm, a PDI of 0.22, and a zeta potential of - 32 mV.

7. Utilizing green synthesis for nanosilver

7.1 Antimicrobial and antibiofilm activity

Green-synthesized nanosilver (AgNPs) possesses strong antibacterial and antibiofilm properties, which make it an excellent option for use in the food, pharmaceutical, medical, and environmental industries. The primary benefit of green-synthesized AgNPs is the combination of silver's inherent antibacterial properties with a coating of active biomolecules, such as proteins, polyphenols, and flavonoids derived from biological extracts, that are affixed to the surface of the nanoparticles as a capping agent. By disrupting the integrity of cell membranes, increasing cell wall permeability, and releasing Ag^+ ions that interact with proteins, enzymes, and microbial DNA, AgNPs can mechanistically inhibit the growth of a variety of pathogenic microorganisms, including Gram-positive and Gram-negative bacteria, fungi, and some viruses [24]. While the biomolecular coating from green synthesis can enhance affinity to the cell membrane and decrease nonspecific toxicity to mammalian cells, AgNPs' nanoscale size enables them to engage with the microbial cell surface and pierce its protective barrier. The AgNPs exhibit strong antibiofilm properties in addition to planktonic antimicrobial activity [50]. These include preventing biofilm development from the first adhesion stage, causing damage to the extracellular polymer matrix (EPS), and interfering with microbial cell communication (quorum sensing) (see **Figure 2**). The AgNPs' capacity to infiltrate and break biofilm structures offers a significant therapeutic benefit because biofilms are known to be highly resistant to traditional antibiotics.

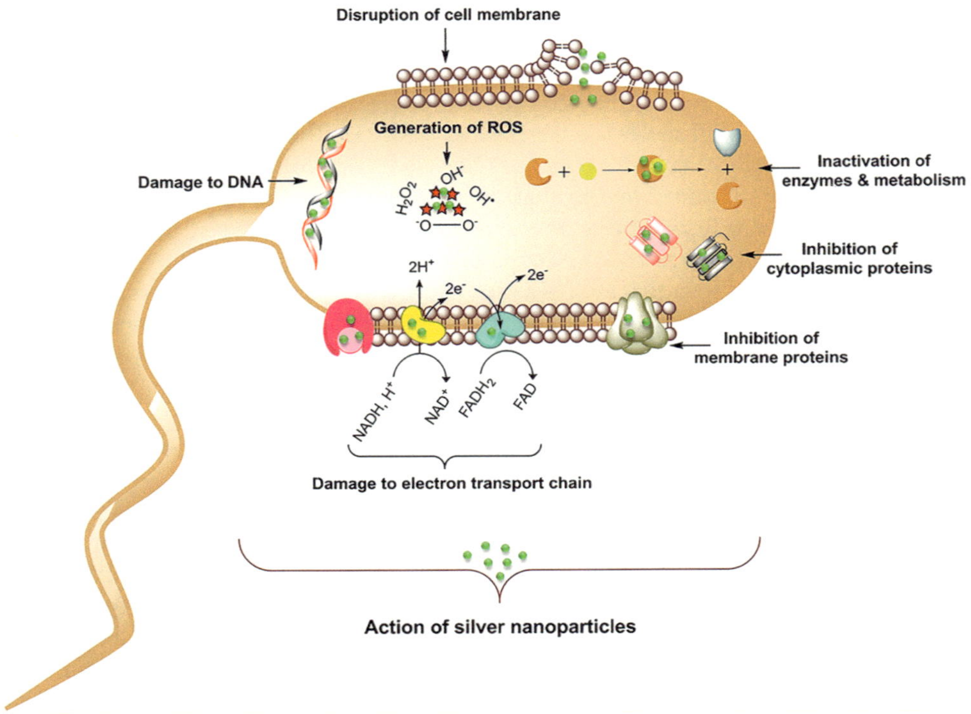

Figure 2.
Antimicrobial and antibiofilm activity of AgNPs.

Green-synthesized AgNPs can lower cell viability in established biofilms and stop the creation of new biofilms on medical surfaces such as catheters, implants, and wound dressings, according to experimental research. Therefore, using green-synthesized nanosilver as an antimicrobial and antibiofilm agent provides an efficient, long-lasting, and possibly multifunctional way to lower microbial resistance; however, dosage optimization and long-term safety assessments are still required for clinical and industrial applications. According to numerous reports, green-synthesized AgNPs have the following properties:

1. The ability to combat both Gram-positive and Gram-negative bacteria.

2. Antiprotozoal and antifungal properties.

3. The capacity to prevent the production of biofilms.

It is interesting to note that the synergistic impact between AgNPs and these phytochemical compounds can be enhanced by the presence of bioactive molecules on the surface (such as terpenoids and flavonoids), leading to stronger antibacterial activity than typical chemical AgNPs. Among the primary mechanisms are the following:

1. Disruption of cell membranes and alterations in permeability.

2. The production of ROS and oxidative damage.

3. Interaction with DNA and microbial proteins.

7.2 Applications for anticancer

Due to its distinct physicochemical characteristics and the contribution of natural biomolecules adhered to its surface, green-synthesized nanosilver (AgNPs) has garnered a lot of interest as a possible anticancer drug. The biomolecular capping layer, which includes proteins, polyphenols, and flavonoids, not only improves stability and biocompatibility but also offers specific cytotoxic effects against cancer cells. The nanoscale size enables AgNPs to interact with cancer cells effectively through enhanced cellular uptake [19]. According to reports, green-produced AgNPs cause oxidative stress by boosting the production of ROS in cancer cells (see **Figure 3**). This leads to damage to the mitochondria, depolarization of the mitochondrial membrane, and activation of the intrinsic death pathway. Lipid peroxidation, protein oxidation, and DNA damage are all brought on by ROS accumulation, which eventually halts the growth of cancer cells and encourages programmed cell death. Additionally, by arresting cells in particular stages [20] (such as G_0/G_1 or G2/M), AgNPs can interfere with the cancer cell cycle and prevent cell division.

The benefits of green synthesis are also evident in biological selectivity, as multiple studies have demonstrated that AgNPs with natural biomolecular capping are less toxic to normal cells than to cancer cells [20]. This may be because the two cell types have different redox and metabolic states. Additionally, because AgNPs' surfaces can be altered or interact with anticancer bioactive substances, they can function as therapeutic adjuvants, either in concert with traditional chemotherapy or as a drug

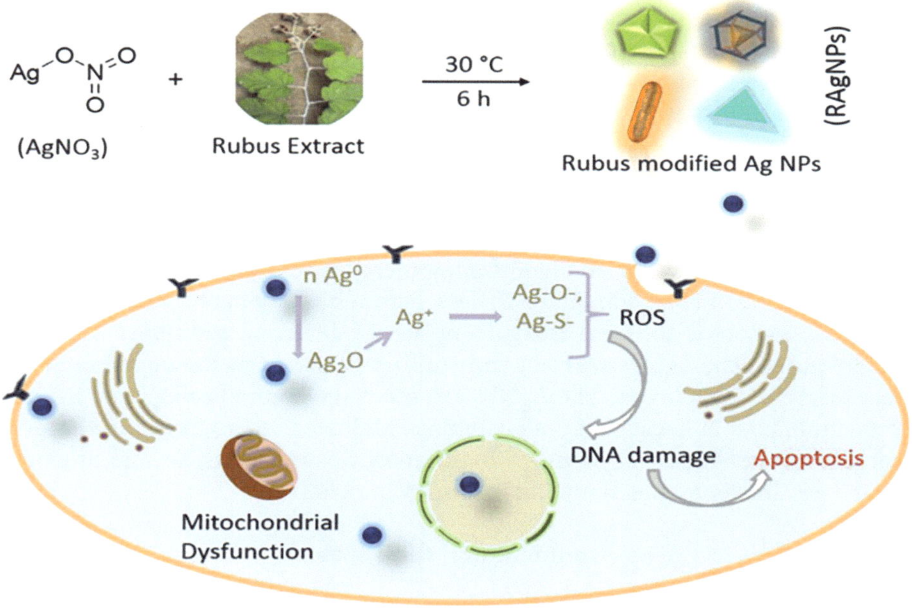

Figure 3.
Anticancer mechanism of AgNPs (ROS, apoptosis, cell cycle).

delivery system. Therefore, using green-synthesized nanosilver as an anticancer agent offers a promising multifunctional approach that combines proliferation inhibition, apoptosis induction, and potential selectivity. However, before widespread therapeutic application, long-term safety evaluation, a deeper understanding of the molecular mechanisms, and preclinical and clinical trials are still required.

7.3 Topical application and wound healing

Due to its superior biocompatibility profile and robust antibacterial activity compared to traditional chemically synthesized nanosilver, green-synthesized nanosilver (AgNPs) has demonstrated significant promise in topical applications and wound healing. Microbial infection and biofilm formation are important aspects of wound healing that impede tissue regeneration, extend the inflammatory phase, and raise the possibility of complications. Through a variety of processes, such as avoiding biofilm development on the wound surface, lowering the local microbial burden, and suppressing the growth of harmful bacteria, green-synthesized AgNPs can address these problems. The AgNPs can directly interact with bacteria in the wound region due to their nanoscale size, and the regulated release of Ag^+ ions offers a long-lasting antimicrobial action without overly harming skin cells. Furthermore, natural biomolecular coatings made from plant or microorganism extracts serve as capping agents that boost the stability of nanoparticles while also offering additional biological activities like anti-inflammatory and antioxidant effects, which are crucial for accelerating the healing process [3].

Green-synthesized AgNPs are commonly used in topical applications such as ointments, creams, gels, hydrogels, wound dressings, and polymer films that are intended to make direct, uniform contact with the skin's surface or wound tissue. By preserving moisture, lowering local inflammation, and promoting the growth of fibroblasts and keratinocytes, AgNP-based topical formulations not only act as a physical barrier against microbial contamination but also establish a microenvironment that promotes tissue regeneration. The AgNPs can speed up wound closure, promote granulation tissue formation, and improve collagen deposition in wound dressings, all of which enhance healing quality and reduce the chance of scarring, according to a number of studies. Because the natural biomolecule residues on the surface of AgNPs might lessen skin irritation and increase tolerance compared to nanoparticles stabilized by synthetic compounds [51], the benefits of green production are especially pertinent in topical applications. Therefore, green synthesized–nanosilver provides an efficient, secure, and long-lasting therapeutic strategy for topical application and wound healing; nevertheless, before being widely used in medical practice, assessment of the ideal dosage, long-term safety, and additional clinical trials are still required. The AgNPs are frequently utilized in bandages for wounds and topical antibacterial gels and creams. The AgNPs can speed up wound healing and prevent subsequent infections because of their combined antibacterial properties and capacity to promote fibroblast/keratinocyte migration. Nanosilver-containing wound dressing products are already on the market in a number of countries.

7.4 Applications in the environment and food sectors

Green-synthesized nanosilver (AgNPs) has broad potential for applications in the food and environmental sectors because it combines high antimicrobial activity with a more environmentally friendly and sustainable production approach. The AgNPs are extensively developed in the food industry as active ingredients in polymer composite materials, surface coatings, and active food packaging to increase product shelf life. The AgNPs' antimicrobial action lowers the risk of contamination and quality loss during storage by preventing the growth of pathogens and microorganisms that cause food spoilage, such as fungi and spoilage bacteria. The natural biomolecule coating that is affixed to the surface of green-synthesized AgNPs is advantageous because it can enhance compatibility with food polymer matrices [26] and lessen the possibility of excess silver ion migration into food [45]. Additionally, as these materials are biodegradable and lessen reliance on traditional plastics, the usage of AgNPs in biopolymer-based packaging like chitosan, starch, or cellulose supports the idea of sustainable food.

Because they can inactivate pathogenic microorganisms and biofilms, which are frequently the source of fouling in water filtration and distribution systems, they are used in the environmental field for water treatment, environmental microbial control, and pollutant remediation. The AgNPs can be incorporated into filter membranes or adsorbent materials to improve the effectiveness of water disinfection, prevent the formation of biofilms, and prolong the life of treatment systems [1]. The AgNPs can also serve as catalysts in the breakdown of harmful chemical compounds, including industrial dyes and persistent organic pollutants, according to a number of studies, which helps to improve the quality of the environment. Because it minimizes the use of hazardous chemicals during the creation of AgNPs, the green synthesis approach is highly relevant in an environmental context, reducing the risk of secondary pollution and the ecological impact. Although ecotoxicological risk assessment and usage regulation are still required to ensure long-term sustainability and safety, the application of green synthesis nanosilver in the food and environmental sectors offers an efficient multifunctional solution to enhance food safety and environmental quality.

The following are some instances of nanosilver uses in the food and environmental industries:

1. Packaging for food

   Adding AgNPs to polymer films (PVA, chitosan, and PLA) enhances antibacterial qualities and extends the shelf life of food products.

2. Treatment of waste

   a. AgNPs serve as catalysts for the breakdown of organic contaminants, such as insecticides and textile dyes.

   b. AgNP-modified filtration membranes have improved anti-biofouling characteristics.

3. Biosensors and sensors

   AgNPs' plasmonic characteristics are used in color sensors (colorimetric sensors) to detect metal ions, insecticides, and specific biomolecules.

8. Safety and toxicity concerns with nanosilver

8.1 Mammalian cell toxicity

Prior to their widespread use in the biomedical, food, and environmental domains, the assessment of the toxicity and safety aspects of green-synthesized nanosilver (AgNPs) to mammalian cells is a key concern, despite the fact that AgNPs provide numerous application advantages, particularly as antimicrobials. A number of physicochemical parameters, such as particle size, shape, concentration, surface charge, type of capping agent, and rate of release of silver ions (Ag^+), have a significant impact on the toxicity of AgNPs to mammalian cells [41]. Reactive oxygen species and oxidative stress are produced when AgNPs interact with cell membranes at the cellular level via endocytosis or adsorption processes. High amounts of ROS can result in apoptosis or cell necrosis by causing membrane lipid peroxidation, protein degradation, mitochondrial malfunction, and DNA damage [19, 20]. Furthermore, it is known that Ag^+ ions produced from the surface of AgNPs have a strong affinity for thiol groups (–SH) in mammalian cell proteins and enzymes, interfering with critical metabolic processes and enzymatic functions.

In contrast to traditionally chemically manufactured AgNPs, the presence of a natural biomolecule layer on the surface of AgNPs, such as polyphenols, flavonoids, proteins, and polysaccharides, is crucial for lowering toxicity levels in the context of green synthesis. This biological capping layer can improve colloidal stability, prevent excessive Ag^+ ion release, and exhibit antioxidant properties that help counteract ROS. Green synthesized–produced AgNPs at low to medium concentrations are comparatively safe for a variety of mammalian cell types, including fibroblasts, keratinocytes, and epithelial cells, and do not significantly alter cell shape, according to numerous *in vitro* investigations. Long-term exposure or high concentrations may have cytotoxic effects; however, toxicity is dose and time-dependent [20]. As a result, thorough cytotoxicity testing, oxidative stress analysis, genotoxicity studies, and *in vivo* research to comprehend the distribution, accumulation, and removal of nanoparticles in the body should all be included in nanosilver safety evaluations. Using this method, green-synthesized nanosilver can be developed as a functional material that is safe, effective, and poses little risk to human health or mammalian cells.

Particularly at high concentrations and very small sizes (<20 nm), AgNPs can result in:

1. Overproduction of ROS leads to oxidative stress.

2. Damage to DNA, mitochondria, and cell membranes.

3. Apoptosis/necrosis and cell signaling pathway disruption.

Numerous *in vitro* and *in vivo* investigations have documented harmful effects on the immune system, liver, and kidney cells at high dosages or with extended exposure. But it is crucial to remember that the toxicity profile depends heavily on:

1. The capping layer, size, and shape.

2. Exposure duration and dosage.

3. Exposure route (oral, inhalation, intravenous, topical).

8.2 Environmental toxicity

Even though nanosilver (AgNPs), especially those made by green synthesis, has many advantages for food, health, and the environment, its existence in ecosystems raises grave concerns about its possible harmful effects. The AgNPs are released into water, soil, and sediment after entering the environment through a variety of routes, including consumer goods, domestic wastewater, industrial waste, and the breakdown of objects containing nanosilver. The AgNPs are known to undergo physicochemical changes in aquatic environments, such as oxidation, agglomeration, or the release of silver ions (Ag^+), which are the most hazardous form for aquatic life. By causing oxidative stress, preventing photosynthesis, and interfering with enzymatic systems and cellular respiration, Ag^+ ions and AgNPs can impair the physiological processes of aquatic microbes, algae, zooplankton, and fish. The food chain and the equilibrium of aquatic ecosystems could be disrupted by these effects.

The AgNPs can interact with organic matter, bacteria, and soil particles in the soil environment, ultimately influencing the composition and activity of soil microbial communities. Disruption of these populations can affect soil fertility and ecosystem production since soil microbes are essential to the cycling of nutrients like carbon and nitrogen. The AgNPs can suppress the activity of nitrogen-fixing bacteria and decomposer microorganisms at high concentrations, according to several studies. However, this effect is highly dependent on particle size, concentration, and environmental factors like pH and organic matter content. When compared to traditional chemically synthesized AgNPs, the presence of a natural biomolecular layer on the surface of AgNPs can modulate their interaction with the environment, slowing the release of Ag^+ ions and lowering toxicity in the context of green synthesis. However, more research is needed to fully understand the long-term consequences and possible bioaccumulation of AgNPs in higher trophic species. Therefore, in order to guarantee that the use of nanosilver occurs safely and sustainably without having a major detrimental effect on the ecosystem, environmental risk evaluation of nanosilver should include thorough ecotoxicological studies, including acute and chronic tests on various organisms, environmental transformation analysis, and the development of appropriate management and regulatory strategies. When introduced into the environment, AgNPs can:

1. Impact aquatic and soil microbes.

2. Build up within the food chain.

3. Modify the makeup of significant microbial populations, such as those found in wastewater treatment facilities' activated sludge.

The necessity of thorough risk assessments, which cover the entire product life cycle – from manufacture to use to disposal – is emphasized by recent evaluations.

8.3 Long-term effects on the environment and toxicology: Conventional versus green-synthesized AgNP

Overview (bottom line): Since environmental transformation (sulfidation, complexation) and Ag^+ release are the primary variables affecting ecological toxicity, both green and chemical AgNPs exhibit major potential environmental concerns. On the other hand, the bioavailability, affinity for organic sediments and micelles, and transformation kinetics can all be altered by the biocapped surface. Depending on the capping properties, this can either increase or decrease biological persistence/uptake. According to traditional environmental studies and evaluations, environmental transformation, such as sulfidation – which can reduce toxicity rather than solely the synthesis technique – is the main factor. Details of ecotoxicology and fate are as follows:

1. Ag^+ release. In general, silver ions (Ag^+) are more hazardous than the nanoparticles themselves. Although organic coatings can prolong the particle lifespan because nanoparticle entities can be transported over long distances, capping biomolecules can decrease Ag^+ release (raising metal retention) and hence lessen acute toxicity.

2. AgNPs quickly change in the environment by mixing with organic matter, sulfidation, and chloridation. Although sulfidized particles might linger longer in sediments, sulfidation usually decreases Ag^+ release and acute toxicity. There is substantial evidence that the end result is determined by the ambient circumstances (pH, S_2, dissolved organic carbon).

3. Microbial interactions and resistance: Long-term exposure to AgNPs can cause cross-resistance mechanisms and disturb soil/sediment microbial populations. Although some studies have demonstrated decreased acute toxicity, it is uncertain whether green AgNPs are "safer" in the long run. Long-term field research is necessary to rule out changes in microbial populations.

4. Bioaccumulation and biomagnification: Long-term data on trophic chains are scarce; however, current research points to a low risk of bioaccumulation for altered AgNPs (such as sulfides). In conclusion, green synthesis lowers some acute toxicity indices, but without thorough fate/tox studies, it does not always ensure decreased long-term environmental consequences.

5. The NPs with various cappings are compared in multi-parameter (field and microcosm) persistence studies that track Ag^+ release, sulfidation kinetics, microbial community alterations, and trophic accumulation. In order to facilitate the comparability of toxicological results, contemporary studies highlight the necessity of uniform reporting of characterization (size, zeta, $\%Ag^+$ dissolved).

8.4 The impact of biological surface ligands (biocapping) on AgNP applications' electronic structure, reactivity, and particular performance

Three important factors are influenced by surface ligands (proteins, phenolics, and polysaccharides): (a) electronic structure and LSPR; (b) chemical reactivity (surface access and ion release); and (c) biological interactions and functional selectivity. These effects are caused by direct attachment to the surface of silver atoms (altering local coordination), modifications to the surrounding dielectric environment (altering the LSPR), and the development of chemical or physical barriers to ionization or substrate access.

1. LSPR
   The charge distribution on the NP surface can be altered by the adsorption of ligands with π/electron donor groups (phenol, thiol, amine), which can impact plasmon dephasing (bandwidth) and shift the SPR peak (red/blue shift). The SPR peak is shifted by polar ligands, which increase the local dielectric constant. These modifications affect optical applications, such as biosensors and SERS; for example, biocapped AgNPs occasionally exhibit a decrease in SPR intensity but an increase in spectral stability.

2. Reactivity and Ag^+ release
   Strong ligands, such as thiols from proteins or peptides, can passivate the surface, which lowers oxidative reactivity and Ag^+ release. On the other hand, loose or readily desorbed ligands increase Ag^+ release and antibacterial action by providing access to water and oxidants. Put differently, the trade-off between stability and catalytic/biological reactivity is determined by ligands.

3. Specific application performance
   Antimicrobials: The Ag can work in concert with ligands that offer extra interactions, such as phenolics with their own antibacterial activity, to reduce the MIC. Catalysis: A porous or semipermeable ligand design can preserve access and stability, while thick ligands prevent substrate adsorption and slow down the rate of catalysis. Biomedical/targeting: By introducing a target moiety (specific peptides, polysaccharides), ligands might improve therapeutic efficacy and lower systemic toxicity by increasing target cell uptake and decreasing nonspecific uptake.

9. Obstacles in the creation of green synthesis nanosilver

9.1 Reproducibility and standardization

Green synthesis of nanosilver (AgNPs) presents a potentially safer and more environmentally friendly approach than traditional chemical synthesis methods, but its application at advanced research and industrial scales still faces significant challenges related to process standardization and reproducibility. The primary obstacles are caused by the biological sources that are utilized, such as plant extracts, microorganisms, or natural biomolecules, whose chemical makeup can change based on the species, organism part, growth conditions, season, location, and extraction technique. The AgNPs with varying size, shape, and stability between batches are the result of these differences in phytochemical composition, which have a direct impact on reduction ability, nucleation rate, growth mechanism, and capping effectiveness. As a result, even though the main methods employed seem to be equivalent, synthesis findings acquired in one laboratory are frequently challenging to replicate exactly in another.

The absence of established process parameters is a significant barrier, in addition to the heterogeneity of biological materials. Comparisons between research are challenging since variables, including the concentration of silver precursor, the extract-to-precursor ratio, pH, temperature, reaction duration, mixing conditions, and lighting, are frequently reported inconsistently or incompletely. Significant variations in the properties of the final AgNPs might result from even minor adjustments to a single parameter. Additionally, there are discrepancies in the reporting of particle size (crystallite size vs. hydrodynamic size), size distribution, and colloidal stability, among other aspects of the characterization of synthesized products. The lack of a thorough mechanistic knowledge of the intricate interactions between silver ions and biomolecules in green synthesis systems exacerbates this problem. Therefore, deliberate attempts to establish standard protocols, use more specified biomolecules or well-characterized extracts, and harmonize characterization methodologies and result reporting are necessary for the progress of green nanosilver production toward practical applications. This approach is expected to improve reproducibility, enable more accurate comparisons between studies, and pave the way for consistent, safe, and reliable production of green synthesis-based nanosilver on an industrial scale. The variability of biological extracts is one of the main obstacles:

1. Season, location, growing technique, and extraction procedure all affect the phytochemical makeup.

2. For industrial and medicinal applications, achieving batch-to-batch uniformity is challenging.

The following is required:

1. Standardization of phytochemical characterization and extraction techniques.

2. Using marker chemicals to regulate the quality of extracts.

9.2 Scalability of the process

Since biologically-based systems are far more complex than traditional chemical synthesis methods, process scalability is one of the most important issues in the development of green synthesis of nanosilver (AgNPs) from laboratory size to pilot and industrial scale. Green synthesis is often performed in small volumes at the laboratory scale with relatively simple parameter control, including pH, temperature, extract-to-precursor ratio, and reaction duration, producing AgNPs with regulated properties. However, mixing inhomogeneity, mass and heat transport constraints, and differences in the rate of silver ion reduction in larger reaction volumes make it challenging to maintain consistent nanoparticle quality when the process is scaled up. Because large-scale extract production frequently encounters limitations on raw material availability, variations in phytochemical composition between batches, and extract stability during storage and repeated use, biological extracts that serve as reducing agents and capping agents also pose special scalability challenges. Additionally, green synthesis's comparatively slower reaction rate than chemical synthesis can lower production efficiency on a large scale. This calls for reactor design optimization, such as the use of bioreactors or continuous-flow reactors that can maintain homogeneity and real-time parameter control.

The procedure of extracting and purifying AgNPs from the reaction media, which can be more complicated, costly, and perhaps produce secondary waste on a large scale, presents another scalability difficulty. Therefore, the ability to create a standardized, effective, and sustainable process while preserving the consistent properties of the nanoparticles and the environmental advantages that form the foundation of the green synthesis approach itself is crucial to the success of green nanosilver synthesis development on an industrial scale. While laboratory scale-up is straightforward, industrial scale-up presents a number of difficulties:

1. Controlling temperature and homogeneity in large quantities.

2. Perishable biological extract handling and storage.

3. Modified large-scale reaction kinetics.

Specialized process engineering approaches and bioreactor designs are required to make green synthesis truly industrially viable.

9.3 Rules and safety considerations

The unique properties of nanoparticles are frequently not adequately accommodated within traditional regulatory frameworks, which poses serious regulatory and safety problems for the creation of nanosilver (AgNPs) through green synthesis methodologies. Regulations in many countries still focus on silver compounds in bulk or ionic form rather than on nanoparticles with fundamentally different physicochemical features, despite the fact that green synthesis offers the advantage of employing natural resources and minimizing hazardous chemical residues. This raises questions about risk classification, safe exposure limits, and the requirements for toxicological testing that must be fulfilled before nanosilver can be widely used, especially in the consumer goods, food, and healthcare industries. Given that the natural biomolecular coating on the surface of green-synthesized nanosilver can modulate biological and environmental interactions differently than chemically synthesized AgNPs, the lack of standardized methods for testing the toxicity and ecotoxicity of green-synthesized nanosilver also presents safety challenges. Furthermore, it is challenging to set universal safety limits for nanosilver products due to differences in size, surface charge, and Ag^+ ion release rates. From the standpoint of occupational safety, the manufacture of nanosilver, even when it is based on green synthesis, still presents the possibility of exposure dangers for workers through skin contact or inhalation, requiring certain safety regulations, which are currently scarce.

The absence of worldwide standards harmonization is another obstacle at the regulatory level, since different agencies' and nations' requirements might impede the licensing and commercialization process. To ensure that the use of nanosilver is safe, responsible, and sustainable, the development of green synthesis nanosilver necessitates an adaptable, science-based regulatory approach that includes the creation of characterization standards, safety testing protocols specific to nanomaterials, and thorough risk assessments throughout the product lifecycle. Nanomaterial regulations are continuously being developed. The following are necessary for food and medical applications:

1. Extensive information on toxicity.
2. Clinical and preclinical studies.
3. Details about biodistribution and degradation profiles.

Although green synthesis facilitates the process and is more ecologically friendly, it does not automatically remove the regulatory requirement to evaluate the dangers associated with nanosilver as a finished product.

10. Conclusion

A possible method for producing AgNPs in a more cost-effective, sustainable, and ecologically friendly way is the green synthesis of nanosilver. The basic idea is to substitute hazardous chemicals used in traditional procedures with plant extracts, microbes, or biomolecules as reducing agents, capping agents, and stabilizers. It has been demonstrated that plant extracts high in phenols, flavonoids, terpenoids, and other redox-active substances can effectively reduce Ag^+ to Ag_2 while also stabilizing nanoparticles. To regulate the size, shape, and stability of AgNPs, parameters like pH, temperature, reaction duration, extract concentration, and $AgNO_3$ concentration must be tuned. To guarantee the quality and functional characteristics of the final nanoparticles, a variety of characterization techniques (UV–vis, TEM/SEM, XRD, FTIR, DLS, zeta potential) are employed. Green synthesis-derived nanosilver exhibits extremely promising antibacterial, anticancer, and wound-healing properties, as well as possible applications in the food, environmental, and sensing domains. However, because nanosilver can release Ag^+ ions and cause oxidative stress, toxicity and environmental safety continue to be key concerns. Among the main obstacles are the standardization and reproducibility of biological extracts, the process's ability to scale to industrial levels, and the necessity of thorough regulation and risk evaluation. In order to create more regulated, effective, and sustainable green synthesis systems, future research aims to integrate omics technologies, machine learning, and biomass waste utilization within a circular economy framework.

Author details

Ally Kafesa* and Farhan Baehaki
Institute Kesehatan Rajawali, Bandung City, Indonesia

*Address all correspondence to: kafeally@gmail.com

References

[1] Eker F, et al. Green synthesis of silver nanoparticles using plant extracts: A comprehensive review of physicochemical properties and multifunctional applications. International Journal of Molecular Sciences. 2025;**26**(1):1–22. DOI: 10.3390/ijms26136222

[2] Michalcová A, Machado L, Marek I, Martinec M, Sluková M, Vojtěch D. Properties of Ag nanoparticles prepared by modified Tollens' process with the use of different saccharide types. Journal of Physics and Chemistry of Solids. 2018;**113**:125–133. DOI: 10.1016/j.jpcs.2017.10.011

[3] Nejad FS, et al. Investigating the effectiveness of iron nanoparticles synthesized by green synthesis method in chemoradiotherapy of colon cancer. Heliyon. 2024;**10**(7). DOI: 10.1016/j.heliyon.2024.e28343

[4] Riswan M, Arifin M, Santoso I, Nawa K, Nakamura K, Suharyadi E. Effect of change in number of electrons to optical properties and surface plasmon resonance of noble metals. Computational Materials Science. 2025;**247**. DOI: 10.1016/j.commatsci.2024.113519

[5] Dutta PP, et al. Antimalarial silver and gold nanoparticles: Green synthesis, characterization and in vitro study. Biomedical Pharmacother. 2017;**91**:567–580

[6] Lestari GAD, Cahyadi KD, Esati NK, Suprihatin IE, Ankamwar B. Karakterisasi Green Synthesis Nanopartikel Emas (Npau) Menggunakan Ekstrak Air Biji Cengkeh. J. Kim. 2022;122. DOI: 10.24843/jchem.2022.v16.i01.p16

[7] Mavani K, Shah M. Synthesis of Silver Nanoparticles by Using Sodium Borohydride as a Reducing Agent. International Journal of Engineering Research & Technology (IJERT). 2013;2:3. DOI: 10.13140/2.1.3116.8648

[8] Oktavia IN, Sutoyo S. Synthesis Of Silver Nanoparticles Using Bioreductor From Plant Extract As An Antioxidant. 2021

[9] Li Z, Liu Z, Chen Y. A green strategy for decoration of silver nanoparticles on MOF for catalytic reduction and cytotoxic activity against colon cancer. Alexandria Engineering Journal. 2024;**93**:67–72. DOI: 10.1016/j.aej.2024.03.002

[10] Gahlawat G, Shikha S, Chaddha BS, Chaudhuri SR, Mayilraj S, Choudhury AR. Microbial glycolipoprotein-capped silver nanoparticles as emerging antibacterial agents against cholera. Microbial Cell Factories. 2016;**15**(1). DOI: 10.1186/s12934-016-0422-x

[11] Gudikandula K, Charya Maringanti S. Synthesis of silver nanoparticles by chemical and biological methods and their antimicrobial properties. Journal of Experimental Nanoscience. 2016;**11**(9):714–721. DOI: 10.1080/17458080.2016.1139196

[12] AbuDalo MA, Al-Mheidat IR, Al-Shurafat AW, Grinham C, Oyanedel-Craver V. Synthesis of silver nanoparticles using a modified Tollens' method in conjunction with phytochemicals and assessment of their

antimicrobial activity. PeerJ. 2019;**2019** (2). DOI: 10.7717/peerj.6413

[13] Sari EM, Ma'ruf WF, Sumardianto, et al. Kajian Senyawa Bioaktif Ekstrak Teripang Hitam (Holothuria Edulis) Basah Dan Kering Sebagai Antibakteri Alami. Journal Pengolahan dan Bioteknologi Hasil Perikanan. 2014;**3** (4):16–24. https://ejournal3.undip.ac.id/index.php/jpbhp/article/view/7771

[14] Yang J, Chen X, Rao S, Li Y, Zang Y, Zhu B. Identification and Quantification of Flavonoids in Okra (Abelmoschus esculentus L. Moench) and Antiproliferative Activity In Vitro of Four Main Components Identified. Metabolites. 2022;**12**(6):483. DOI: 10.3390/metabo12060483

[15] Abdel-Razek MAM, Abdelwahab MF, Abdelmohsen UR, Hamed ANE. A review: Pharmacological activity and phytochemical profile of Abelmoschus esculentus (2010-2022). Royal Society of Chemistry. 2023;**13** (22):15280–15294. DOI: 10.1039/d3ra01367g

[16] Wahab S, Alshahrani MY, Ahmad MF, Abbas H. Current trends and future perspectives of nanomedicine for the management of colon cancer. European Journal of Pharmacology. 2021;**910**. DOI: 10.1016/j.ejphar.2021.174464

[17] Okur EE, Eker F, Akdaşçi E, Bechelany M, Karav S. Comprehensive review of silver nanoparticles in food packaging applications. International Journal of Molecular Sciences. 2025;**26**(20):9842. DOI: 10.3390/ijms26209842

[18] Khan K, et al. Mitochondria-derived reactive oxygen species are the likely primary trigger of mitochondrial retrograde signaling in Arabidopsis. Current Biology Journal. 2024;**34** (2):327–342.e4. DOI: 10.1016/j.cub.2023.12.005

[19] Kafesa A, Darmanto W, Wahyuningsih SPW. Silver nanoparticles with modified synthesis use sodium borohydride reducing agent and okra (Abelmoschus Esculentus L.) raw polysaccharide extract as an anti-colon cancer. Trends Science. 2024;**21**. DOI: 10.48048/tis.2024.8149

[20] Kasefa A, Wahyuningsih SPA, Darmanto W, Zulkufli NS. New development of Nanosilver manufacture using Abelmoschus esculentus l. pods extract as Bioreductor by Tollen's method and anti-colon cancer ability. International Journal of Innovative Research in Science Studies. 2025;**8**(3):3140–3154. DOI: 10.53894/ijirss.v8i3.7215

[21] Prasetyaningtyas T, Prasetya AT, Widiarti N, et al. Sintesis Nanopartikel Perak Termodifikasi Kitosan dengan Bioreduktor Ekstrak Daun Kemangi (Ocimum basilicum L.) dan Uji Aktivitasnya sebagai Antibakteri. Indonesian Journal of Chemical Science Sintesis. 2020;**9**(1). http://journal.unnes.ac.id/sju/index.php/ijcs

[22] Tamuly C, Hazarika M, Borah SC, Das MR, Boruah MP. In situ biosynthesis of Ag, Au and bimetallic nanoparticles using Piper pedicellatum C.DC: Green chemistry approach. Colloids Surfaces B Biointerfaces. 2013;**102**:627–634. DOI: 10.1016/j.colsurfb.2012.09.007

[23] Akdaşçi E, Eker F, Duman H, Bechelany M, Karav S. Microbial-based green synthesis of silver nanoparticles: A comparative review of bacteria- and

fungi-mediated approaches. International Journal of Molecular Sciences 2025;**26**(20):10163. DOI: 10.3390/ijms262010163

[24] Anees Ahmad S, et al. Bactericidal activity of silver nanoparticles: A mechanistic review. Materials Science for Energy Technologies. 2020;**3**:756–769

[25] Fabiani VA, Silvia D, Liyana D, Akbar H. Sintesis Nanopartikel Perak Menggunakan Bioreduktor Ekstrak Daun Pucuk Idat (Cratoxlum glaucum) melalui Iradiasi Microwave serta Uji Aktivitasnya sebagai Antibakteri. Fullerenes, Nanotubes and Carbon Nanostructures Journal. 2019;**4**(2):96–101

[26] Karahmet Sher E, et al. Nanotechnology in medicine revolutionizing drug delivery for cancer and viral infection treatments. International Journal of Pharmaceutics. 2024;**660**:124345

[27] Tiyaboonchai W. Chitosan Nanoparticles : A Promising System for Drug Delivery. 2003

[28] Lestari GAD, Ratnasari PMD, Sibarani J. Aplikasi Antibakteri Nanopartikel Perak (NPAg) Hasil Biosintesis dengan Ekstrak Air Daun Kemangi. KOVALEN: Jurnal Riset Kimia. 2022;**8**(1):17–24. DOI: 10.22487/kovalen.2022.v8.i1.15771

[29] Wu D, et al. Utilizing nanotechnology and advanced machine learning for early detection of gastric cancer surgery. Environmental Research. 2024;**245**. DOI: 10.1016/j.envres.2023.117784

[30] Chen Y, et al. Enhancing cancer immunotherapy: Nanotechnology-mediated immunotherapy overcoming immunosuppression. Acta Pharmaceutica Sinica B. 2024;**14**(9):3834–3854. DOI: 10.1016/j.apsb.2024.05.032

[31] Li Q, et al. Structural characterisation and anti-colon cancer activity of an arabinogalactan RSA-1 from Raphani semen. Carbohydrate Polymers. 2024;**342**. DOI: 10.1016/j.carbpol.2024.122417

[32] Chavada VD, Bhatt NM, Sanyal M, Shrivastav PS. Pyrophosphate functionalized silver nanoparticles for colorimetric determination of deferiprone via competitive binding to Fe(III). Microchimica Acta. 2017;**184**(10):4203–4208. DOI: 10.1007/s00604-017-2417-7

[33] Al Sufyani NM, Hussien NA, Hawsawi YM. Characterization and Anticancer Potential of Silver Nanoparticles Biosynthesized from Olea chrysophylla and Lavandula dentata Leaf Extracts on HCT116 Colon Cancer Cells. Journal of Nanomaterials. 2019;**2019**. DOI: 10.1155/2019/7361695

[34] Slepička P, Kasálková NS, Siegel J, Kolská Z, Švorčík V. Methods of gold and silver nanoparticles preparation. Materials. 2020;**13**(1):1. DOI: 10.3390/ma13010001

[35] Abdelghany AM, Elamin NY, Younis S, Ayaad DM. Polyvinyl pyrrolidone/carboxymethyl cellulose (PVP/CMC) polymer composites containing CuO nanoparticles synthesized via laser ablation in liquids. Journal of Molecular Liquids. 2024;**403**. DOI: 10.1016/j.molliq.2024.124857

[36] Ghazal S, Khandannasab N, Hosseini HA, Sabouri Z, Rangrazi A, Darroudi M. Green synthesis of

copper-doped nickel oxide nanoparticles using okra plant extract for the evaluation of their cytotoxicity and photocatalytic properties. Ceramics International. 2021;**47**(19):27165–27176. DOI: 10.1016/j.ceramint.2021.06.135

[37] Jalali MR, Kaushik D, Verma S, Singh H. A coupled model of finite element method and Mie theory for heat transfer inside expanded perlite vacuum insulation panels (VIPs) at high temperatures. International Journal of Heat and Mass Transfer. 2024;**219**. DOI: 10.1016/j.ijheatmasstransfer.2023.124885

[38] Oliveira AEF, Pereira AC, Resende MAC, Ferreira LF. Gold Nanoparticles: A Didactic Step-by-Step of the Synthesis Using the Turkevich Method, Mechanisms, and Characterizations. Analytica. 2023;**4**(2):250–263. DOI: 10.3390/analytica4020020

[39] Booth SG, et al. The significance of bromide in the Brust-Schiffrin synthesis of thiol protected gold nanoparticles. Chemical Science 2017;**8**(12):7954–7962. DOI: 10.1039/c7sc03266h

[40] Ozaki K, et al. Hydrogen-bonded structure of hydrated water in polyvinyl pyrrolidone aqueous solution investigated by X-ray absorption and emission spectroscopy. Journal of Molecular Liquids. 2024;**403**. DOI: 10.1016/j.molliq.2024.124822

[41] Strużyńska L. Dual implications of nanosilver-induced autophagy: Nanotoxicity and anti-cancer effects. International Journal of Molecular Sciences. 2023;**24**(20):15386. DOI: 10.3390/ijms242015386

[42] Ul Q, Naqvi A, et al. Size-dependent inhibition of bacterial growth by chemically engineered spherical ZnO nanoparticles. Journal of Biological Physics. 2019;**45**(2):147–159. DOI: 10.1007/s10867-019-9520-4

[43] De Rosa IM, Kenny JM, Puglia D, Santulli C, Sarasini F. Morphological, thermal and mechanical characterization of okra (Abelmoschus esculentus) fibres as potential reinforcement in polymer composites. Composites Science and Technology. 2010;**70**(1):116–122. DOI: 10.1016/j.compscitech.2009.09.013

[44] Ariani Edityaningrum C, Oktafiani AT, Widiyastuti L, Arimurni DA. Formulation and Evaluation of Silver Nanoparticles Gel. Indonesian Journal of Pharmaceutical Science and Technology. 2022;**9**(2): 126–139. http://jurnal.unpad.ac.id/ijpst/

[45] Jiang S, Shi Y, Li M, Xiong L, Sun Q. Characterization of Maillard reaction products micro/nano-particles present in fermented soybean sauce and vinegar. Scientific Reports. 2019;**9**(1):11285. DOI: 10.1038/s41598-019-47800-6

[46] Sumadiyasa M, Manuaba IBS. Penentuan Ukuran Kristal Menggunakan Formula Scherrer, Williamson-Hull Plot, dan Ukuran Partikel dengan SEM. Boletín de Física. 2018;**19**(1):28–35

[47] Abdullah A, Mohammed A. Scanning Electron Microscopy (SEM): A Review. In: Proceedings of 2018 International Conference on Hydraulics and Pneumatics. Băile Govora, Romania: HERVEX; 2018. p. 77–85

[48] Anggraini R, Surini S, Saputri FC. Formulation and characterization of bitter melon (Momordica charantia Linn.) fruit fraction loaded solid lipid nanoparticles. Pharmacognosy Journal.

2021;**13**(6):1347–1354. DOI: 10.5530/PJ.2021.13.170

[49] Kolb CG, Lehmann M, Kulmer D, Zaeh MF. Modeling of the stability of water-based graphite dispersions using polyvinylpyrrolidone on the basis of the DLVO theory. Heliyon. 2022;**8**(12). DOI: 10.1016/j.heliyon.2022.e11988

[50] Almatroudi A. Silver nanoparticles: Synthesis, characterisation and biomedical applications. Open Life Sciences. 2020;**15**(1):819–839. DOI: 10.1515/biol-2020-0094

[51] Sakore P, Bhattacharya S, Belemkar S, Prajapati BG, Elossaily GM. The theranostic potential of green nanotechnology-enabled gold nanoparticles in cancer: A paradigm shift on diagnosis and treatment approaches. Results in Chemistry. 2024;7. DOI: 10.1016/j.rechem.2023.101264

Chapter 3

Green/Biological Synthesis of Silver Nanoparticles

Yasmina Khane, Fares Fenniche, Sofiane Khane and Hafsi Zoulikha

Abstract

In nanotechnology, the establishment of dependable and environmentally sustainable techniques for nanoparticle synthesis is essential. Traditional methods for nanoparticle synthesis are expensive, toxic, and environmentally unfriendly. To address these challenges, natural sources have been utilized for the synthesis of silver nanoparticles (AgNPs). This chapter will examine the green and biological synthesis of AgNPs as an environmentally friendly and sustainable alternative to conventional physical and chemical synthesis methods by elucidating the specific biochemical reduction pathways and evaluating the effectiveness of different flora and microbiota in regulating nanoparticle architecture. The emphasis will be on biogenic methods utilizing plants, microorganisms (bacteria, fungi, algae), and biomolecules as natural reducing and stabilizing agents, thereby eliminating the need for hazardous chemicals and energy-intensive processes. The discussion will begin with an overview of green nanotechnology principles, followed by the processes of nanoparticle synthesis through the biological reduction of silver ions (Ag^{+}) utilizing various biological sources. It will examine their roles in regulating nanoparticle size, morphology, stability, properties, and applications, supported by a discussion of the characterization techniques used to confirm nanoparticle formation, assess the structural integrity of these particles, and explore the influence of biosynthetic methods on the physicochemical and functional properties of AgNPs, establishing a foundation for quality control. Specific focus will be placed on their prospective applications in the biomedical field and environmental uses of green-synthesized AgNPs, as well as their utilization across various domains, including biosensors, catalysis, and magnetic resonance imaging, in conjunction with the challenges of reproducibility and the potential for industrial scalability.

Keywords: green synthesis, biological synthesis, silver nanoparticles, eco-friendly nanotechnology, biomolecules, sustainable chemistry, biocompatibility

1. Introduction

Nanotechnology has become a multidisciplinary field in which understanding nanostructures' electrical, optical, magnetic, and mechanical properties promises to lead to the next generation of functional materials with diverse applications. Surface effects, quantum size effects, and macro-quantum tunneling provide nanoscale materials with better physical and chemical properties than bulk materials [1–3]. Nanoparticles (NPs)' increased reactivity and utility depend on chemical diversity, particle size, shape, surface area, and charge [4, 5]. These materials have 35–45% higher surface-to-volume ratios than larger particles [6]. The form and distribution of nanoscale particles are novel and enhanced [7, 8], assisting them in resolving ecological and technical issues related to healing, solar power, water purification, and catalysis [9, 10].

Among these, different types of metallic nanoparticles (MNPs) are distinguished by their unique physicochemical properties. Metal-based NPs, particularly silver NPs (AgNPs), are among the most widely considered due to their demand in various fields, primarily biomedical sciences [11, 12]. The interest in studying this type of metal lies primarily in its antimicrobial properties. Silver (Ag) is a natural transition metal known to be one of the most universal antimicrobial substances in history [13, 14]. Its antimicrobial qualities were recognized as early as the identification of bacteria as agents responsible for infections [15–17]. One of the most important and exciting nanomaterials is silver nanoparticles (AgNPs), which are molecules with a size of 1–100 nm and consist of 80% silver atoms and 20% silver ions [6, 18]. They sell well compared to carbon nanotubes and titanium nanoparticles for environmental release [19, 20]. AgNPs have unique physicochemical properties. They are known for their strong electrical and thermal conductivity, chemical stability, catalytic activity, and non-linear optical properties, including light absorption and scattering [21]. AgNPs also have different mechanical and magnetic characteristics than bulk materials [22]. These qualities are greatly influenced by nanoparticle size and form. Thus, these two parameters should be controlled during synthesis to suit photocatalysis (plasmonic effects in photocatalytic processes) and biology (antibacterial activities).

Three main ways for generating silver nanoparticles in clouding are chemical, physical, and biological. Each has benefits in scalability, affordability, and repeatability [23]. In the past, antimicrobial applications have relied on physical and mechanical processes like evaporation-condensation, laser ablation, or grinding. However, these approaches frequently come with downsides, such as aggregate formation and ineffective bactericidal effects [24]. A greater control over the size and shape of nanoparticles, which can increase their antibacterial activity, is achieved using chemical approaches, such as chemical reduction or the silver salt technique [25]. Even the usage of toxic chemicals is prevalent in these procedures, and which may also be more costly. Due to their user-friendliness and superior control over nanoparticle dimensions and morphology, chemical techniques, particularly chemical reduction in organic or aqueous solutions, are the predominant strategy used. This process employs three agents: a reducing agent, a stabilizing agent, and a metallic precursor [6, 26]. Nonetheless, the use of potentially dangerous and costlier chemicals has prompted the development of alternative biological methods. The principal advantage of these green technologies is their reduced environmental impact. These techniques, using microbes or plant extracts, provide mild synthesis conditions and superior nanoparticle dispersion. Compared to traditional methods of nanoparticle production, these biological methodologies provide enhanced environmental sustainability and economic efficiency [27]. Biological products find imperative use in the synthesis of agent nanoparticles

(Ag NPs) without the use of harsh [28], toxic, and dangerous chemicals. This is mainly achieved through the use of plant or fruit extracts [29] and bioorganisms [30]. Biological synthesis uses biochemicals from nature, such as alcohols, flavonoids, alkaloids, quinones, terpenoids, phenolic compounds, cellulose, enzymes, and exopolysaccharides, as bioreduction and stabilizing agents [19, 31, 32], to form crystalline nanoparticles of various shapes and sizes. Bio-corona prevents nanoparticle aggregation, modulates surface charge, and improves stability and dispersion in physiological settings [33]. These properties mostly rely on process factors such as the plant extract's type, its relative concentrations, pH, temperature, reaction duration, and mixing rate [34]. Repeatable synthesis and detailed toxicity evaluation techniques are currently missing.

This chapter investigates the entire lifecycle of green/biological production of silver nanoparticles (AgNPs), including synthesis methods and various applications, while critically analyzing structure-function correlations and addressing contemporary issues for future advancement. It seeks to bridge the significant divide between biogenic synthesis theory and the practical use of functionalized silver nanoparticles (AgNPs). First, we analyze the metabolic processes by which natural extracts and microorganisms reduce silver ions, providing a mechanistic foundation for green synthesis and emphasizing phytochemicals' significance in nanoparticle formation, shape, and stability. In terms of synthesis variables, AgNPs' size, shape, and surface properties are carefully examined, along with the plant species and reaction conditions. We evaluate the efficacy of floral extracts and microbial cultures as biological mediators and how source selection affects particle properties. This chapter establishes a rigorous characterization approach to ensure AgNP crystalline and surface integrity. We aim to demonstrate how biologically produced surfaces enhance biocompatibility and therapeutic efficacy in clinical applications using these "green" capping agents. Plant-based AgNPs have several applications in antimicrobial, agricultural, environmental, industrial, and medicinal fields, including drug delivery and wound healing. It addresses present issues, including toxicity, and discusses future standards, mechanistic knowledge, and translational possibilities across varied applications.

2. What the advantages are of investigated the green synthesis silver nanoparticles?

We decided to explore biologically-mediated green synthesis of AgNPs because of the available recorded evidence and community interest in this topic. However, metal nanoparticles (MNPs), particularly silver nanoparticles (Ag NPs), are entities comprised of several hundred to several thousand atoms with zero-dimensional characteristics. They possess exceptional properties that augment their appeal to society and have been the primary focus of research; their field of development is remarkably extensive due to their size (1 to 100 nanometers), which distinguishes these nano-objects from classical materials, endowing them with extraordinary physical, chemical, and biological properties [35]. The predominant cause of their prevalence is the discrepancy between the bulk structure and the morphology, composition, crystallinity, and architecture of Ag NPs [36].

Recent publications indicate that AgNPs are mostly identified in research focused on green synthesis among metal-based nanoparticles. **Figure 1** demonstrates that

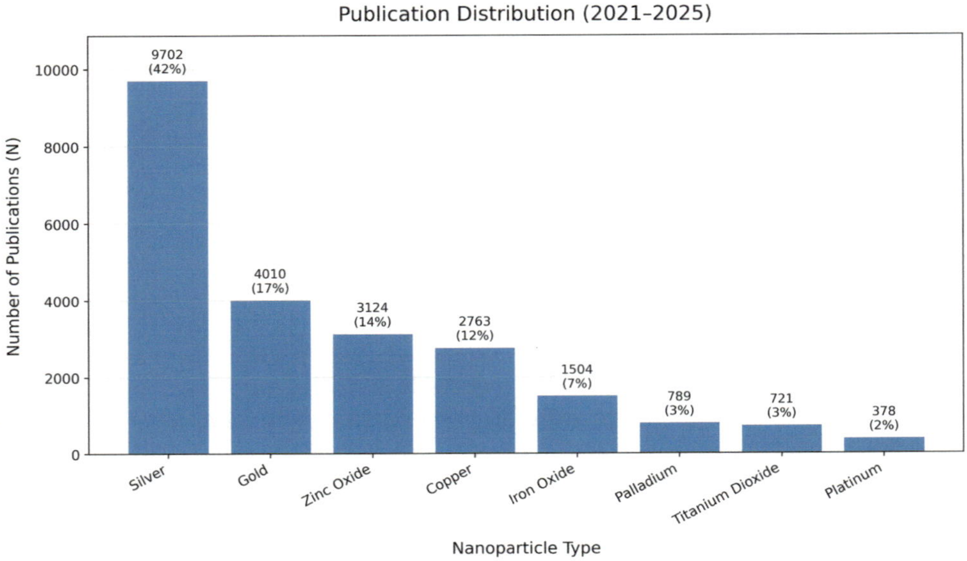

Figure 1.
Histogram showing the distribution of publications on green synthesis of nanoparticles by nanoparticle type over the last five years (2021–2025). Data were retrieved from the web of science core collection using a topic search for "green synthesis of [nanoparticle type]". Values above bars represent the number of publications (N) and their percentage contribution to the total dataset [37].

AgNPs account for 42% of all publications during the last five years concerning the green production of nanoparticles. AgNPs predominantly feature in the green production of nanoparticles, with roughly 10,000 documented instances, effectively tripling the publications concerning gold nanoparticles. This is primarily ascribed to the advantageous physicochemical properties of silver nanoparticles (AgNPs), which include robust optical characteristics stemming from their plasmonic behavior (surface plasmon resonance (SPR)), a high surface-area-to-volume ratio, and notable electrical conductivity [35, 38]. My review of scholarly literature revealed several bibliometric studies analyzing publication patterns in the green synthesis of nanoparticles (**Figure 2**). Oliveira and Lenzi [39] conducted a bibliometric analysis utilizing multiple databases with the search phrases "green chemistry," "synthesis," and "nanomaterials." Their data revealed a significant increase in publishing over time, with 456 articles published from 2014 to 2023 compared to only 50 from 2007 to 2013. Over the past decade, the volume of publications has surged ninefold.

Morenko *et al.* [40] analyzed 6,062 academic publications on green nanoparticle production from 2001 to October 2023 using bibliometrics. The study found that this research subject became prominent after 2010 [41].

Employing the term "Green synthesis of silver nanoparticles," Degrecci and Parreira [42] performed a Web of Science search and identified 1,140 publications published from 2011 to 2021. We decided to explore biologically mediated green synthesis of AgNPs because of the available recorded evidence and community interest in this topic. However, metal nanoparticles (MNPs), particularly silver nanoparticles (Ag NPs), are entities comprised of several hundred to several thousand atoms with zero-dimensional characteristics. They possess exceptional properties that augment their appeal to society and have been the primary focus of

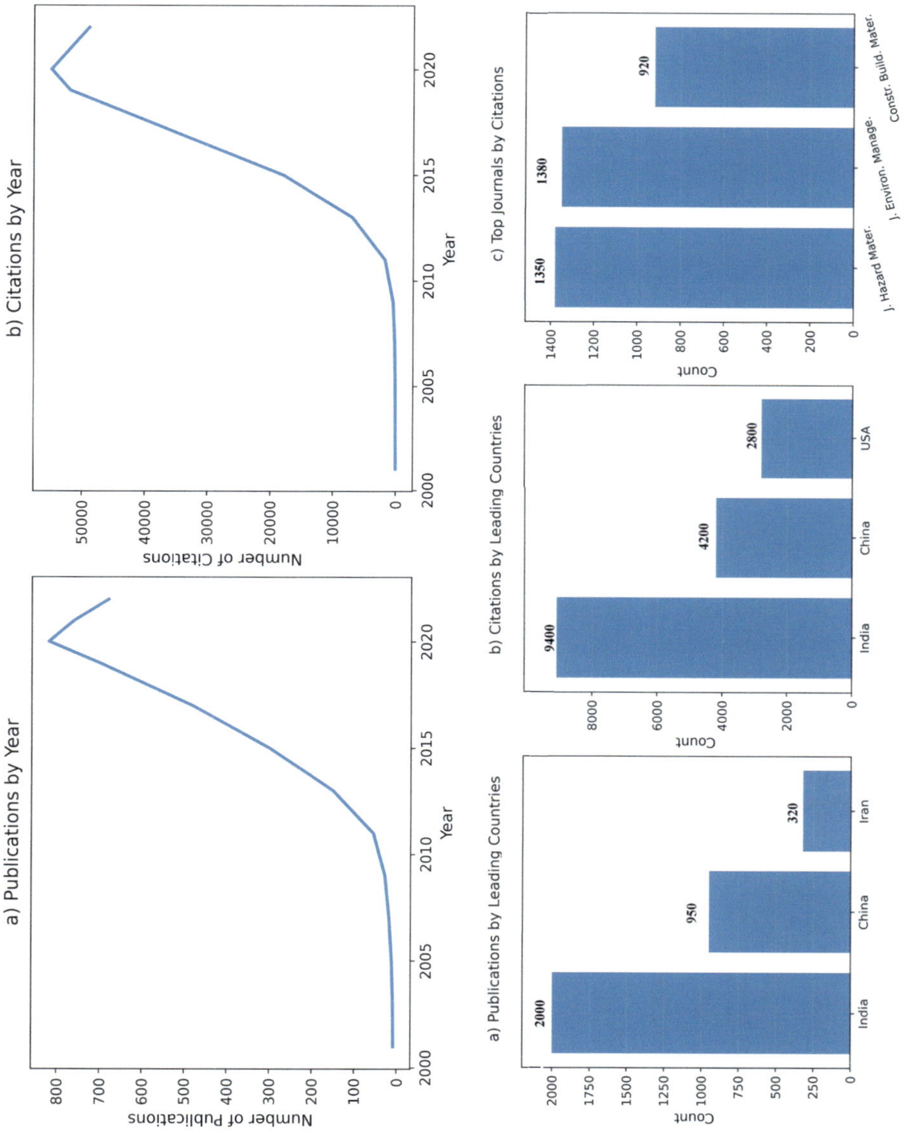

Figure 2.
(a) The annual number of papers on environmentally friendly nanoparticle production. (b) Citations to papers on green production of nanoparticles by Morenko et al. *[40].*

research; their field of development is remarkably extensive due to their size (1 to 100 nanometers), which distinguishes these nano-objects from classical materials, endowing them with extraordinary physical, chemical, and biological properties [35]. The predominant cause of their prevalence is the discrepancy between the bulk structure and the morphology, composition, crystallinity, and architecture of Ag NPs [36].

3. Biosynthesis strategies of silver nanoparticles

The main aim of formulating methodologies for AgNPs is to exploit their distinctive characteristics, including optical, electrical, high surface-to-volume ratio, and biological functionalities [43]. Consequently, attaining the requisite characteristics of nanoparticles necessitates the regulated synthesis of nanocrystals. The exceptional properties of nanomaterials are contingent upon their size, structure, interactions with stabilizers and surrounding environments, and production techniques [6]. However, a notable drawback of AgNP use is their potential toxicity, which arises from the synthesis methods employed. Recently, green synthesis methods have predominantly been employed in the fabrication of metal nanoparticles as an environmentally friendly and sustainable alternative [44]. The synthesis of Ag NPs typically employs two primary methodologies: "top-down" or "bottom-up" (**Figure 3**), including various physical and chemical techniques. These methodologies vary in their production processes and the corresponding quality, efficiency, and expense [45].

Physical synthesis can produce silver nanoparticles instead of relying on chemical bottom-up methods (**Figure 4**). High purity, without the use of chemical stabilizers or reductants, is achieved through heat, electromagnetic radiation, and plasma [47–51]. Laser ablation generates Ag nanoparticles in solvents including N, N-dimethylformamide, acetonitrile, and tetrahydrofuran without reductants or stabilizers, ensuring both purity and sustainability [52, 53]. In the absence of capping agents, compounds often exhibit enhanced microbial activities [54–56]. Techniques such as evaporation-condensation in a gas tube, physical vapor deposition via electron beam PVD [57], sputtering under high pressure to produce nanocrystalline

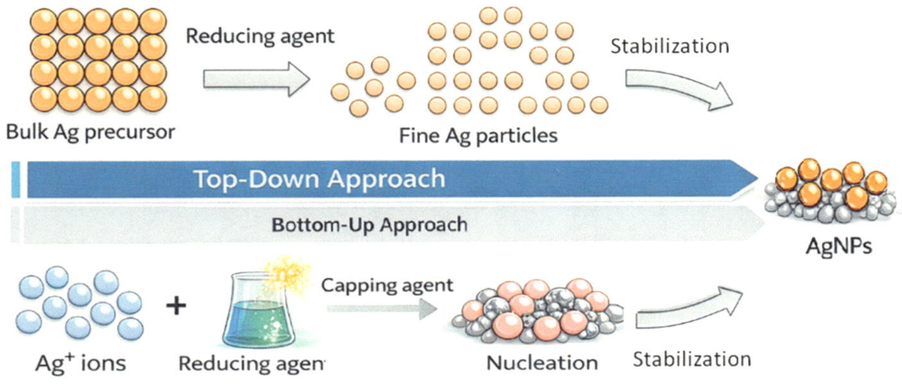

Figure 3.
Schematic representation of the synthesis of AgNPs by top-down and bottom-up processes [46].

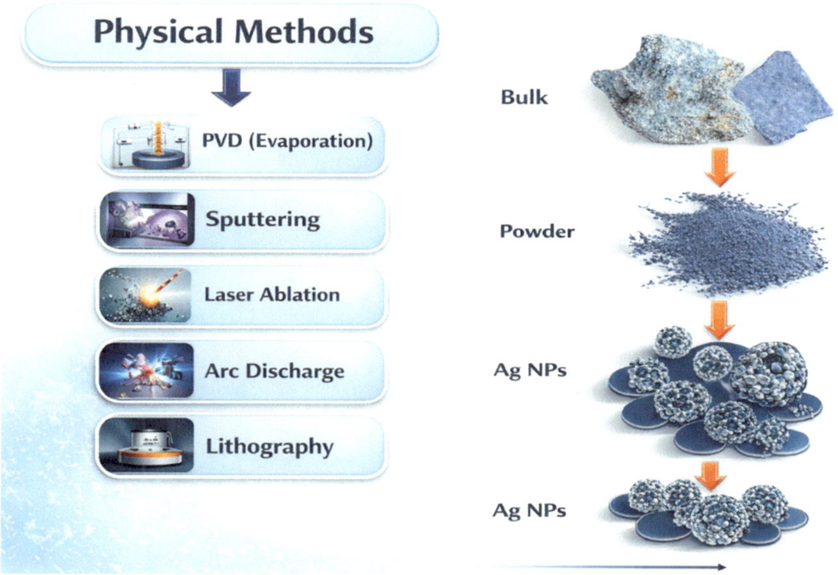

Figure 4.
Schematic diagram for physical synthesis of AgNPs.

thin films and powders [58], arc discharge using titanium electrodes to extinguish an arc in a mixture, and reducing the current by applying a 15 A current [59, 60] are also employed. Lithography, a plasmonic method, enables precise control of Ag NP dimensions but is both time-consuming and expensive [61, 62].

Chemical synthesis is the most common way to make silver nanoparticles (**Figure 5**) due to its simplicity, affordability, and scalability [63, 64]. A metal precursor, a reducing agent like sodium borohydride or trisodium citrate, a stabilizing agent like polyvinylpyrrolidone or cetyltrimethylammonium bromide to inhibit aggregation and regulate nanoparticle size distribution, and a solvent are the usual components of chemical reduction [65]. Chemical reduction, including borohydride reduction, the citrate method, the polyol process, and the Tollens reaction, is the most common method for converting silver precursors into nanoparticles [66–72]. Chemical reduction [73], polyol [74, 75], sol-gel [76], hydrothermal methods [77], chemical vapor deposition [78], microemulsion [79, 80], and electrochemical methods [81–83] are widely used.

The bottom-up method uses growth and nucleation to assemble molecular elements into complex aggregates, which regulate nanoparticle size, shape, and other properties for many applications. It is efficient and cost-effective. Chemical and biological syntheses from precursor salts are common bottom-up nanoparticle production strategies. Integrating light, electricity, microwaves, or sound waves improves chemical synthesis efficiency. Chemical nanoparticles may be produced quickly and used in many designs [84, 85]. Although these chemical methods are effective and cost-efficient, the use of hazardous reagents like sodium borohydride and hydrazine raises environmental and safety concerns, increasing interest in green synthesis alternatives that use biocompatible reducing agents [63].

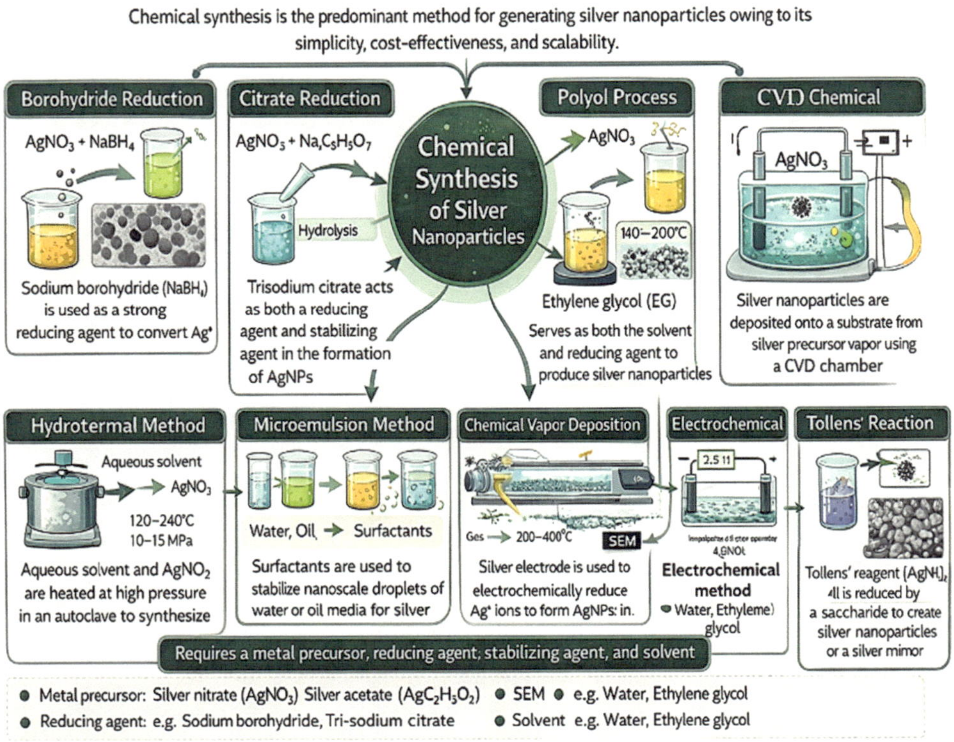

Figure 5.
Schematic diagram for chemical synthesis of AgNPs.

Green synthesis, which is superior to standard chemical synthesis, is attracting attention. Biological or "green" silver nanoparticle (AgNP) production is a sustainable alternative to chemical and physical methods.

3.1 Biological methods

In recent years, several efforts have been made to develop environmentally sustainable and cost-effective nanoparticle synthesis processes. Biological methods can generate silver nanoparticles without the use of costly and dangerous chemicals. The environmentally friendly synthesis of AgNPs requires silver nitrate and a natural reducing agent [86]. Natural reducing agents or cellular components stabilize or encapsulate cells, eliminating the need for exogenous agents [87].

Numerous biological synthesis techniques use fungi, yeasts, bacteria, and actinomycetes. An alternative method encompasses plants, plant extracts, viruses, membranes, and diatoms (**Figure 6**). Food waste, including shells and peels, contains several physiologically active compounds. The production of silver nanoparticles (Ag NPs) from food waste is superior to chemical synthesis [88].

The bio-reduction of metal ions using enzymes, proteins, amino acids, polysaccharides, and vitamins from organism extracts is environmentally sustainable and effective (**Figure 7**). The production of plant-based nanoparticles is a burgeoning field of study [89].

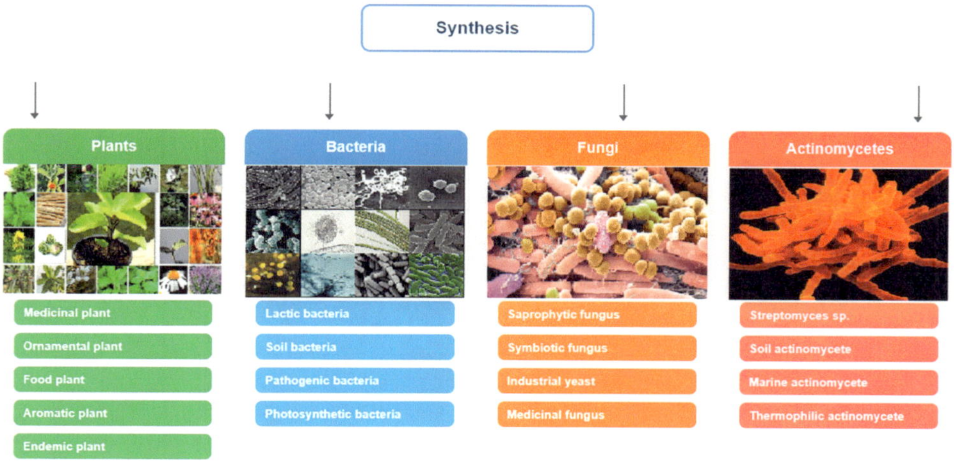

Figure 6.
The types of different biological resource for biosynthesis of AgNPs.

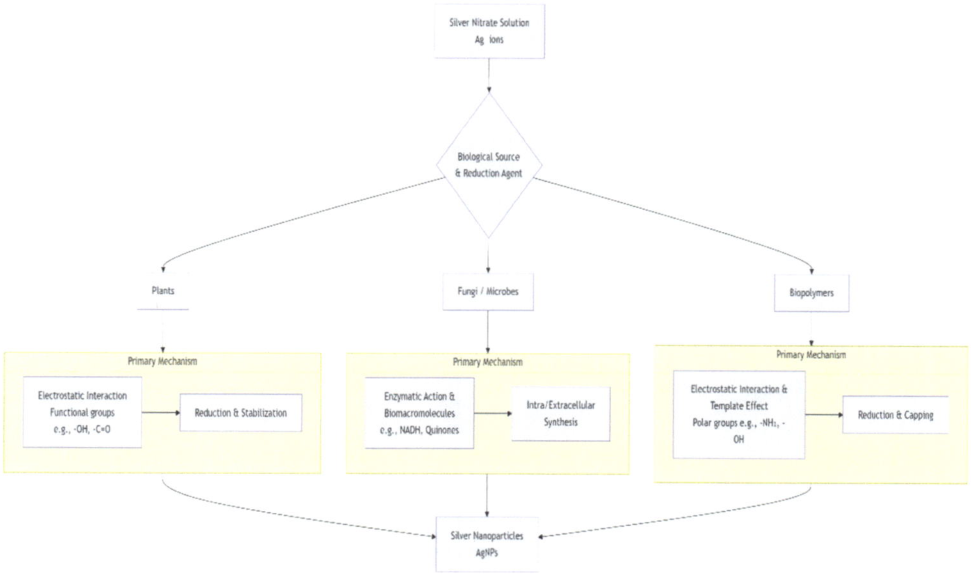

Figure 7.
Various constituents of plant, bacteria, fungi, and biopolymer responsible for the reduction of silver nitrate to AgNPs.

The quantity of the biological extract, silver nitrate ($AgNO_3$), pH of the reaction medium, reaction temperature and duration, mixing speed, and light exposure all influence the efficacy and repeatability of the green synthesis of silver nanoparticles (AgNPs). The yield, morphology, size distribution, and surface chemistry of AgNPs are contingent upon these characteristics. Acidic environments accelerate nucleation

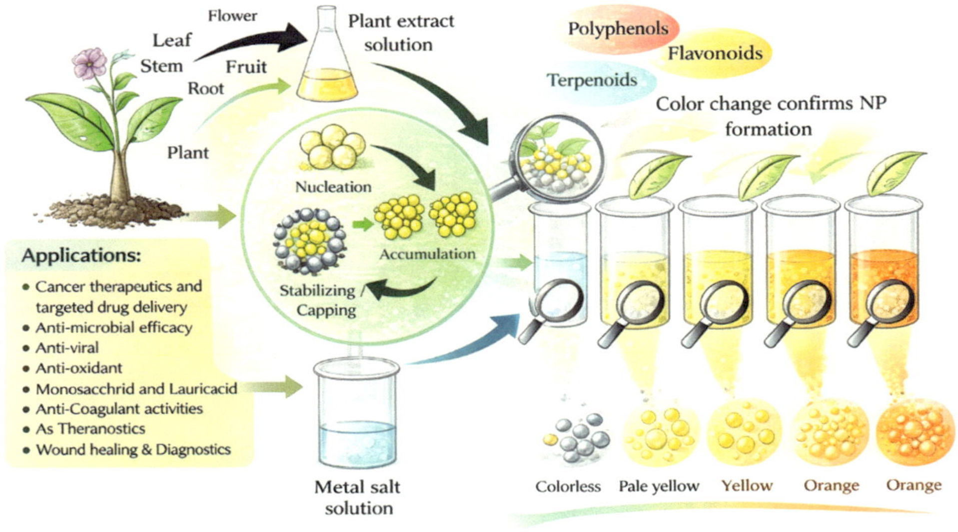

Figure 8.
Impact of plant extract phytochemicals in green synthesis of AgNPs [91].

and produce smaller nanoparticles, whereas alkaline conditions may lead to aggregation and polydispersity. Biologically synthesized AgNPs may exhibit spherical, rod-like, triangular, or quasi-spherical morphologies, contingent upon the biological agent used. Extracts from the Azadirachta indica plant provide homogeneous spherical nanoparticles, whereas synthesis by Aspergillus niger results in anisotropic forms. Pseudomonas aeruginosa synthesizes polydispersed quasi-spherical AgNPs. Morphologies influence antibiotic efficacy, cellular absorption, and optical properties [90].

Various reducing agents have been used in the synthesis of AgNPs, as shown in **Figure 8,** along with the overall process for AgNP production.

3.1.1 Plant-mediated synthesis

Plant-based AgNP production is more scalable, biohazard-free, and cell culture-free than microorganism-based methods [92–94]. Plant extracts are better than microbial nanoparticles because of their low synthesis rate, speed, efficiency, and viability. Use the entire plant – leaves, fruits, roots, seeds, stems. Enzymes, alkaloids, polysaccharides, tannins, terpenoids, phenols, and vitamins convert silver ions (Ag^{+}) to metallic silver (Ag^{0}) in plant-mediated synthesis and are medicinal and environmentally friendly despite their complex structures (**Figure 8**) [95]. Nanoparticle synthesis and waste management benefit from banana, orange, and potato peels. Polyols and water-soluble heterocyclic chemicals reduce silver ions, while leaf broth flavonoids and terpenoids stabilize nanoparticles [95].

Nanoparticle stabilization and capping chemicals must differ from microbial chemicals because plant cells contain a complex of antioxidant metabolites that prevent oxidation and biological damage [96, 97]. Enzymes, glycosides, and saponins stabilize nanoparticles [98, 99]. In plant extracts with metal salts, silver ions attach to water-soluble compounds via –OH and –COOH groups [100, 101]. This simplifies the conversion of metal ions to silver nanoparticles by restructuring the protein complex. Protein amino groups and cysteine residues reduce silver and produce AgNPs [102]. AgNPs are "capped" by alkanes, amines, phenols, polyphenols, alkaloids, lignans, terpenoids, and flavonoids [103, 104]. The antioxidant power of flavonoids is particularly important. Hydrophilic functional groups in surrounding molecules make nanoparticles colloid-stable in water [105]. Garlic extract reduces and stabilizes sucrose and fructose [74]. Ag^{+} ions are converted into silver nanoparticles via polyols [75]. AgNPs should be stabilized and kept clump-free by surface-active terpenoids [106].

By changing terpene C–O groups to –C–O, terpenoids may help convert Ag^{+} ions into Ag0 nanoparticles [107]. Terpenoids convert aldehydic groups into carboxylic acids, thereby reducing metal ions [108].

Nanoparticle manufacturing may use plant extracts as capping agents, and current research reveals that the plant source strongly affects nanoparticle form and bioactivity. Due to their phytochemical content and therapeutic benefits, *Tephrosia purpurea*, *Sesbania grandiflora*, and *Morinda citrifolia* are the main AgNP producers [109–111] capped spherical silver nanoparticles (10–30 nm) with *Alternanthera dentata*, *Premna herbacea*, and *Parthenium hysterophorus* leaf extracts, which inhibited *Shigella dysenteriae* and *Escherichia coli*.

Ultrasound-assisted synthesis using weed plants enhances nanoparticle properties. Manjamadha and MuthuKumar noted that ultrasound enhanced the reaction rate in the ultrasonic-assisted synthesis of AgNPs from weed plants [112, 113]. *Capsicum annuum* extract may be used to produce AgNPs. Understanding AgNP formation requires comprehension of recognition-reduction-limited nucleation, as shown by the study. The manufacture of plant-based silver nanoparticles (AgNPs) is more scalable, devoid of biohazards, and independent of cell cultures compared to microorganism-based methods [92–94]. The entire plant – foliage, fruits, roots, seeds, and stems. Enzymes, alkaloids, polysaccharides, tannins, terpenoids, phenols, and vitamins that facilitate the reduction of silver ions (Ag^{+}) to metallic silver (Ag^{0}) and provide therapeutic and ecological benefits despite their complex structures [114].

The production of environmentally friendly silver nanoparticles utilizes agricultural waste and byproducts from the food industry for reduction and stabilization purposes. Banana, orange, and potato peels contain polyphenols and flavonoids. These compounds facilitate the reduction of silver ions to nanoparticles and stabilize them, thereby minimizing the need for hazardous chemicals. This nanoparticle synthesis method is sustainable and enhances the value of agricultural and industrial waste, contributing to waste management and environmental preservation [115]. Hui Xu *et al.* synthesized silver nanoparticles (AgNPs) from food waste in 10 minutes using grape seed extract for reduction and stabilization. Silver nanoparticles (AgNPs) with sizes ranging from 25 to 35 nm and a zeta potential of -28.4 mV exhibited stability and antibacterial effectiveness against eight Gram-negative and Gram-positive pathogens [116].

3.1.2 Bacterial-mediated synthesis

Metal nanoparticle-making microbes should be considered. Organic matter's unique properties enable complex NPs. Microbes produce intracellular and extracellular AgNPs [117]. Both intracellular and extracellular techniques can synthesize microbial silver nanoparticles [118].

During intracellular synthesis, enzymes and biomolecules in microbial cells convert Ag ions to NPs [119]. AgNP nucleation occurs due to cellular Ag concentration. Intracellular silver buildup produces nanoparticles and boosts bacterial activity. Intracellular interactions diminish metal ions [120, 121]. After optimal bacterial growth, viable cell nanoparticles are extracted. The gathered cells must precisely release nanoparticles. Live cells are collected when fully mature. Additionally, special handling methods are used to release NPs from recovered cells [117]. Noble metal NPs like Ag and gold are synthesized from bacteria-generated intra or extracellular inorganic elements.

Extracting bacteria's extracellular secretions is used to make extracellular goods. Microorganisms that resist silver ions create good AgNPs. Enzymes/proteins, amino acids, polysaccharides, vitamins, etc., in extracts may lower Ag^{+} ions. The procedure is unknown. Lihong Liu *et al.* invented a microbe-free Ag NP synthesis in microorganism culture broth. This study reveals how pH, light, and broth content affect pure Ag NP production. This study created Ag NPs without living microorganisms under optimal light and pH conditions for nanomaterial synthesis [122]. Some organisms cannot nanoform metals.

Karthik *et al.* [123] synthesized AgNPs outside cells using Streptomyces sp. LK3. Some Ag-resistant bacteria are biocompatible with AgNPs, according to Slawson *et al.* [124]. Pooley [125] suggested employing bacteria to industrially recover Ag from ore because their cell walls agglomerate Ag.

3.1.3 Fungi-mediated synthesis

Compared to bacteria, fungi contain more enzymes, proteins, and reducing components on their cell surfaces, making them a better nanoparticle manufacturing option. Fungi outperform yeast, bacteria, plants, and physicochemical methods in nanoparticle production [126]. Fungi reduce metal salts with enzymes and proteins to make metal nanoparticles. In vitro cultivation allows many fungal strains to produce nanoparticles [127]. Fungal enzymes, such as naphthoquinones and anthraquinones, can reduce Ag^{+} on cell surfaces [120, 128]. Xue B. *et al.* [129] synthesized AgNPs using *Arthroderma fulvum* at 1.5 mM substrate concentration, alkaline pH, 55°C reaction temperature, and 10 hours of reaction time. Crystalline AgNPs were synthesized with an ideal particle size of 15.5 ± 2.5 nm, exhibiting antifungal activity against Candida, Fusarium, and Aspergillus. Honary S. *et al.* evaluated soil-derived *Penicillium citrinum*-based green extracellular silver nanoparticle (AgNP) synthesis. Nanoparticles with an average diameter of 109 nm were synthesized [130]. The first controlled and scalable green method produced well-shaped silver nanoparticles (AgNPs) from the cell-free aqueous filtrate of the non-pathogenic biocontrol agent Trichoderma asperellum [131]. Birla SS *et al.* [131] optimized *Fusarium oxysporum* AgNPs by manipulating medium, pH, temperature, light intensity, filtrate volume, salt concentration, and biomass. Ishida K. *et al.* [132] synthesized and tested Fusarium oxysporum aqueous extract AgNPs for antifungal activity.

3.1.4 Bio-polymer-mediate

Polymers and biopolymers facilitate the reduction and capping of AgNPs, rendering the process more environmentally friendly and biocompatible [63]. Water serves as a ubiquitous solvent. Starch, chitosan, PVA, and PEG reduce Ag^{+} ions.

These polymers initiate and stabilize silver ions through amino and hydroxyl groups [133]. Temperature, polymer molecular weight, and concentration influence particle morphology and dimensions.

Silver nanoparticles (AgNPs) made from polymers and polysaccharides are biocompatible and ecologically friendly. Natural and manmade polymers reduce and stabilize AgNPs, improving colloidal stability and reducing chemical use. This strategy is safer and more sustainable, especially for health and the environment. Polymeric components in the final product may affect nanoparticle performance in electronics and catalysis, where purity and surface accessibility are crucial. Polyquaternary phosphonium oligochitosan (PQPOCs–AgNPs) demonstrated better antiviral efficacy than other formulations [134]. Most biopolymers used to synthesize AgNPs reduce and stabilize, except starch, which acts as a capping agent [135]. Leung TC *et al.* [136] produced AgNPs in 10–15 minutes using carboxymethylated-curdlan or fucoidan as reducing and stabilizing agents. Heating the reaction mixture to 100°C resulted in 40–80 nm AgNPs.

The proposed mechanism (**Figure 9**) for the biosynthesis of AgNPs utilizing *Phyllanthus amarus* leaf extract, as described by Ajitha *et al.* [137], indicates that the leaf extract acted as both a capping and reducing agent, substantially influencing the size and morphology of the nanoparticles. This mechanism suggests that the antioxidant phytochemicals in the leaf extract converted positively charged metal ions (Ag^+ ions) to a neutral state (Ag^0 atoms) and subsequently to small Ag^0 particles through electron donation, facilitated by bioreducing agents in the initial stage. The Ag^0 atoms subsequently expanded into petal formations, which, aided by

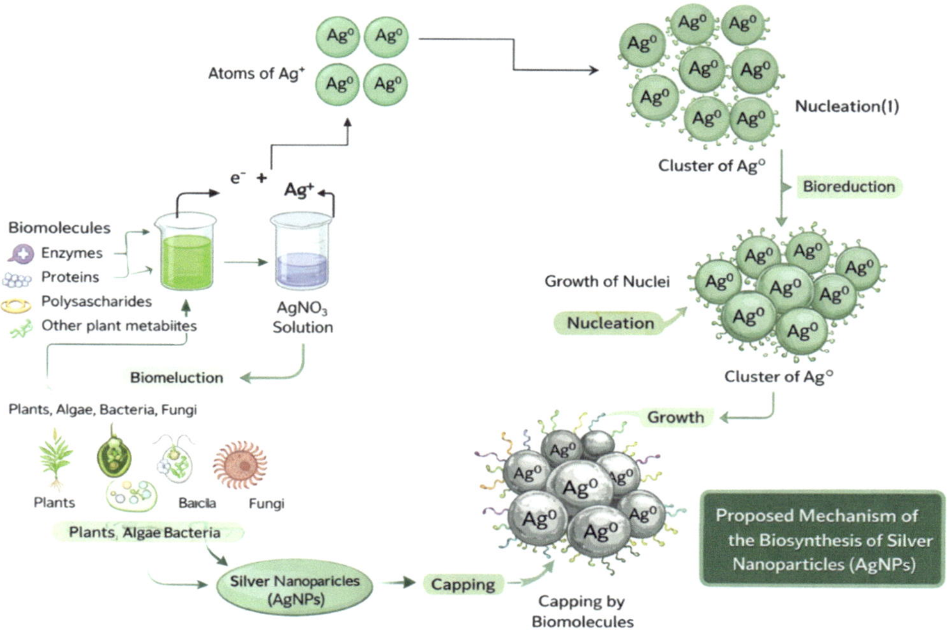

Figure 9.
Green synthesis may be categorized into: (a) plant and extract-based methods, (b) bacterial methods, (c) fungal methods, and (d) biopolymer-based methods.

bio-capping agents, amalgamated into nanostructures that resembled flowers. Reduced metals typically undergo nucleation and cluster formation, leading to the assembly of particles that are stabilized through complexation or electrostatic interactions with oxygen-containing groups such as –COO and –OH [138]. The phytochemicals responsible for the formation of AgNPs also inhibited further aggregation, consistent with our proposed mechanism, by coating the particles and supplying capping agents abundant in electron-rich hydroxyl groups [100, 139].

4. Characterization of biogenic silver nanoparticles

Various characterization (**Figure 10**) approaches are employed to verify nanoparticle production, evaluate the structural integrity of these particles, and investigate the impact of biosynthetic processes on the physicochemical and functional aspects of AgNPs, thereby establishing a basis for quality control. The behavior, biodistribution, safety, and efficacy of nanoparticles are predominantly influenced by their physicochemical properties. Shin *et al*. [140] suggest that size distribution, electrostatics, surface area, morphology, and aggregation may affect nanomaterial-biotarget interactions.

Identifying AgNPs is critical to assessing particle performance. UV spectroscopy can verify AgNP synthesis, FTIR can identify bio-capping agent functional groups (e.g., –OH, C=O, N–H) responsible for reduction and stability, and XRD can confirm the crystalline nature and phase (face-centered cubic structure). Silver is confirmed by EDX elemental analysis. DLS evaluates suspension hydrodynamic size and stability. Zeta potential measures surface

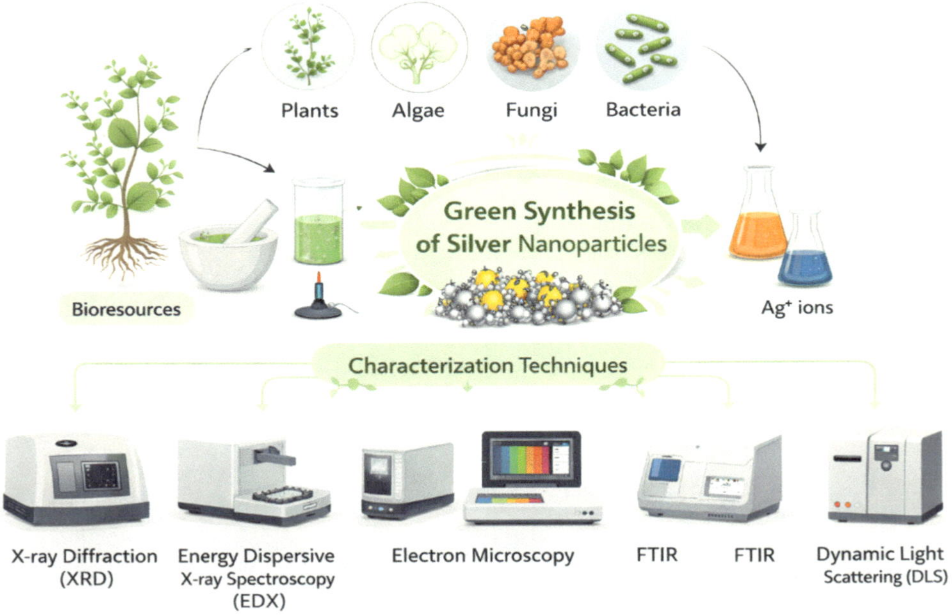

Figure 10.
The different characterization analysis to confirm the formation of AgNPs.

charge and colloidal stability. SEM and TEM measurements have determined size, shape (spheres, rods, triangles), and distribution [141]. However, the basic techniques' key concepts are summarized below to optimize synthesis processes, including precursor amount, temperature, pH, and reducing and stabilizing agents [142].

UV-visible spectroscopy is the primary method for measuring AgNP production and stability. Due to their unique optical characteristics, silver nanoparticles (AgNPs) interact with light at specific wavelengths. AgNPs' low band gap allows electrons to move freely, creating a surface plasmon resonance (SPR) band from their collective oscillation with light [143–145]. Numerous studies indicate that green-synthesized silver nanoparticles (AgNPs) have Surface Plasmon Resonance (SPR) peaks ranging from 380 to 450 nm. The SPR width correlates with NP size distributions; smaller nanoparticles exhibit peaks at lower wavelengths, whereas larger particles generate peaks at higher wavelengths. AgNPs with atypical geometries may exhibit two or more plasmon resonances, depending on nanoparticle composition [146, 147].

Zeta potentials indicate stability, and DLS evaluated AgNP suspension particle size and dispersion. Dubey *et al.* [148] found that AgNPs had a lower zeta potential at acidic pH and a higher one at basic pH. Dynamic Light Scattering revealed a hydrodynamic diameter of 45 ± 8 nm and a zeta potential of -28.5 mV [3, 149], indicating improved colloidal stability above the -25 mV threshold for repulsion-dominated suspensions [150] and -27.9 mV from E. hirta. Zetasizer Nano Series analyzers measure AgNP size.

XRD determines material crystallinity. XRD peaks at 38.13°, 44.21°, 64.47°, 77.37°, 81.47°, 98.01°, 110.56°, and 114.80° confirm AgNPs' FCC structure. Silver face-centered cubic planes had X-ray diffraction peaks at 38.1°, 44.3°, 64.4°, and 77.4°. XRD and EDAX confirm the purity of the produced silver nanoparticles (AgNPs) [136, 137, 151].

TEM and SEM analyze AgNPs' morphology. SEM, TEM, and AFM examined AgNP surface morphology (**Figure 11**), dimensions, and forms [152, 153]. Transmission Electron Microscopy pictures showed spherical, evenly dispersed Ag nanoparticles with an average size of 25 ± 5 nm [154, 155], smaller than

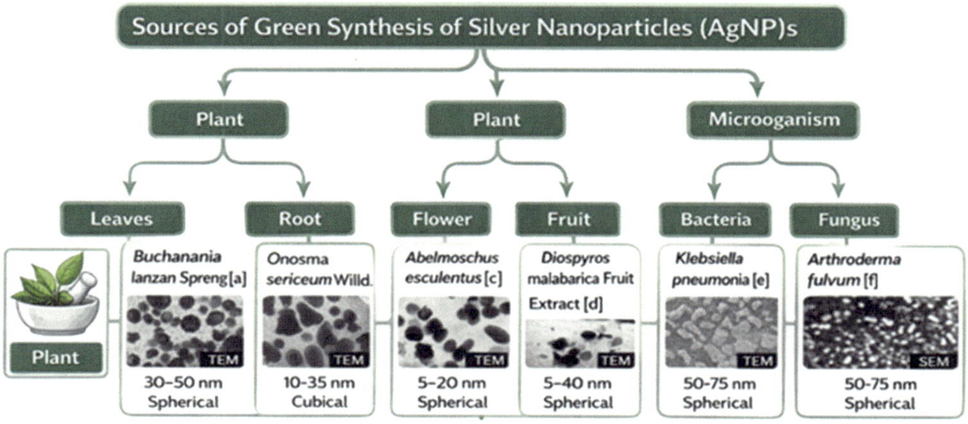

Figure 11.
TEM and SEM images of green synthesis AgNPs using different biological components [161].

those produced via microbial methods (e.g., 74 nm from Ehretia rigida leaves), yet suitable for biomedical applications. Koyla *et al.* [156] produced spinach leaf extract-based globular polycrystalline silver nanoparticles. Hamelian M. *et al.* [153] synthesized plate or rod-shaped AgNPs from Thymus kotschyanus extract.

Madhusudanan [157] and Nasruddin *et al.* [158] found capping biomolecules with peaks at 3,420 cm^{-1}, 1,635 cm^{-1} (C=O amide), and 1,045 cm^{-1} from phytochemicals in vegetable peels, with similar shifts in citrus peels (3,380–1,641 cm^{-1}) as reported by Jalani *et al.* [159], improving biocompatibility over chemical methods. Functional groups that reduce and stabilize AgNPs are shown by FTIR [130, 151, 160].

5. Unique physicochemical properties and enhanced biocompatibility

Nanoscale materials differ greatly from bulk materials. In nanotechnology, numerous nanomaterials exhibit different characteristics based on the fundamental and physical properties of metallic molecules and the surface-to-volume ratio [41]. The chemical composition, shape, size, and controlled dispersity of nanoparticles affect their physicochemical properties [162]. Modern nanoparticle manufacturing emphasizes nanoscale dimensions, simplicity, cost-effectiveness, environmental friendliness, and application adaptability [163].

5.1 Physical properties

Quantum size effect refers to the difference in behavior of atoms or molecules in small clusters compared to bulk material, which represents the average of all quantum forces. When item sizes become so small that averaging fails, nanoscale materials are significantly affected by surface energy. The specific surface area increases dramatically with smaller particles [164]. It is essential to assess product dimensions before synthesis [164]. Sotiriou *et al.* [165] observed that smaller nanosilver particles release silver ions more quickly, thereby increasing potential risks. Synthesizes include spherical, rod-like, triangular, cubic, wire-like, and star-shaped AgNPs [166]. Morphology plays a crucial role in influencing the optical, catalytic, and electrical characteristics of AgNPs. Recent research has revealed that the forms of AgNPs alter their antibacterial, catalytic, penetrating, and optical capabilities. Therefore, selecting specific AgNP morphologies during synthesis may optimize their applications [167, 168]. NP stability must be carefully controlled throughout synthesis. Plant extracts containing flavonoids and sugars are often used to cap nanoparticles [169]. The surface charge can be adjusted to influence nanoparticle aggregation, solubility, and reactivity. Silver exhibits excellent heat conductivity, and nanoscale thermal conductivity is enhanced for AgNPs due to their high surface-area-to-volume ratio [170–174].

5.2 Chemical properties

Numerous studies have revealed that AgNPs' surface chemistry, oxidation states, and ion release potential boost their antibacterial properties. Nanoparticle toxicity may be affected by the release of silver ions from their surfaces, which can cause oxidative stress and kill bacteria.

Surface functionalization with synthetic or biological substances may alter nanoparticle stability, solubility, and cellular interaction. These chemical features demonstrate how AgNPs function in complex contexts and how to manage them for safer, more efficient uses [175]. In the complex physicochemical process of nanoparticle formation, bioextracts reduce greenness and modify nanomaterial structure and surface properties. Adjustments provide a novel property tweaking method beyond synthesis [176]. Crystallinity depends on reduction kinetics and biomolecule-inorganic precursor interactions. Soft reductants create ordered atomic arrangements and highly crystalline structures, whereas strong reductants like polyphenols nucleate quickly and form semicrystalline or polycrystalline phases [177]. Plant extracts may influence thermodynamically stable crystalline phases. Toton Sarkar and colleagues found that the concentration of Tabernaemontana divaricata flower extract controls crystallite size, defect density, and surface chemistry in ZnO nanoparticle formation. A higher extract amount enhanced crystallite diameters and decreased dislocation density, improving crystallinity. They demonstrated that biosynthesis may alter surface function and cause defects [178].

5.3 Optical properties

Localized surface plasmon resonance (LSPR) causes the collective oscillation of conduction electrons at the metal surface when excited by light at specific wavelengths, resulting in AgNPs' unique optical properties. The surface plasmon resonance of AgNPs makes them excellent light scatterers and absorbers [179]. This property makes silver nanoparticles (AgNPs) more effective at absorbing and scattering light than many visible spectrum dyes and pigments [180].

5.4 Biological properties

Silver compounds precipitate bacterial cellular proteins and block the respiratory systems of aerobic and anaerobic bacteria. Research shows that AgNPs, ranging from 1 to 10 nm, may cling to bacterial membranes and impair critical functions, ultimately killing the cells [181].

AgNPs' positive surface charge increases membrane permeability, cell leakage, and lysis by electrostatically adhering to negatively charged bacterial membranes. Due to their small size and tailored surface functionalization, AgNPs may enter bacterial cells and directly interact with internal proteins and nucleic acids, altering protein synthesis and causing genotoxicity [182]. Size-dependent and surface chemistry-dependent toxicity of biogenic AgNPs are related, with surface chemistry likely determining toxicity [183]. Nanoparticle size impacts toxicity, with smaller particles ($\leq$10 nm or $\leq$ 45 nm for nanoprisms) having higher surface-area-to-volume ratios, leading to faster dissolution, increased Ag^+ ion release, and enhanced cellular uptake and internalization [184–186]. Surface chemistry affects toxicity, but capping agents like proteins, polyphenols, or synthetic coatings such as citrate, PVP, or BPEI regulate the nanoparticle's surface charge (zeta potential), stability, and interaction with biological membranes. However, the texts emphasize the difficulty of distinguishing these processes because geometry-related dissolving rates are inconsistent [187], and small size directly facilitates dissolution, correlating size with ionic toxicity. Additionally, leaching may produce ultra-small nanoparticles (1–30 nm) with distinct quantum properties, obscuring the differences between these groups. In conclusion, nanoparticle toxicity research has focused on size, but purified suspensions show that surface charge is a key determinant of AgNP toxicity [188].

6. Application of biological synthesis silver nanoparticles

Owing to their distinctive characteristics, eco-friendly AgNPs are intriguing candidates for photocatalytic and sensor applications [189, 190]. Due to their wide antibacterial activity against diverse pathogens, cost-effectiveness, and easy chemical and sustainable synthesis, the biomedical, pharmaceutical, environmental, and industrial sectors have adopted them.

6.1 Biomedical application

Silver nanoparticles (AgNPs), which are 1–100 nm in diameter, possess higher surface energy, catalytic characteristics, and antibacterial activity compared to bulk silver [191, 192]. Due to their powerful antibacterial capabilities, biocompatibility, and ease of surface functionalization, AgNPs are widely studied in drug delivery, wound healing, diagnostics, vaccine adjuvants, and biosensors [31, 64, 193, 194]. Customizing AgNPs' size, shape, and surface chemistry allows for application-specific features [195, 196]. Silver nanoparticles (AgNPs) are extensively used in biomedicine owing to their potent antibacterial, antiviral, and anticancer effects, as well as their size-dependent biocompatibility. Their microscopic size enables direct contact with microbial membranes, leading to membrane rupture, ROS production, DNA damage, and enzyme inhibition. These mechanisms enhance the inhibition of microbial growth and reproduction. Functionalized silver nanoparticles (Ag@PTX) with paclitaxel demonstrated improved cytotoxicity against A549 lung cancer cells and inhibited tumor growth, as reported by Li *et al.* [192]. Moradi *et al.* [240] developed green AgNPs using Taxus leaf extract, which exhibited an IC50 of 1.5 mM, lower than Taxol's 3.1 mM, indicating superior anticancer efficacy with fewer adverse effects. Wound dressings incorporate silver nanoparticles (AgNPs) to prevent infection and promote tissue regeneration [195]. Noble metal nanoparticles, particularly silver, are widely utilized as antibacterials [196–201]. Le *et al.* [202, 203] and Shrivastava *et al.* [204] tested silver nanoparticles against Escherichia coli and Staphylococcus aureus.

To reduce microbiological contamination and metallic taste in food packaging, researchers have added silver nanoparticles to filter paper, LDPE, and PMMA. Smith and Zysk reported that wastewater treatment facilities use silver nanoparticles for their antibacterial activity [205]. Silver dispersion on membrane surfaces increased *E. coli* and *S. aureus* antibacterial efficiency. The Ag-PES membrane may suppress bacterial growth by 100%, according to this study [206]. They prevent wound infections and fight cancer as antibacterial dressings [207]. PMMA bone cements include antimicrobial silver nanoparticles [208]. Implanted devices and surgical masks containing silver are antimicrobial [209]. Silver protects HIV-1-infected cells [210].

6.2 Environmental applications

Water filtration systems employ silver nanoparticles (AgNPs) to prevent biofouling and microbiological contamination. Their antibacterial effect comes from the release of silver ions (Ag^+), which damage cell membranes and hinder respiration and reproduction [211]. AgNPs are immobilized on ceramic membranes to extend activity and reduce leaching [212].

Despite these advantages, AgNPs' ecological effects in aquatic habitats remain a concern. Prolonged operation might cause slow leaching of immobilized Ag^+ ions or unattached nanoparticles, leading to waste stream accumulation. The accidental discharge of AgNPs highlights the necessity of assessing their life cycle dynamics beyond filtering device operation [213, 214].

Thus, AgNPs' waste stream disposal must be assessed to balance their technological benefits with environmental safety. Nanoparticle immobilization, AgNP recovery, and biodegradable stabilizers are proposed as ecological mitigation measures. In addition, risk assessment frameworks increasingly monitor dissolved Ag^+ and particulate AgNPs in treated effluents to understand their long-term ecological effects. Incorporating these environmental parameters into AgNP-based water purification systems would ensure that biofouling resistance increases are not offset by downstream pollution, thereby boosting their sustainability [215].

6.3 Industrial applications

AgNPs are valued in electronics for their electrical conductivity and shapeability. Ultrafine silver powders, precursors for AgNPs, are nanometer to submicron-sized spheres, rods, and plates. Surface area and performance are greatly influenced by nanostructure architecture, notably in antibacterial coatings and conductive inks [216]. Progress in physical and chemical synthesis has allowed for nanoparticle size and shape control. In the recent decade, silver nanocubes, nanowires, and nanospheres have been developed for printed electronics, sensors, and optoelectronic devices.

The biological production of silver nanoparticles advances sustainable nanotechnology by eliminating the main drawbacks of traditional technologies: high prices, toxicity, and environmental impact. The switch to biological synthesis utilizes microbes, plants, and food waste as sustainable reducing and stabilizing agents. This discussion emphasizes how phytochemicals, microbial activities, and biopolymers improve Ag NPs [217].

7. Conclusion

This research addresses green synthesis processes and applications. It critically analyzes structure-function correlations and explores future research challenges while addressing the whole lifecycle of green/biological synthesis of AgNPs.

This debate emphasizes green nanotechnology by examining nanoparticle manufacturing employing plants, microorganisms (bacteria, fungi, algae), and particular biomolecules as natural reducing and stabilizing agents to reduce silver ions. These biogenic methods govern nanoparticle size, shape, and stability without toxic chemicals or significant energy consumption. Green synthesis provides a sustainable alternative to chemical reduction; however, linking raw biogenic production and functional application is difficult. This chapter shows that biogenic routes improve biocompatibility and functionality, making this Green/Biological Synthesis of Silver Nanoparticles useful for biomedical, antibacterial, and environmental applications, and paving the way for sustainable nanotechnology.

Recent research has focused on large-scale, high-yield Ag NP manufacturing with adjustable particle shapes and sizes. However, uniform NP sizes and forms remain a worldwide challenge [218]. Repeatable synthesis and detailed toxicity evaluation techniques are currently lacking. AI integration into AgNP research provides new insights into synthesis optimization, toxicity prediction, and application design. Machine learning methods and predictive modeling are used to optimize synthesis conditions, forecast particle behavior, evaluate toxicity, and create application-specific nanostructures [53]. Multiple machine learning (ML) models have been utilized to enhance silver nanoparticle (AgNP) synthesis and forecast toxicity. Algorithms, including Artificial Neural Networks (ANN), Support Vector Regression (SVR), and Random Forest models, have been employed for synthesis optimization to correlate experimental parameters (e.g., precursor concentration, pH, temperature, and reaction time) with nanoparticle size and stability [219, 220]. Classification models, including Support Vector Machines (SVM), k-Nearest Neighbors (k-NN), and Random Forest classifiers, have been utilized for toxicity prediction to forecast cytotoxic or ecotoxic effects based on physicochemical descriptors such as particle size, surface charge, and surface functionalization. These machine learning methodologies reduce experimental burdens and facilitate more efficient and predictive nanoparticle design [221, 222].

AI tools

The author acknowledges the use of ChatGPT for translating some text from French into English, for language polishing of the manuscript, and for assisting in the design of certain figures.

Conflict of Interest

The authors declare no conflict of Interest.

Author details

Yasmina Khane[1,2]*, Fares Fenniche[1,2], Sofiane Khane[1] and Hafsi Zoulikha[1,2]

1 Faculty of Sciences and Technology, University of Ghardaïa, Ghardaïa, Algeria

2 Materials, Energy Systems Technology and Environment Laboratory, Faculty of Sciences and Technology, University of Ghardaïa, Ghardaïa, Algeria

*Address all correspondence to: yasminekhane@yahoo.fr; khane.yasmina@univ-ghardaia.edu.dz

References

[1] Nie P, Zhao Y, Xu H. Synthesis, applications, toxicity and toxicity mechanisms of silver nanoparticles: A review, Ecotoxicol. Ecotoxicology & Environmental Safety. 2023; **253**:114636

[2] Patil SP, Burungale VV. Physical and Chemical Properties of Nanomaterials. Elsevier: Nanomedicines for Breast Cancer Theranostics; 2020. p. 17–31

[3] Hosny S, Abu-Dief AM, Farhan N, Shehata MR, Abd El-Baki RF. Green Synthesis of Nanoscale Mixed-Ligand Complexes: DNA Interaction, Biomedical Applications, and Computational Studies. Applied Organometallic Chemistry. 2025;**39**: e70230

[4] Medici S, Peana M, Pelucelli A, Zoroddu MA. In: An updated overview on metal nanoparticles toxicity. 2021;**76**:17–26

[5] Ali SH, Emran MY, Gomaa H. Rice husk-derived nanomaterials for potential applications. In Waste Recycling Technologies for Nanomaterials Manufacturing. Springer; 2021. p. 541–588

[6] Abbas R, Luo J, Qi X, Naz A, Khan IA, Liu H, Yu S, Wei J. Silver nanoparticles: Synthesis, structure, properties and applications. Nanomaterials (Basel, Switzerland). 2024;**14**:1425

[7] Auffan M, Rose J, Wiesner MR, Bottero J-Y. Chemical stability of metallic nanoparticles: A parameter controlling their potential cellular toxicity in vitro. Environmental Pollution. 2009;**157**:1127–1133

[8] Jamkhande PG, Ghule NW, Bamer AH, Kalaskar MG. Metal nanoparticles synthesis: An overview on methods of preparation, advantages and disadvantages, and applications. Journal of Drug Delivery Science and Technology. 2019;**53**:101174

[9] Hoang V-T, Mai M, Tam LT, Vu NP, Khi NT, Tam PD, Huy TQ, Le A-T, Dinh NX, Tran V-H. Functionalized-AgNPs for Long-Term Stability and Its Applicability in the Detection of Manganese Ions. Advances in Polymer Technology. 2020;**2020**:9437108

[10] Altammar KA. A review on nanoparticles: Characteristics, synthesis, applications, and challenges, Front. Microbiology. 2023;**14**:1155622

[11] Shanmuganathan R, Karuppusamy I, Saravanan M, Muthukumar H, Ponnuchamy K, Ramkumar VS, Pugazhendhi A. Synthesis of silver nanoparticles and their biomedical applications-a comprehensive review. Current Pharmaceutical Design. 2019;**25**:2650–2660

[12] Zargar M, Hamid AA, Bakar FA, Shamsudin MN, Shameli K, Jahanshiri F, Farahani F. Green synthesis and antibacterial effect of silver nanoparticles using Vitex negundo L. Molecules. 2011;**16**:6667–6676

[13] Alexander JW. History of the medical use of silver. Surgical Infections. 2009;**10**:289–292

[14] Spear M. Silver: An age-old treatment modality in modern times.

Plastic and Aesthetic Nursing. 2010;**30**:90–93

[15] Gerba CP. Silver as disinfectant. Encyclopedia of Metalloproteins. New york, Ny: Springer;2013

[16] Barillo DJ, Marx DE. Silver in medicine: A brief history BC 335 to present. Burns. 2014;**40**:S3–S8

[17] Sim W, Barnard RT, Blaskovich MAT, Ziora ZM. Antimicrobial silver in medicinal and consumer applications: A patent review of the past decade (2007–2017). Antibiotics. 2018;**7**:93

[18] Nasrollahzadeh M, Mahmoudi-Gom Yek S, Motahharifar N, Gorab MG. Recent developments in the plant-mediated green synthesis of Ag-based nanoparticles for environmental and catalytic applications. The Chemical Record. 2019;**19**:2436–2479

[19] Bouafia A, Laouini SE, Ahmed ASA, Soldatov AV, Algarni H, Chong KF, Ali GAM. The recent progress on silver nanoparticles: Synthesis and electronic applications. Nanomaterials. 2021;**11**:2318

[20] Sintubin L, De Windt W, Dick J, Mast J, Ha DVD, Verstraete W, Boon N. Lactic acid bacteria as reducing and capping agent for the fast and efficient production of silver nanoparticles. Applied Microbiology and Biotechnology. 2009;**84**:741–749

[21] Tran QH, Nguyen VQ, Le A-T. Silver nanoparticles: Synthesis, properties, toxicology, applications and perspectives. Advances in Natural Sciences: Nanoscience and Nanotechnology. 2013;**4**:33001

[22] de Lima R, Seabra AB, Durán N. Silver nanoparticles: A brief review of cytotoxicity and genotoxicity of chemically and biogenically synthesized nanoparticles. Journal of Applied Toxicology. 2012;**32**:867–879

[23] Burdușel A-C, Gherasim O, Grumezescu AM, Mogoantă L, Ficai A, Andronescu E. Biomedical applications of silver nanoparticles: An up-to-date overview. Nanomaterials. 2018;**8**:681

[24] Meher A, Tandi A, Moharana S, Chakroborty S, Mohapatra SS, Mondal A, Dey S, Chandra P. Silver nanoparticle for biomedical applications: A review. Hybrid Advances. 2024;**6**:100184

[25] Ismail M, Jabra R. Investigation the parameters affecting on the synthesis of silver nanoparticles by chemical reduction method and printing a conductive pattern. Journal of Materials and Environmental Science. 2017;**8**:4152–4159

[26] Shayo GM, Elimbinzi E, Shao GN. Preparation methods, applications, toxicity and mechanisms of silver nanoparticles as bactericidal agent and superiority of green synthesis method. Heliyon. 2024;**10**:e36539

[27] Parida VK, Das S, Maity S, Mahanty A, Datta D, Pradhan A. Green Synthesis of Nanocatalysts and Nanomaterials for Effluent Treatment. In Sustainable Effluent Treatment and Resource Recovery. Vol. **1**. ACS Publications; 2025. p. 1–44

[28] Sankar R, Karthik A, Prabu A, Karthik S, Shivashangari KS, Ravikumar V. Origanum vulgare mediated biosynthesis of silver nanoparticles for its antibacterial

and anticancer activity. Colloids and Surfaces B: Biointerfaces. 2013;**108**:80–84

[29] Ghodake GS, Deshpande NG, Lee YP, Jin ES. Pear fruit extract-assisted room-temperature biosynthesis of gold nanoplates. Colloids and Surfaces B: Biointerfaces. 2010;**75**:584–589

[30] Sanghi R, Verma P. Biomimetic synthesis and characterisation of protein capped silver nanoparticles. Bioresource Technology. 2009;**100**:501–504

[31] Xu L, Wang -Y-Y, Huang J, Chen C-Y, Wang Z-X, Xie H. Silver nanoparticles: Synthesis, medical applications and biosafety. Theranostics. 2020;**10**:8996

[32] Alharbi NS, Alsubhi NS, Felimban AI. Green synthesis of silver nanoparticles using medicinal plants: Characterization and application. Journal of Radiation Research and Applied Sciences. 2022;**15**:109–124

[33] Al-Khayri JM, Alnaddaf LM, Jain SM. Nanomaterial Interactions with Plant Cellular Mechanisms and Macromolecules and Agricultural Implications. Springer; 2023

[34] Noruzi M, Zare D, Khoshnevisan K, Davoodi D. Rapid green synthesis of gold nanoparticles using Rosa hybrida petal extract at room temperature, Spectrochim. Spectrochimica Acta, Part A: Molecular and Biomolecular Spectroscopy. 2011;**79**:1461–1465

[35] Eker F, Duman H, Akdaşçi E, Bolat E, Sarıtaş S, Karav S, Witkowska AM. A comprehensive review of nanoparticles: From classification to application and toxicity. Molecules (Basel, Switzerland). 2024;**29**:3482

[36] Sudarman F, Shiddiq M, Armynah B, Tahir D. Silver nanoparticles (AgNPs) synthesis methods as heavy-metal sensors: A review. International Journal of Environmental Science & Technology. 2023;**20**:9351–9368

[37] Eker F, Akdaşçi E, Duman H, Bechelany M, Karav S. Green synthesis of silver nanoparticles using plant extracts: A comprehensive review of physicochemical properties and multifunctional applications. International Journal of Molecular Sciences. 2025;**26**:6222

[38] Dawadi S, Katuwal S, Gupta A, Lamichhane U, Thapa R, Jaisi S, Lamichhane G, Bhattarai DP, Parajuli N. Current research on silver nanoparticles: Synthesis, characterization, and applications. Journal of Nanomaterials. 2021;**2021**:6687290

[39] Oliveira JRP, Lenzi GG. Advances in the use of green and sustainable synthesis to obtain nanomaterials. In Green Chem. Environ. Sustain. Approach. IntechOpen; 2023

[40] Morenko I, Isaeva I, Odinokova I, Ostaeva G. Green synthesis of metal nanoparticles and application of nanoadditives in diesel fuel: Bibliometric analysis. In E3S Web of Conferences. EDP Sciences; 2024. p. 4015

[41] Sharma VK, Yngard RA, Lin Y. Silver nanoparticles: Green synthesis and their antimicrobial activities. Advances in Colloid and Interface Science. 2009;**145**:83–96

[42] Degrecci MG, Parreira AG. Estudo exploratório de patentes de produção de AgNPs por rotas de síntese verde. Research, Society and Development. 2022;**11**:e57111932239

[43] Khodashenas B, Ghorbani HR. Synthesis of silver nanoparticles with different shapes. Arabian Journal of Chemistry. 2019;**12**:1823–1838

[44] El-Seedi HR, Omara MS, Omar AH, Elakshar MM, Shoukhba YM, Duman H, Karav S, Rashwan AK, El-Seedi AH, Altaleb HA. Updated review of metal nanoparticles fabricated by green chemistry using natural extracts: Biosynthesis, mechanisms, and applications. Bioengineering. 2024;**11**:1095

[45] Pryshchepa O, Pomastowski P, Buszewski B. Silver nanoparticles: Synthesis, investigation techniques, and properties. Advances in Colloid and Interface Science. 2020;**284**:102246

[46] Braydich-Stolle L, Hussain S, Schlager JJ, Hofmann M-C. In vitro cytotoxicity of nanoparticles in mammalian germline stem cells. Toxicological Sciences. 2005;**88**:412–419

[47] Wei L, Lu J, Xu H, Patel A, Chen Z-S, Chen G. Silver nanoparticles: Synthesis, properties, and therapeutic applications. Drug Discovery Today. 2015;**20**:595–601

[48] Kang W, Cheng C, Li Z, Feng Y, Shen G, Du X. Ultrafine Ag nanoparticles as active catalyst for electrocatalytic hydrogen production. ChemCatChem. 2019;**11**:5976–5981

[49] Kibis LS, Stadnichenko AI, Pajetnov EM, Koscheev SV, Zaykovskii VI, Boronin AI. The investigation of oxidized silver nanoparticles prepared by thermal evaporation and radio-frequency sputtering of metallic silver under oxygen. Applied Surface Science. 2010;**257**:404–413

[50] Miranzadeh M, Kassaee MZ. Solvent effects on arc discharge fabrication of durable silver nanopowder and its application as a recyclable catalyst for elimination of toxic p-nitrophenol. Chemical Engineering Journal. 2014;**257**:105–111

[51] Kylián O, Kuzminova A, Štefaníková R, Hanuš J, Solař P, Kúš P, Cieslar M, Choukourov A, Biederman H. Silver/plasma polymer strawberry-like nanoparticles produced by gas-phase synthesis. Materials Letters. 2019;**253**:238–241

[52] Amendola V, Polizzi S, Meneghetti M. Free silver nanoparticles synthesized by laser ablation in organic solvents and their easy functionalization. Langmuir. 2007;**23**:6766–6770

[53] Menazea AA. Femtosecond laser ablation-assisted synthesis of silver nanoparticles in organic and inorganic liquids medium and their antibacterial efficiency. Radiation Physics and Chemistry. 2020;**168**:108616

[54] Amendola V, Meneghetti M. Laser ablation synthesis in solution and size manipulation of noble metal nanoparticles. Physical Chemistry Chemical Physics. 2009;**11**:3805–3821

[55] Szegedi Á, Popova M, Valyon J, Guarnaccio A, De Stefanis A, De Bonis A, Orlando S, Sansone M, Teghil R, Santagata A. Comparison of silver nanoparticles confined in nanoporous silica prepared by chemical synthesis and by ultra-short pulsed laser

ablation in liquid. Applied Physics A. 2014;**117**:55–62

[56] Zhang J, Chaker M, Ma D. Pulsed laser ablation based synthesis of colloidal metal nanoparticles for catalytic applications. Journal of Colloid and Interface Science. 2017;**489**:138–149

[57] Lytvyn SY, Kurapov YA, Ruban NM, Churkina LN, Andrusyshyna IM, Didikin GG, Boretskyi VV. Influence of temperature on the physical properties and bio-activity of pure (ligand-free) EB PVD silver nanoparticles. Applied Nanoscience. 2023;**13**:5171–5183

[58] Chandra R, Taneja P, John J, Ayyub P, Dey GK, Kulshreshtha SK. Synthesis and TEM study of nanoparticles and nanocrystalline thin films of silver by high pressure sputtering. Nanostructured Materials. 1999;**11**:1171–1179

[59] El-Khatib AM, Badawi MS, Ghatass ZF, Mohamed MM, Elkhatib M. Synthesize of silver nanoparticles by arc discharge method using two different rotational electrode shapes. Journal of Cluster Science. 2018;**29**:1169–1175

[60] Apriliani AI, Berliana JD, Putri RAP, Rohilah S, Thifalizalfa V, Guniawaty Y, Nandiyanto ABD. Synthesis of silver nanoparticles in several methods. Maghrebian Journal of Pure & Applied Sciences. 2020;**6**:91–110

[61] Tien D-C, Tseng K-H, Liao C-Y, Huang J-C, Tsung -T-T. Discovery of ionic silver in silver nanoparticle suspension fabricated by arc discharge method. Journal of Alloys and Compounds. 2008;**463**:408–411

[62] Biswas A, Bayer IS, Biris AS, Wang T, Dervishi E, Faupel F. Advances in top–down and bottom–up surface nanofabrication: Techniques, applications & future prospects. Advances in Colloid and Interface Science. 2012;**170**:2–27

[63] Rafique M, Sadaf I, Rafique MS, Tahir MB. A review on green synthesis of silver nanoparticles and their applications, Artif. Artificial Cells, Nanomedicine, and Biotechnology. 2017;**45**:1272–1291

[64] Hosny S, Gaber GA, Ragab MS, Ragheb MA, Anter M, Mohamed LZ. A comprehensive review of silver nanoparticles (AgNPs): Synthesis strategies, toxicity concerns, biomedical applications, AI-driven advancements, challenges, and future perspectives. Arabian Journal for Science and Engineering. 2025;1–48

[65] Chugh D, Viswamalya VS, Das B. Green synthesis of silver nanoparticles with algae and the importance of capping agents in the process. Journal of Genetic Engineering and Biotechnology. 2021;**19**:126

[66] Polte J, Tuaev X, Wuithschick M, Fischer A, Thuenemann AF, Rademann K, Kraehnert R, Emmerling F. Formation mechanism of colloidal silver nanoparticles: Analogies and differences to the growth of gold nanoparticles. ACS Nano. 2012;**6**:5791–5802

[67] Liu T, Yin B, He T, Guo N, Dong L, Yin Y. Complementary effects of nanosilver and superhydrophobic coatings on the prevention of marine bacterial adhesion. ACS Applied Materials & Interfaces. 2012;**4**:4683–4690

[68] Desai R, Mankad V, Gupta SK, Jha PK. Size distribution of silver

nanoparticles: UV-visible spectroscopic assessment. Nanoscience and Nanotechnology Letters. 2012;**4**:30–34

[69] Blommaerts N, Vanrompay H, Nuti S, Lenaerts S, Bals S, Verbruggen SW. Unraveling structural information of Turkevich synthesized plasmonic gold–silver bimetallic nanoparticles. Small. 2019;**15**:1902791

[70] Chen Z, Balankura T, Fichthorn KA, Rioux RM. Revisiting the polyol synthesis of silver nanostructures: Role of chloride in nanocube formation. ACS Nano. 2019;**13**:1849–1860

[71] Silva RRD, Yang M, Choi S-I, Chi M, Luo M, Zhang C, Li Z-Y, Camargo PHC, Ribeiro SJL, Xia Y. Facile synthesis of sub-20 nm silver nanowires through a bromide-mediated polyol method. ACS Nano. 2016;**10**:7892–7900

[72] Eby DM, Luckarift HR, Johnson GR. Hybrid antimicrobial enzyme and silver nanoparticle coatings for medical instruments. ACS Applied Materials & Interfaces. 2009;**1**:1553–1560

[73] Alqadi MK, Abo Noqtah OA, Alzoubi FY, Alzouby J, Aljarrah K. pH effect on the aggregation of silver nanoparticles synthesized by chemical reduction. Materials Science. 2014;**32**:107–111

[74] Von White G, Kerscher P, Brown RM, Morella JD, Mc Allister W, Dean D, Kitchens CL. Green synthesis of robust, biocompatible silver nanoparticles using garlic extract. Journal of Nanomaterials. 2012; **2012**:730746

[75] Ghosh S, Patil S, Ahire M, Kitture R, Kale S, Pardesi K, Cameotra SS, Bellare J, Dhavale DD, Jabgunde A. Synthesis of silver nanoparticles using Dioscorea bulbifera tuber extract and evaluation of its synergistic potential in combination with antimicrobial agents. International Journal of Nanomedicine. 2012;7:483–496

[76] Guo X, Zhang Q, Ding X, Shen Q, Wu C, Zhang L, Yang H. Synthesis and application of several sol–gel-derived materials via sol–gel process combining with other technologies: A review. Journal of Sol-Gel Science and Technology. 2016;**79**:328–358

[77] Rajeswari R, Prabu HG, Amutha DM. One Pot Hydrothermal synthesis characterizations of silver nanoparticles on reduced graphene oxide for its enhanced antibacterial and antioxidant properties. IOSR Journal of Applied Chemistry. 2017;**10**:64–69

[78] Maślak E, Arendowski A, Złoch M, Walczak-Skierska J, Radtke A, Piszczek P, Pomastowski P. Silver nanoparticle targets fabricated using chemical vapor deposition method for differentiation of bacteria based on lipidomic profiles in laser desorption/ionization mass spectrometry. Antibiotics. 2023;**12**:874

[79] Ayhan ME. A single-step fabrication of Ag nanoparticles and CVD graphene hybrid nanostructure as SERS substrate. Microelectronic Engineering. 2020;**233**:111421

[80] Petica A, Buruntea N, Nistor C, Ionescu CR. Antimicrobial colloidal silver solutions. Preparation and characterization. Journal of Optoelectronics and Advanced Materials. 2007;**9**:3435

[81] Dobre N, Petică A, Buda M, Anicăi L, Vişan T. Electrochemical

synthesis of silver nanoparticles in aqueous electrolytes. UPB Science Bulletin. 2014;**76**:127–136

[82] Singaravelan R, Alwar SBS. Electrochemical synthesis, characterisation and phytogenic properties of silver nanoparticles. Applied Nanoscience. 2015;**5**:983–991

[83] Kuntyi OI, Kytsya AR, Mertsalo IP, Mazur AS, Zozula GI, Bazylyak LI, Topchak RV. Electrochemical synthesis of silver nanoparticles by reversible current in solutions of sodium polyacrylate. Colloid and Polymer Science. 2019;**297**:689–695

[84] Raut SS, Sharma A, Yadav R, Kumar S, Shukla A, Singh A, Mishra A. Biological Green Synthesis of Nanoparticles: A Sustainable Platform for Agricultural, Environmental, Food, and Cosmetic Applications. Journal of Environmental Chemical Engineering. 2026:121102

[85] Kumar S, Bhushan P, Bhattacharya S. Fabrication of nanostructures with bottom-up approach and their utility in diagnostics, therapeutics, and others. In Environ. Chem. Med. Sensors. Springer; 2017. p. 167–198

[86] Chaloupka K, Malam Y, Seifalian AM. Nanosilver as a new generation of nanoproduct in biomedical applications. Trends in Biotechnology. 2010;**28**:580–588

[87] Srikar SK, Giri DD, Pal DB, Mishra PK, Upadhyay SN. Green synthesis of silver nanoparticles: A review. Green and Sustainable Chemistry. 2016;**6**:34

[88] Das G, Shin H-S, Patra JK. Comparative assessment of antioxidant, anti-diabetic and cytotoxic effects of three peel/shell food waste extract-mediated silver nanoparticles. International Journal of Nanomedicine. 2020;Vol **15**:9075–9088

[89] Rauwel P, Küünal S, Ferdov S, Rauwel E. A review on the green synthesis of silver nanoparticles and their morphologies studied via TEM. Advances in Materials Science and Engineering. 2015; **2015**:682749

[90] Bhat RS, Al-Dbass AM, Khayyat AIA, Al-Daihan S. Utilizing biomolecule-rich citrus fruit waste as a medium for the eco-friendly preparation of silver nanoparticles with antimicrobial properties. Inorganics. 2024;**12**:180

[91] Shreyash N, Bajpai S, Khan MA, Vijay Y, Tiwary SK, Sonker M. Green synthesis of nanoparticles and their biomedical applications: A review. ACS Applied Nano Materials. 2021;**4**:11428–11457

[92] Ghaffari-Moghaddam M, Hadi-Dabanlou R, Khajeh M, Rakhshanipour M, Shameli K. Green synthesis of silver nanoparticles using plant extracts. Korean Journal of Chemical Engineering. 2014;**31**:548–557

[93] Sastry M, Ahmad A, Khan MI, Kumar R. Biosynthesis of metal nanoparticles using fungi and actinomycete. Current Science. 2003;**85** (2):162–170

[94] Durán N, Marcato PD, Alves OL, De Souza GIH, Esposito E. Mechanistic aspects of biosynthesis of silver nanoparticles by several Fusarium oxysporum strains. Journal of Nanobiotechnology. 2005;**3**:8

[95] Logeswari P, Silambarasan S, Abraham J. Synthesis of silver nanoparticles using plants extract and analysis of their antimicrobial property. Journal of Saudi Chemical Society. 2015;**19**:311–317

[96] Jain AS, Pawar PS, Sarkar A, Junnuthula V, Dyawanapelly S. Bionanofactories for green synthesis of silver nanoparticles: Toward antimicrobial applications. International Journal of Molecular Sciences. 2021;**22**:11993. 11993

[97] Al-Musawi MH, Ibrahim KM, Albukhaty S. In vitro study of antioxidant, antibacterial, and cytotoxicity properties of Cordia myxa fruit extract. Iranian Journal of Microbiology. 2022;**14**:97

[98] Shang Y, Hasan MK, Ahammed GJ, Li M, Yin H, Zhou J. Applications of nanotechnology in plant growth and crop protection: A review. Molecules. 2019;**24**:2558

[99] Tripathi RM, Ranac D, Shrivastav A, Singh RP, Shrivastav BR. Biogenic Synthesis of Silver Nanoparticles Using Saraca indica Leaf Extract and Evaluation of Their Antibacterial Activity. Nano Biomedicine and Engineering. 2013;**5**

[100] Huang J, Lin L, Sun D, Chen H, Yang D, Li Q. Bio-inspired synthesis of metal nanomaterials and applications. Chemical Society Reviews. 2015;**44**:6330–6374

[101] Mikhailova EO. Silver nanoparticles: Mechanism of action and probable bio-application. Journal of Functional Biomaterials. 2020;**11**:84

[102] Kasithevar M, Saravanan M, Prakash P, Kumar H, Ovais M, Barabadi H, Shinwari ZK. Green synthesis of silver nanoparticles using Alysicarpus monilifer leaf extract and its antibacterial activity against MRSA and CoNS isolates in HIV patients. Journal of Interdisciplinary Nanomedicine. 2017;**2**:131–141

[103] Mondal AH, Yadav D, Mitra S, Mukhopadhyay K. Biosynthesis of silver nanoparticles using culture supernatant of Shewanella sp. ARY1 and their antibacterial activity. International Journal of Nanomedicine. 2020;Vol **15**:8295–8310

[104] Arunachalam KD, Arunachalam KD. Memecylon edule leaf extract mediated green synthesis of silver and gold nanoparticles. International Journal of Nanomedicine. 2011;1265–1278

[105] Ebrahiminezhad A, Taghizadeh S, Ghasemi Y. Green Synthesis of Silver Nanoparticles using Mediterranean Cypress (Cupressus sempervirens) Leaf Extract. American Journal of Biochemistry and Biotechnology. 2017;**13**:1–6

[106] Parlinska-Wojtan M, Kus-Liskiewicz M, Depciuch J, Sadik O. Green synthesis and antibacterial effects of aqueous colloidal solutions of silver nanoparticles using camomile terpenoids as a combined reducing and capping agent. Bioprocess and Biosystems Engineering. 2016;**39**:1213–1223

[107] Shankar SS, Rai A, Ahmad A, Sastry M. Rapid synthesis of Au, Ag, and bimetallic Au core–Ag shell nanoparticles using Neem (Azadirachta indica) leaf broth. Journal of Colloid and Interface Science. 2004;**275**:496–502

[108] Mahendia S, Tomar AK, Goyal PK, Kumar S. Tuning of refractive

index of poly (vinyl alcohol): Effect of embedding Cu and Ag nanoparticles. Journal of Applied Physics. 2013;113

[109] Shyam A, S SC, George B, E S. Plant mediated synthesis of AgNPs and its applications: An overview. Inorganic and Nano-Metal Chemistry. 2021;**51**:1646–1662

[110] Kumar DA, Palanichamy V, Roopan SM. Green synthesis of silver nanoparticles using Alternanthera dentata leaf extract at room temperature and their antimicrobial activity, Spectrochim. Spectrochimica Acta, Part A: Molecular and Biomolecular Spectroscopy. 2014;**127**:168–171

[111] SantoshKumar SK, Daimary RM, Mwkthang Swargiary MS, Anjali Brahma AB, Sudhesh Kumar SK, Mukesh Singh MS, Biosynthesis of silver nanoparticles using Premna herbacea leaf extract and evaluation of its antimicrobial activity against bacteria causing dysentery., (2013).

[112] Rodríguez-Félix F, Graciano-Verdugo AZ, Moreno-Vásquez MJ, Lagarda-Díaz I, Barreras-Urbina CG, Armenta-Villegas L, Olguín-Moreno A, Tapia-Hernández JA. Trends in Sustainable Green Synthesis of Silver Nanoparticles Using Agri-Food Waste Extracts and Their Applications in Health. Journal of Nanomaterials. 2022;**2022**:8874003

[113] Xu H, Wang L, Su H, Gu L, Han T, Meng F, Liu C. Making good use of food wastes: Green synthesis of highly stabilized silver nanoparticles from grape seed extract and their antimicrobial activity. Food Biophysics. 2015;**10**:12–18

[114] Roy S, Das TK. Plant mediated green synthesis of silver nanoparticles-A. International Journal of Plant Biology & Research. 2015;**3**:1044–1055

[115] Sharma K, Kaushik S, Jyoti A. Green synthesis of silver nanoparticles by using waste vegetable peel and its antibacterial activities. Journal of Pharmaceutical Sciences and Research. 2016;**8**:313

[116] Vasyliev G, Vorobyova V, Skiba M, Khrokalo L. Green synthesis of silver nanoparticles using waste products (apricot and black currant pomace) aqueous extracts and their characterization. Advances in Materials Science and Engineering. 2020;**2020**:4505787

[117] Roy A, Bulut O, Some S, Mandal AK, Yilmaz MD. Green synthesis of silver nanoparticles: Biomolecule-nanoparticle organizations targeting antimicrobial activity. RSC Advances. 2019;**9**:2673–2702

[118] Das VL, Thomas R, Varghese RT, Soniya EV, Mathew J, Radhakrishnan EK. Extracellular synthesis of silver nanoparticles by the Bacillus strain CS 11 isolated from industrialized area. 3 Biotechnology. 2014;**4**:121–126

[119] Rana A, Yadav K, Jagadevan S. A comprehensive review on green synthesis of nature-inspired metal nanoparticles: Mechanism, application and toxicity. Journal of Cleaner Production. 2020;**272**:122880

[120] Mukherjee P, Ahmad A, Mandal D, Senapati S, Sainkar SR, Khan MI, Parishcha R, Ajaykumar PV, Alam M, Kumar R. Fungus-mediated synthesis of silver nanoparticles and their immobilization in the mycelial matrix: A novel biological approach to

nanoparticle synthesis. Nano Letters. 2001;**1**:515–519

[121] Narayanan KB, Sakthivel N. Synthesis and characterization of nano-gold composite using Cylindrocladium floridanum and its heterogeneous catalysis in the degradation of 4-nitrophenol. Journal of Hazardous Materials. 2011;**189**:519–525

[122] Liu L, Liu T, Tade M, Wang S, Li X, Liu S. Less is more, greener microbial synthesis of silver nanoparticles. Enzyme and Microbial Technology. 2014;**67**:53–58

[123] Karthik L, Kumar G, Kirthi AV, Rahuman AA, Rao KVB. Streptomyces sp. LK3 mediated synthesis of silver nanoparticles and its biomedical application. Bioprocess and Biosystems Engineering. 2014;**37**:261–267

[124] Slawson RM, Trevors JT, Lee H. Silver accumulation and resistance in Pseudomonas stutzeri. Archives of Microbiology. 1992;**158**:398–404

[125] Pooley FD. Bacteria accumulate silver during leaching of sulphide ore minerals. Nature. 1982;**296**:642–643

[126] Dhillon GS, Brar SK, Kaur S, Verma M. Green approach for nanoparticle biosynthesis by fungi: Current trends and applications. Critical Reviews in Biotechnology. 2012;**32**:49–73

[127] Siddiqi KS, Husen A. Fabrication of metal and metal oxide nanoparticles by algae and their toxic effects. Nanoscale Research Letters. 2016;**11**:363

[128] Slawson RM, Van Dyke MI, Lee H, Trevors JT. Germanium and silver resistance, accumulation, and toxicity in microorganisms. Plasmid. 1992;**27**:72–79

[129] Xue B, He D, Gao S, Wang D, Yokoyama K, Wang L. Biosynthesis of silver nanoparticles by the fungus Arthroderma fulvum and its antifungal activity against genera of Candida, Aspergillus and Fusarium. International Journal of Nanomedicine. 2016;**11**:1899–1906

[130] Honary S, Barabadi H, Gharaei-Fathabad E, Naghibi F. Green synthesis of silver nanoparticles induced by the fungus Penicillium citrinum. Tropical Journal of Pharmaceutical Research. 2013;**12**:7–11

[131] Birla SS, Gaikwad SC, Gade AK, Rai MK. Rapid Synthesis of Silver Nanoparticles from Fusarium oxysporum by Optimizing Physicocultural Conditions. The Scientific World Journal. 2013;**2013**:796018

[132] Ishida K, Cipriano TF, Rocha GM, Weissmüller G, Gomes F, Miranda K, Rozental S. Silver nanoparticle production by the fungus Fusarium oxysporum: Nanoparticle characterisation and analysis of antifungal activity against pathogenic yeasts. Memórias Do Instituto Oswaldo Cruz. 2013;**109**:220–228

[133] Pinzaru I, Coricovac D, Dehelean C, Moacă E-A, Mioc M, Baderca F, Sizemore I, Brittle S, Marti D, Calina CD. Stable PEG-coated silver nanoparticles–A comprehensive toxicological profile. Food and Chemical Toxicology. 2018;**111**:546–556

[134] Sofy AR, Hmed AA, Haliem NFAE, Zein MA-E, Elshaarawy RFM. Polyphosphonium-oligochitosans decorated with nanosilver as new prospective inhibitors for common human enteric viruses.

Carbohydrate Polymers. 2019;**226**:115261

[135] Elgorban AM, Al-Rahmah AN, Sayed SR, Hirad A, Mostafa AA-F, Bahkali AH. Antimicrobial activity and green synthesis of silver nanoparticles using Trichoderma viride. Biotechnology and Biotechnological Equipment. 2016;**30**:299–304

[136] Leung TC-Y, Wong CK, Xie Y. Green synthesis of silver nanoparticles using biopolymers, carboxymethylated-curdlan and fucoidan. Materials Chemistry and Physics. 2010;**121**:402–405

[137] Ajitha B, Reddy YAK, Jeon H-J, Ahn CW. Synthesis of silver nanoparticles in an eco-friendly way using Phyllanthus amarus leaf extract: Antimicrobial and catalytic activity. Advanced Powder Technology. 2018;**29**:86–93

[138] Wu X, Lu C, Zhou Z, Yuan G, Xiong R, Zhang X. Green synthesis and formation mechanism of cellulose nanocrystal-supported gold nanoparticles with enhanced catalytic performance. Environmental Science Nano. 2014;**1**:71–79

[139] Singh P, Pandit S, Beshay M, Mokkapati V, Garnaes J, Olsson ME, Sultan A, Mackevica A, Mateiu RV, Lütken H. Anti-biofilm effects of gold and silver nanoparticles synthesized by the Rhodiola rosea rhizome extracts. Artificial Cells, Nanomedicine, and Biotechnology. 2018;**46**:886–899

[140] Regiel-Futyra A, Kus-Liśkiewicz M, Sebastian V, Irusta S, Arruebo M, Kyzioł A, Stochel G. Development of noncytotoxic silver–chitosan nanocomposites for efficient control of biofilm forming microbes. RSC Advances. 2017;**7**:52398–52413

[141] Shin SW, Song IH, Um SH. Role of physicochemical properties in nanoparticle toxicity. Nanomaterials. 2015;**5**:1351–1365

[142] Kareem EAA, Sultan AE, Oraibi HM. Synthesis and characterization of silver nanoparticles: A review. Ibn AL-Haitham Journal for Pure and Applied Sciences. 2023;**36**:177–200

[143] Zhang X-F, Liu Z-G, Shen W, Gurunathan S. Silver nanoparticles: Synthesis, characterization, properties, applications, and therapeutic approaches. International Journal of Molecular Sciences. 2016;**17**:1534

[144] Das R, Nath SS, Chakdar D, Gope G, Bhattacharjee R. Preparation of silver nanoparticles and their characterization. Journal of Nanotechnology. 2009;**5**:1–6

[145] Link S, El-Sayed MA. Optical properties and ultrafast dynamics of metallic nanocrystals. Annual Review of Physical Chemistry. 2003;**54**:331–366

[146] Salem JK, El-Nahhal IM, Najri BA, Hammad TM. Utilization of surface Plasmon resonance band of silver nanoparticles for determination of critical micelle concentration of cationic surfactants. Chemical Physics Letters. 2016;**664**:154–158

[147] Shrivas K, Sahu S, Patra GK, Jaiswal NK, Shankar R.Localized surface plasmon resonance of silver nanoparticles for sensitive colorimetric detection of chromium in surface water, industrial waste water and vegetable samples. Analytical Methods. 2016;**8** (9):2088–2096

[148] Dubey SP, Lahtinen M, Sillanpää M. Tansy fruit mediated greener synthesis of silver and gold nanoparticles. Process Biochemistry. 2010;**45**:1065–1071

[149] Surega R, Anita B, Ramakrishnan S, Gunasekaran K, Nakkeran S. Synthesis and characterization of AgNps using plant extracts. International Journal of Current Microbiology and Applied Sciences. 2020;**9**:1939–1947

[150] Vinay CH, Goudanavar P, Acharya A. Development and characterization of pomegranate and Orange fruit peel extract based silver nanoparticles. Journal of Manmohan Memorial Institute of Health Sciences. 2018;**4**:72–85

[151] Tho NTM, An TNM, Tri MD, Sreekanth TVM, Lee J-S, Nagajyothi PC, Lee KD. Green synthesis of silver nanoparticles using Nelumbo nucifera seed extract and its antibacterial activity. Acta Chimica Slovenica. 2013;**60**:673–678

[152] Vankar PS, Shukla D. Biosynthesis of silver nanoparticles using lemon leaves extract and its application for antimicrobial finish on fabric. Applied Nanoscience. 2012;**2**:163–168

[153] Hamelian M, Zangeneh MM, Amisama A, Varmira K, Veisi H. Green synthesis of silver nanoparticles using Thymus kotschyanus extract and evaluation of their antioxidant, antibacterial and cytotoxic effects. Applied Organometallic Chemistry. 2018;**32**:e4458

[154] Ahamed MS, Ali MS, Ahmed S, Sadia SI, Islam MR, Rahaman MA, Alam MA. Synthesis of silver nanomaterials capping by fruit-mediated extracts and antimicrobial activity: A critical review. International Research Journal of Pure & Applied Chemistry. 2024;**25**:45–60

[155] Deepa RD, Kamble SS, Kamble SS, Kamakshi SG. Biosynthesis and characterization of silver nanoparticles generated from peels of Solanum tuberosum (potato) and their antibacterial and wastewater treatment potential. Frontiers in Nanotechnology. 2023;**5**:1213160

[156] Kolya H, Maiti P, Pandey A, Tripathy T. Green synthesis of silver nanoparticles with antimicrobial and azo dye (Congo red) degradation properties using Amaranthus gangeticus Linn leaf extract. Journal of Analytical Science and Technology. 2015;**6**:33

[157] Madhusudanan M, Green Synthesis of Silver Nanoparticles and their Polymer Composites for Enhanced Antimicrobial and Anticancer Applications, (2024)

[158] Srivastava V, Pandey S, Mishra A, Choubey AK. Green synthesis of biogenic silver particles, process parameter optimization and application as photocatalyst in dye degradation, SN Appl. Science. 2019;**1**:1722

[159] Joshi SJ, Mamari SA, Azkawi AA, Green synthesis of silver nanoparticles using pomegranate peel extracts and its application in photocatalytic degradation of methylene blue, (2018)

[160] Ali ZA, Yahya R, Sekaran SD, Puteh R. Green synthesis of silver nanoparticles using apple extract and its antibacterial properties. Advances in Materials Science and Engineering. 2016;**2016**:4102196

[161] Albukhaty S, Tomah AA, Khane Y, Al-Karagoly H, Kadhim NJ, Fenniche F, Aouf D. Eco-Friendly Synthesis of Silver Nanoparticles: Principles and Their Antimicrobial Characteristics. In Sustain. Nanoremediation, Apple Academic Press; 2024. p. 253–295

[162] Eker F, Akdaşçi E, Duman H, Bechelany M, Karav S. Gold nanoparticles in nanomedicine: Unique properties and therapeutic potential. Nanomaterials. 2024;**14**:1854

[163] Natsuki J, Natsuki T, Hashimoto Y. A review of silver nanoparticles: Synthesis methods, properties and applications. International Journal of Materials Science and Applications. 2015;**4**:325–332

[164] Beger M. The Future is Flat—Two-Dimensional Nanomaterials. Nanotechnology Future. 2016

[165] Duman H, Eker F, Akdaşçi E, Witkowska AM, Bechelany M, Karav S. Silver nanoparticles: A comprehensive review of synthesis methods and chemical and physical properties. Nanomaterials. 2024;**14**:1527

[166] Sotiriou GA, Pratsinis SE. Engineering nanosilver as an antibacterial, biosensor and bioimaging material. Current Opinion in Chemical Engineering. 2011;**1**:3–10

[167] Zhang J, Ahmadi M, Fargas G, Perinka N, Reguera J, Lanceros-Méndez S, Llanes L, Jiménez-Piqué E. Silver nanoparticles for conductive inks: From synthesis and ink formulation to their use in printing technologies. Metals (Basel). 2022;**12**:234

[168] Pal S, Tak YK, Song JM. Does the Antibacterial Activity of Silver Nanoparticles Depend on the Shape of the Nanoparticle? A Study of the Gram-Negative Bacterium Escherichia coli. Applied and Environmental Microbiology. 2007;**73**:1712–1720

[169] Xu R, Wang D, Zhang J, Li Y. Shape-dependent catalytic activity of silver nanoparticles for the oxidation of styrene. Chemistry – An Asian Journal. 2006;**1**:888–893

[170] Malik AQ, Ul G. Mir T, Kumar D, Mir IA, Rashid A, Ayoub M, Shukla S. A review on the green synthesis of nanoparticles, their biological applications, and photocatalytic efficiency against environmental toxins. Environmental Science and Pollution Research. 2023;**30**:69796–69823

[171] Iyahraja S, Rajadurai JS. Study of thermal conductivity enhancement of aqueous suspensions containing silver nanoparticles. AIP Advances. 2015;**5**

[172] Zhou Y, Zhuang X, Wu F, Liu F. High-performance thermal management nanocomposites: Silver functionalized graphene nanosheets and multiwalled carbon nanotube. Crystals. 2018;**8**:398

[173] Sun Z, Li J, Yu M, Kathaperumal M, Wong C-P. A review of the thermal conductivity of silver-epoxy nanocomposites as encapsulation material for packaging applications. Chemical Engineering Journal. 2022;**446**:137319. 137319

[174] Wu H, Carrete J, Zhang Z, Qu Y, Shen X, Wang Z, Zhao L-D, He J. Strong enhancement of phonon scattering through nanoscale grains in lead sulfide thermoelectrics. NPG Asia Materials. 2014;**6**:e108

[175] Lok C-N, Ho C-M, Chen R, He Q-Y, Yu W-Y, Sun H, Tam PK-H, Chiu J-F, Che C-M. Proteomic analysis of the mode of antibacterial action of silver nanoparticles. Journal of Proteome Research. 2006;**5**:916–924

[176] Labanni A, Nasir M, Arief S. Research progress and prospect of copper oxide nanoparticles with controllable nanostructure, morphology, and function via green synthesis. Mater Today Sustain. 2023;**24**:100526

[177] Sarkar T, Kundu S, Ghorai G, Sahoo PK, Bhattacharjee A. Structural, spectroscopic and morphology studies on green synthesized ZnO nanoparticles. Advances in Natural Sciences: Nanoscience and Nanotechnology. 2023;**14**:035001

[178] Mahiuddin M, Ochiai B. Green synthesis of crystalline bismuth nanoparticles using lemon juice. RSC Advances. 2021;11,26683–26686

[179] Yusuf M. Silver Nanoparticles: Synthesis and Applications. Ecomater H, Springer; 2017. 1–14 p

[180] Duman H, Akdaşçi E, Eker F, Bechelany M, Karav S. Gold nanoparticles: Multifunctional properties, synthesis, and future prospects. Nanomaterials (Basel, Switzerland). 2024;**14**:1805

[181] Syafiuddin A, Salmiati MRS, Kueh ABH, Hadibarata T, Nur H. A review of silver nanoparticles: Research trends, global consumption, synthesis, properties, and future challenges. Journal of the Chinese Chemical Society. 2017;**64**:732–756

[182] Aggarwal R, Sheikh A, Akhtar M, Ghazwani M, Hani U, Sahebkar A, Kesharwani P. Understanding gold nanoparticles and their attributes in ovarian cancer therapy, Mol. Cancer. 2025;**24**:88

[183] El Badawy AM, Silva RG, Morris B, Scheckel KG, Suidan MT, Tolaymat TM. Surface Charge-Dependent toxicity of silver nanoparticles. Environmental Science and Technology. 2010;**45**(1):283–287. DOI: 10.1021/es1034188

[184] Helmlinger J, Sengstock C, Groß-Heitfeld C, Mayer C, Schildhauer TA, Köller M, Epple M. Silver Nanoparticles with Different Size and Shape: Equal Cytotoxicity, but Different Antibacterial Effects. RSC Advances. 2016;**6**:18490–18501. DOI: 10.1039/C5RA27836H

[185] Dobias J, Bernier-Latmani R. Silver Release from Silver Nanoparticles in Natural Waters. Environmental Science and Technology. 2013;**47**:4140–4146. DOI: 10.1021/es304023p

[186] Buchman JT, Hudson-Smith NV, Landy KM, Haynes CL. Understanding Nanoparticle Toxicity Mechanisms To Inform Redesign Strategies To Reduce Environmental Impact. Accounts of Chemical Research. 2019;**52**:1632–1642. DOI: 10.1021/acs.accounts.9b00053

[187] Zhang C, Hu Z, Deng B. Silver Nanoparticles in Aquatic Environments: Physiochemical Behavior and Antimicrobial Mechanisms. Water Research. 2016;**88**:403–427. DOI: 10.1016/j.watres.2015.10.025

[188] Silva T, Pokhrel LR, Dubey B, Tolaymat TM, Maier KJ, Liu X. Particle size, surface charge and concentration dependent ecotoxicity of three organo-coated silver nanoparticles: Comparison between general linear

model-predicted and observed toxicity. The Science of the Total Environment. 2013;468–469,968–976. DOI: 10.1016/j.scitotenv.2013.09.006

[189] Rupanshi VK, Yadav N, Singh D, Beniwal V, Chhabra J, Singh B. Biogenic Silver Nanoparticles as Next-Generation Green Catalysts for Multifaceted Applications. Transactions of Tianjin University. 2025;1–34

[190] Saravanan M, Barik SK, MubarakAli D, Prakash P, Pugazhendhi A. Synthesis of silver nanoparticles from Bacillus brevis (NCIM 2533) and their antibacterial activity against pathogenic bacteria. Microbial Pathogenesis. 2018;**116**:221–226

[191] Beyene HD, Werkneh AA, Bezabh HK, Ambaye TG. Synthesis paradigm and applications of silver nanoparticles (AgNPs), a review, Sustain. Mater Technology. 2017;**13**:18–23

[192] Li X, Li B, Liu R, Dong Y, Zhao Y, Wu Y. Development of pH-responsive nanocomposites with remarkably synergistic antibiofilm activities based on ultrasmall silver nanoparticles in combination with aminoglycoside antibiotics. Colloids and Surfaces B: Biointerfaces. 2021;**208**:112112

[193] Chung I-M, Park I, Seung-Hyun K, Thiruvengadam M, Rajakumar G. Plant-mediated synthesis of silver nanoparticles: Their characteristic properties and therapeutic applications. Nanoscale Research Letters. 2016;**11**:40

[194] Maghimaa M, Alharbi SA. Green synthesis of silver nanoparticles from Curcuma longa L. and coating on the cotton fabrics for antimicrobial applications and wound healing activity. Journal of Photochemistry & Photobiology, B: Biology. 2020;**204**:111806

[195] Madani M, Hosny S, Alshangiti DM, Nady N, Alkhursani SA, Alkhaldi H, Al-Gahtany SA, Ghobashy MM, Gaber GA. Green synthesis of nano particles for varied applications: Green renewable resources and energy-efficient synthetic routes. Nanotechnology Reviews. 2022;**11**:731–759

[196] Taha BA, Addie AJ, Kadhim AC, Azzahran AS, Haider AJ, Chaudhary V, Arsad N. Photonics-powered augmented reality skin electronics for proactive healthcare: Multifaceted opportunities. Microchimica Acta. 2024;**191**:250

[197] Taha BA, Addie AJ, Kadhim AC, Azzahrani AS, Ahmed NM, Haider AJ, Chaudhary V, Arsad N. Plasmonic-enabled nanostructures for designing the next generation of silicon photodetectors: Trends, engineering and opportunities. Surfaces and Interfaces. 2024;**48**:104334

[198] Taha BA, Chaudhary V, Rustagi S, Sonu PS. Fate of sniff-the-diseases through nanomaterials-supported optical biochip sensors. ECS Journal of Solid State Science and Technology. 2024;**13**:47004

[199] Saklani S, Barsola B, Pathania D, Sonu S, Kumari P, Singh P, Taha BA, Rustagi S, Thakur P, Narayan M. Nanomaterials-integrated electrochemical biosensors as pioneering solutions for zoonotic disease diagnosis. Journal of the Electrochemical Society. 2024;**171**:87502

[200] Taha BA, Al-Tahar IA, Addie AJ, Mahdi AB, Haider AJ, Mashhadany YA, Chaudhary V, Arsad N. Nanophotonic

catheters: A lens into the body for biosensing and biomedical imaging. Applied Materials Today. 2024;**38**:102229

[201] Textor T, Fouda MMG, Mahltig B. Deposition of durable thin silver layers onto polyamides employing a heterogeneous Tollens' reaction, Appl. Applied Surface Science. 2010;**256**:2337–2342

[202] Le A-T, Huy PT, Tam PD, Huy TQ, Cam PD, Kudrinskiy AA, Krutyakov YA. Green synthesis of finely-dispersed highly bactericidal silver nanoparticles via modified Tollens technique. Current Applied Physics. 2010;**10**:910–916

[203] Shrivastava S, Dash D. Label-free colorimetric estimation of proteins using nanoparticles of silver. Nano-Micro Letters. 2010;**2**:164–168

[204] Le A-T, Tam PD, Huy PT, Huy TQ, Van Hieu N, Kudrinskiy AA, Krutyakov YA. Synthesis of oleic acid-stabilized silver nanoparticles and analysis of their antibacterial activity, Mater. Materials Science and Engineering. 2010;**30**:910–916

[205] Smith MW, Ambudkar IS, Phelps PC, Regec AL, Trump BF. HgCl2-induced changes in cytosolic Ca^{2+} of cultured rabbit renal tubular cells. Biochimica Et Biophysica Acta (BBA)-Molecular Cell Research. 1987;**931**:130–142

[206] Jain P, Pradeep T. Potential of silver nanoparticle-coated polyurethane foam as an antibacterial water filter, Biotechnol. Biotechnology and Bioengineering. 2005;**90**:59–63

[207] Firdhouse MJ, Lalitha P. Biosynthesis of silver nanoparticles and its applications. Journal of Nanotechnology. 2015;**2015**:829526

[208] Alt V, Bechert T, Steinrücke P, Wagener M, Seidel P, Dingeldein E, Domann E, Schnettler R. An in vitro assessment of the antibacterial properties and cytotoxicity of nanoparticulate silver bone cement. Biomaterials. 2004;**25**:4383–4391

[209] Fu J, Ji J, Fan D, Shen J. Construction of antibacterial multilayer films containing nanosilver via layer-by-layer assembly of heparin and chitosan-silver ions complex. Journal of Biomedical Materials Research Part A Australasian Society for Biomaterials and Tissue Engineering; Korean Society for Biomaterials. Our Society. 2006;**79**:665–674

[210] Elechiguerra JL, Burt JL, Morones JR, Camacho-Bragado A, Gao X, Lara HH, Yacaman MJ. Interaction of silver nanoparticles with HIV-1. Journal of Nanobiotechnology. 2005;**3**:6

[211] Maharubin S, Zhou Y, Tan GZ. Integration of silver nanoparticles and microcurrent for water filtration, Sep. Separation and Purification Technology. 2019;**212**:57–64

[212] Lv G, Li K, Shi Y, Zhang R, Tang H, Tang C. Effect of aminosilane coupling agent-modified nano-SiO2 particles on thermodynamic properties of epoxy resin composites. Processes. 2021;**9**:771

[213] Jiang HS, Yin L, Ren NN, Xian L, Zhao S, Li W, Gontero B. The effect of chronic silver nanoparticles on aquatic system in microcosms. Environmental Pollution. 2017;**223**:395–402

[214] López-Herrera A, Gómez-Merino FC, Zavaleta-Mancera HA,

Avalos-Borja M, García-Nava JR, Trejo-Téllez LI. Effect of silver nanoparticles (AgNPs) on aquatic and wetland plants. Environments. 2024;**11**:297

[215] Onotu OP, Samuel HS, Undie DA, Akinpelu OO, Ibekwe FA, Etim EE. Nanoparticles for targeted removal of emerging contaminants in wastewater: Mechanisms and sustainable practices. Discovery Nano. 2025;**20**:1–33

[216] Sati A, Ranade TN, Mali SN, Yasin HKA, Pratap A. Silver nanoparticles (AgNPs): Comprehensive insights into bio/synthesis, key influencing factors, multifaceted applications, and toxicity— a 2024 update. ACS Omega. 2025;**10**:7549–7582

[217] Pushpa M, Pushpa B, Soumya SB, NR MB. Silver Nanoparticles for the Energy Future: Synthesis, Smart Properties and Disruptive Applications. Journal of the Electrochemical Society. 2025;**172**:90530

[218] Shenashen MA, El-Safty SA, Elshehy EA. Synthesis, morphological control, and properties of silver nanoparticles in potential applications, Part. Particle & Particle Systems Characterization. 2014;**31**:293–316

[219] Prasad A, Santra TS, Jayaganthan R. A Study on Prediction of Size and Morphology of Ag Nanoparticles Using Machine Learning Models for Biomedical Applications. Metals. 2024;**14**(5):539. DOI: 10.3390/met14050539

[220] Lafi Z, Matalqah S, Asha S, Asha N, Mhaidat H, Asha SY. Advenced fabrication and characterization of silver nanoparticle using AI techniques. International Journal of Applied Pharmaceutics. 2025;42–51. DOI: 10.22159/ijap.2025v17i5.55011

[221] Li M, Li QZ, Zhao Y, Gao X. Recent Advances in Machine Learning Models for Predicting Toxicity of Inorganic Nanoparticles. Chem & Bio Engineering. 2025;**2**(11):647–680. DOI: 10.1021/cbe.5c00048

[222] Bilgi E, Karakus CO. Machine learning-assisted prediction of the toxicity of silver nanoparticles: A meta-analysis. Journal of Nanoparticle Research. 2023;**25**(8). DOI: 10.1007/s11051-023-05806-2

Chapter 4

Characteristics and Use of Silver Nanoparticles

Davi L. Rios, Vitória O. M. Duval and Luiz P. da Costa

Abstract

Silver nanoparticles (AgNPs) have unique physical, chemical, and optical properties that are being used in a wide variety of applications. Interest in the usefulness of nanometric silver as a broad-spectrum antimicrobial agent has led to the development of several products that incorporate silver nanoparticles to prevent bacterial growth on various surfaces (such as clothing, sheets, hospital equipment, and others). The optical properties of AgNPs are of interest due to the strong interaction of these metallic nanoparticles (MNPs) at specific wavelengths of incident light. This gives them an adjustable optical response and can develop bright molecules, more efficient thermal absorbers, and nanometric "antennas" that amplify local electromagnetic field activity to detect changes in the nanoparticle environment. This work will emphasize the main properties of AgNPs and how engineering manipulates the size and morphology of these nanoparticles for a wide variety of applications.

Keywords: metals nanoparticles, Properties, antimicrobial agent, optical materials, silver nanoparticles, medical applications

1. Introduction

In recent years, research involving materials at the nanoscale (nanomaterials) has become a subject of great interest in various fields of chemistry, physics, and materials science. Because of their finite size, new electronic, optical, transport, photochemical, magnetic, electrochemical, and catalytic properties are expected. Therefore, the physical and chemical properties of a nanomaterial differ drastically from those of the same material in bulk form, enabling potential applications in various technological fields, often distinct from the bulk. Nanotechnology addresses the control of the physical size of materials on the 0.1 to 100 nm scale, characterizing the nanometric size regime, to modulate mesoscopic phenomena characteristic of this condition of matter. The optical and electronic properties of metals and semiconductors strongly depend on the crystallite size in the nanometric regime.

Metallic nanoparticles exhibit optical properties resulting from their interaction with electromagnetic radiation. This property is most observed in certain metals such as copper, silver, and gold, due to the presence of free electrons in the

 IntechOpen

conduction band. The relationship between the morphology, size, and dielectric medium of nanoparticle (NP) preparation, among other factors, is well established in studies involving the monitoring of these variations through plasmon resonance bands, observed via absorption spectroscopy in the ultraviolet-visible region. By understanding the factors that affect the intensity, the position of the absorption wavelength, the width at half-height (WHT), and the formation of multipoles, it is possible to infer variations in the morphology and size of nanoparticles.

Among the most studied nanoparticles are those made of gold and silver, with silver nanoparticles (AgNPs) offering the greatest versatility in applications, ease of preparation, and passivation, although they are not as inert as gold-based nanoparticles (AuNPs). Before the 1980s, special interest in AgNPs was related to the possibility of signal enhancement of organic molecules in Raman spectroscopy, called the SERS effect. Based on the electronic and optical properties of AgNPs associated with RPS, it was possible to carry out application studies in catalysis and high electrical capacitance devices, which made them special in the development of electronic devices and sensors.

Over the past decade, numerous studies have been conducted investigating the bactericidal properties of AgNPs and their nanocomposites. These properties have been known for centuries, and it is understood that, in small concentrations, silver is safe for human cells but lethal to most bacteria and viruses. Consequently, it is widely used in water and food disinfection, as well as infection control in the medical field. This property has become particularly valuable in modern times due to the number of pathogenic bacteria species that are resistant to the narrow range of broad-spectrum antibiotics and represent a danger to human health. These bactericidal properties of AgNPs are associated with the slow oxidation and release of Ag^+ ions into the environment, as well as their ability to penetrate the cell membrane and affect processes within the body of these microorganisms.

In the literature, several methods for synthesizing AgNPs are described, such as biological, physical, and chemical syntheses. Due to the simplicity and effective occurrence of the reaction, chemical reduction in the liquid phase by reducing a salt of the metal is very common, using different reducing agents. Several studies also provide information on environmentally friendly synthesis methods, using polysaccharides as agents that promote the reduction of Ag^+ ions and passivate the obtained AgNPs. From this green chemistry perspective, the use of plant extracts has been reported as efficient in reducing Ag^+ ions and stabilizing the obtained AgNPs, in addition to other studies in which microorganisms, such as fungi and bacteria, have been used for the preparation of AgNPs.

Regarding the methods for preparing AgNPs, the great advantage of chemical methods lies in the possibility of effectively controlling the size and morphology of the nanoparticles obtained; however, the substances used are, in most cases, toxic to humans and highly contaminating to the environment. On the other hand, green synthesis (using plant extracts or even non-pathogenic microorganisms such as some fungi, bacteria, yeasts, etc.), although not harmful to humans or the environment, does not allow effective control of size and shape because various biological components are present, generating variables that cannot be controlled.

The high surface area-to-volume ratio exhibited by nanoparticles makes agglomeration inevitable in the absence of stabilizers. These stabilizers function to prevent, reduce, or organize interactions between particles, thus preventing their agglomeration. In general, passivating organic substances used in the synthesis of AgNPs are molecules possessing electron-donating groups such as phosphorus (P), nitrogen (N), oxygen (O), or sulfur (S), which can act as reducing agents for Ag^+ ions and stabilizers of the formed AgNPs. Despite the versatility of colloidal AgNPs synthesis, one of the difficulties still encountered is controlling the size and morphology using an organic molecule whose coordination to the metal ions (Ag^+) is sufficient to control the reduction and formation process of AgNPs.

2. Properties of silver nanoparticles

Silver nanoparticles (AgNPs) exhibit a set of intrinsic properties that differ significantly from those observed in silver at the macroscopic scale. These properties arise mainly from the small particle size, the high surface area-to-volume ratio, and effects associated with the nanoscale, which influence their physical, chemical, and biological behavior. Consequently, AgNPs display unique characteristics that broaden their potential for application across various fields. The main intrinsic properties associated with these nanoparticles will be presented.

2.1 Electrical/electronic conductivity

Electrical conductivity is not a property exclusive to bulk conductive materials; it is also preserved when a material is reduced to the nanoscale. In bulk silver, electrical conduction occurs due to the movement of free electrons in the metal conduction band that travel through the crystalline lattice. However, as the size of the metal is reduced to the nanoscale, electron transport becomes significantly influenced by quantum effects associated with the reduced dimensions of the system [1]. One of the most important factors is the increase in the surface area-to-volume ratio, which causes a significant fraction of the electrons to interact with the nanoparticle surface [2]. Under these conditions, electrons undergo additional surface scattering, which can reduce their mobility and modify the conductivity compared with macroscopic metal.

2.2 Quantum confinement

Another relevant factor is the nanoparticle size, which may lead to partial quantum confinement of electrons when the particle dimensions approach the electronic mean free path. This confinement modifies the distribution of electronic states near the Fermi level and can influence transport charge [3]. In addition, in systems composed of assemblies of nanoparticles, conductivity also depends on the electronic coupling between neighboring particles. When particles are separated by small distances or by stabilizing molecules, charge transport may occur through electron tunneling or electron hopping mechanisms between adjacent nanoparticles [4].

It is important to emphasize that these properties are interrelated and depend on several factors, such as particle size, the dielectric constant of the surrounding medium, the concentration and average distance between the particles, and the material to which they are coupled [5]. There are still limited detailed studies on how variations in each of these factors influence one another and affect the properties of nanoparticles. Studies have shown that silver nanoparticles incorporated into synthetic rubber nanocomposites (BN/EPDM) exhibited non-linear electrical conductivity; that is, at higher concentrations of AgNPs, electrical conductivity increased significantly due to the reduction in the average distance between particles, which facilitates the electron tunneling effect.

2.3 Optical properties

Surface plasmon resonance (SPR) is a physical phenomenon characteristic of metallic nanomaterials, resulting from the interaction between incident electromagnetic radiation and the free electrons present in the metal conduction band. Noble metals (Ag, Au, Pt, Pd) exhibit a high density of conduction electrons that can collectively respond to perturbations caused by an external electromagnetic field. When light strikes the nanoparticle, the electric field of the radiation induces polarization of the electron cloud relative to the positive cores of the metallic lattice, producing a temporary charge separation. When the oscillation frequency of the incident radiation matches the frequency of the collective electronic oscillations of the metal nanoparticles (MNPs), surface plasmon resonance occurs, leading to an amplification of the local electromagnetic field at the nanoparticle surface and an intensification of light absorption and scattering processes, which can be measured using spectroscopic methods [6-8].

The position of the spectral band and the intensity of the plasmon resonance depend on several structural and electronic factors of the nanoparticles. Among the most relevant parameters are the particle size, shape, and spatial distribution, as well as the dielectric constant of the surrounding medium [9]. The size of the nanoparticles directly influences the regime of electronic confinement and the relationship between radiation absorption and scattering, while the particle geometry determines the possible plasmon oscillation modes.

2.4 Optoelectronic properties

Optoelectronic properties are those that result from the interaction between electromagnetic radiation and conduction electrons, simultaneously producing optical and electronic effects such as photoelectron emission and photocatalysis.

a) **Photoelectron emission**
   Photoelectron emission is the phenomenon in which electrons are ejected from the surface of a material after the absorption of photons with sufficient energy. When a photon strikes a material, its energy may be transferred to an electron. If this energy is greater than the work function of the material (the energy required to remove an electron from the surface), the electron can escape from the solid. In metallic nanoparticles, photoemission can be intensified due to effects such as the increased surface area-to-volume ratio – since more surface atoms and electrons are exposed to electromagnetic radiation – plasmonic excitation, and local concentration of the electromagnetic field, which favors electron ejection. This phenomenon is important in analytical techniques such as photoelectron spectroscopy, as well as in applications involving sensors and photoelectronic devices that

utilize localized surface plasmon resonance (LSPR) to enhance the electric field and ultimately obtain a stronger emission signal [10, 11].

b) **Photocatalysis**

Silver nanoparticles can promote chemical reactions through pathways different from those observed in traditional catalysts, such as semiconductors. This occurs because, in these materials, the interaction with light can control the selectivity of the chemical products formed, which is a highly attractive feature in photocatalytic processes. In other words, by adjusting the wavelength of the incident light, it is possible to favor the formation of specific reaction products. The mechanism responsible for this behavior is related to the excitation of surface plasmon resonance (SPR). This collective oscillation generates a strong electric field localized in specific regions of the nanoparticle surface, known as hotspots [11]. In these regions, the intensity of the electromagnetic field is much greater than that of the incident field.

After plasmon excitation, the plasmon may decay, and the collective plasmon energy is transferred to individual electrons. As a result, highly energetic electron–hole pairs, known as hot carriers, are formed, which can then participate in chemical reactions at the nanoparticle surface by transferring energy or charge to adsorbed molecules [10, 11]. These hot electrons and holes can, therefore, promote light-induced catalytic processes. An example of this mechanism was demonstrated by Mincheol Kang, Beomjoon Jeon, and Jeong Young Park, who investigated the relationship between plasmon-generated hot electrons and catalytic activity in hydrogen oxidation. For this purpose, they used Schottky-type catalytic nanodiodes ($Pt/Ag/TiO_2$) in a plasmonic photocatalysis antenna–reactor system, in which silver acts as the plasmonic antenna responsible for light absorption and plasmon generation, while platinum functions as the catalytic site where the reaction occurs [11].

The authors compared the behavior of the Pt/TiO_2 and $Pt/Ag/TiO_2$ systems under irradiation with orange-red light (650 nm) and blue-violet light (450 nm). In the case of Pt/TiO_2, the photon-induced enhancement ratio of catalytic activity remained close to zero under both irradiations, indicating the absence of a significant plasmonic effect. In contrast, the $Pt/Ag/TiO_2$ system showed a significant increase in catalytic activity under blue-violet light due to the excitation of surface plasmon resonance in silver. Under these conditions, the electric field is intensified, and hot electrons are generated in the Ag nanoparticle, which can be transferred to platinum, thereby promoting the catalytic reaction.

2.5 Antimicrobial activity

Silver nanoparticles (AgNPs) exhibit broad antimicrobial activity and are effective against different types of microorganisms. Among their main biological actions are antibacterial activity, characterized by the ability to inhibit bacterial growth or cause bacterial cell death; antifungal activity, which interferes with the development and cellular integrity of fungi; and antiviral activity, related to the ability to interact with viral structures and inhibit important stages of the infection process. Due to this variety of mechanisms of action, AgNPs have been widely investigated for applications in areas such as medicine, biomaterials, food preservation, and microbiological control. Some of these mechanisms include the release of Ag^+ ions, the interaction of AgNPs with the plasma membrane, and the generation of reactive oxygen species (ROS).

a) **Release of silver ions**
   In aqueous media, AgNPs may undergo transformation processes and release silver ions (Ag^+), thus coexisting with their ionic species in the surrounding medium. The Ag^+ ions mainly act through chemical interactions with cellular biomolecules. They can bind to thiol groups (–SH) of proteins, resulting in the inactivation of essential enzymes. In addition, they can compromise the integrity of the bacterial membrane and interfere with important cellular processes such as DNA replication, thereby enhancing the toxic effects on microorganisms [12–14].

b) **Interaction of AgNPs with cellular membranes**
   AgNPs may adhere to bacterial cell membranes and cause structural damage by perforating or destabilizing the membrane and increasing its permeability. This effect may lead to the loss of control over the transport of substances between the interior and exterior of the cell, contributing to cell death, and is also influenced by the size and shape of the nanoparticles. Nanoparticles of different sizes and shapes tend to accumulate in mitochondria, causing mitochondrial dysfunction [12, 14].

c) **Reactive oxygen species: ROS**
   Many studies attribute the toxicity of AgNPs to the generation of reactive oxygen species (ROS) [15]. These reactive species include highly oxidizing molecules such as the superoxide radical ($O_2^{\bullet -}$), hydrogen peroxide (H_2O_2), and the hydroxyl radical ($\bullet OH$), which compromise antioxidant defense mechanisms, resulting in the structural disruption of viral envelopes and cellular membranes [14]. The process usually begins when AgNPs encounter the cell surface or penetrate the interior of the cell. During this process, the nanoparticles may release silver ions (Ag^+) and interact with cellular components such as proteins and enzymes of the respiratory chain. These interactions disturb the cellular redox balance and favor electron transfer reactions that lead to the formation of ROS. The increase in these reactive species generates a condition known as oxidative stress, in which the quantity of free radicals exceeds the capacity of the cellular antioxidant systems to neutralize them [15].

3. Applications of silver nanoparticles

3.1 Electronic applications

Due to the world's increasing demand for electronics, printed devices have become very popular compared to traditional methods, primarily because of their low cost and high-volume production capabilities, while also offering the possibility of a high degree of customization and optimization. Some printing technologies include inkjet printing, gravure printing, aerosol jet printing, and screen printing – which are more modern approaches – and stereolithography, which was the pioneer of 3D printing, with each method being more efficient depending on the target application [16]. Since the type and composition of ink used play a significant role in application efficiency, many materials are utilized for these purposes. Functional particles are regularly incorporated into the ink composition, and silver nanoparticles have gained great interest for such applications because of their superior electrical properties. Silver-based inks are currently being used to manufacture micro and nano-electronic-grade devices such as photovoltaic cells, sensors, and supercapacitors [16]. AgNPs conductive ink was made through chemical reduction by $Na_3C_6H_5O_7$ and the final material has shown great results, with electrical resistivity measured at 9.9 μΩ [17]. A water-based silver nanoparticle conductive ink study was conducted using a polyvinylpyrrolidone (PVP) surfactant to stabilize AgNP aggregation, demonstrating excellent electrical conductance of $2.50 \times 10^{\circ}$ S/m, which can be improved up to $2.69 \times 10^{\circ}$ S/m with concentration and temperature control, and indicating high stability due to the PVP action for up to 45 days [18].

Wearable electronics are also receiving great attention in current research toward conductive inks as part of technological development. Silver ink can be used in inkjet printing for knit, woven, and nonwoven fabrics, which cover the fibers without modifying their intrinsic properties [19]. AgNPs synthesized from starch were dispersed in different commercial conductive inks, and their electrical efficacy when deposited on cotton fabric was tested, showing improved conductivity of the original product [20].

Besides the already known high conductivity of silver, common to noble metals, several studies indicate the important factor of particle size, shape, and overall morphology in the resulting material. Silver nanoparticle synthesis methods are vast and simple, and their morphology is well-known in the scientific community, along with ways to influence that morphology through changes in synthesis parameters like temperature and precursor salt (usually $AgNO_3$) concentration, which makes them well-suited to those applications.

Light-emitting diodes (LEDs) are electronic devices that produce light when an electric current is passed through them and are the most common type of light-providing devices for popular equipment. Metal-Insulator-Semiconductor (MIS) Schottky barrier diodes with nanostructured porous silicon layers can be impregnated with silver nanoparticles to overcome conduction losses at the material interface, thereby increasing the overall conductive properties of the material [21]. Solar cells' electronic parameters (diode-ideality factor, series-resistance, energy-barrier height, and shunt-resistance) can be significantly enhanced by the addition of an AgNPs-polyaniline buffer layer due to a reduction in interfacial trap states and an increase in interfacial dipole-moment, thereby improving photovoltaic response [22]. The addition of an intermediary layer of AgNPs at the silicon-organic interface of silicon nanowires (SiNWs) and a highly conductive polymer hybrid solar cell increases power conversion efficiency, photocurrent, and overall photovoltaic performance due to improvements in the polymer layer properties related to morphology while preventing interface recombination [23].

3.2 Sensors applications

With a constant increase in materials and product usage, along with technological advancements overall, the necessity for better sensing is crucial to maintain the infrastructure needed to support the rate at which this growth is occurring. Applications related to sensors can be used to detect and quantify contaminants, highlighting their leading importance in environmental remediation, as well as being utilized to increase the sensitivity and reliability of already existing detectors. The properties of nanoparticles present highly effective grounds for sensor applications. As a noble metal, the conductivity of silver nanoparticles acts efficiently in electronic sensors, and its plasmonic-resonance effect plays a significant role in its optical properties, enhancing the efficacy of colorimetric sensors.

Colorimetric methods are simple, low-cost, and provide a fast and easy way to obtain results. Many applications involve a color change that can be easily observed by the method's "operator" and are important for point-of-care testing [24]. For metal ion detection, dithizone (a selective dye that forms lead complexes) was tested for Pb^{2+} sensing with AgNPs that covalently bond with sulfide groups. The aggregated silver nanoparticles caused a decrease in absorbance for the complex at the 421 nm band due to surface plasmon resonance, and that variation is applied to detect the presence of lead attached to the dithizone [25]. When molecules (analytes) bind to receptors immobilized on the metal surface, they increase the local mass and refractive index, shifting the SPR resonance angle or wavelength. This allows for label-free, real-time monitoring of association/dissociation rates, binding affinity, and concentration, which is widely used in clinical diagnostics, food safety, and environmental monitoring [24].

Water analysis is extremely important due to contamination that can lead to health hazards for humans, animals, and plants. Some of the commonly used analysis methods include inductively coupled plasma mass spectrometry (ICP-MS), high-performance liquid chromatography (HPLC), and atomic fluorescence spectroscopy (AFS), which involve sample preparation, expensive equipment, and other time-consuming resources. Colorimetric sensors can be used to provide fast and on-site readings for the quick determination of contaminants [24]. Lead concentrations in wastewater can be measured using AgNPs-modified polyvinyl alcohol and PADS (paper-based analytical devices). The presence of Pb (II) ions on the AgNPs-polyvinyl disrupts nanoparticle stability and causes aggregation, which changes the characteristic absorption band with a red shift from 410 nm to 550 nm that can be analyzed by a spectrophotometer [26]. Mercury is among the most dangerous metals in terms of toxicity because it accumulates rapidly, as there is no easy way to eliminate it from the body, and it gets passed along the food chain. A redox reaction can be used to rapidly detect Hg (II) with mercaptobenzene heterocyclic compounds attached to silver nanoparticles that form a nanoalloy (Ag–Hg) and provoke a color change that can be observed [27]. Nickel can be detected by its addition to a triangular silver nanoprism solution, where the Ni (II) acts as a catalyst for the formation of H_2O_2 in the solution, sharpening the nanoparticles and changing the system's color from blue to yellow with high linearity to Ni (II) ion concentration [28]. Iron can be detected by chitosan-capped AgNPs, which decrease the surface plasmon resonance band intensity linearly with Fe (III) ions, provoking a color change from brownish yellow to colorless [29]. Many other metals, such as manganese, cobalt, cadmium, and lead, can also be detected using the same principle.

Colorimetric sensors can also be used to detect more complex compounds, such as pesticides. Paraquat is an herbicide commonly used for weed control but is highly toxic to animals and humans. Citrate-capped silver nanoparticles can be used to develop a paper-based colorimetric sensor, enabling on-site detection with a limit of detection of 10 μM [30]. Imidacloprid, a pesticide used to control insects in agricultural fields, can be detected using carbon-dot-stabilized silver nanoparticles, which exhibit great sensitivity and selectivity toward the detection of this pesticide [31]. CPS (Chlorpyrifos), an organophosphate pesticide used for insect control, can be absorbed by fruits and vegetables in agricultural fields as it is applied as a disinfectant in the late stages of the harvesting process, representing a potential risk to human health. An alpha-cyclodextrin-capped silver nanoparticle colorimetric sensor can be used to detect it based on the occurrence of a redshift movement of the surface plasmon resonance band from 410 to 570 nm, showing a color change from yellow to dark red in the presence of the pesticide [32]. Colorimetric sensing can also be applied to detect species such as hydrogen peroxide and ammonia, both of which are important chemicals for industrial production [33].

In the electrochemical field, silver nanoparticles are also well-established due to their great efficacy in the matter. Electrochemical sensors consist of a receptor, which is the detection agent that can detect a substance or compound of interest, and a transducer, which turns the chemical response of the system into an electrical response that can be read by certain equipment [34]. Many times, the usage of an unmodified working electrode can lead to slow reaction kinetics, which is why modification with nanomaterials, like nanoparticles, and other electrocatalysts can improve electrode performance [34]. Silver nanoparticles can be used as a modification to electrodes for heavy metal ion tracking. An AgNPs-modified carbon paste electrode showed low charge transfer resistance (electrochemical reaction energy barrier), high metal recovery levels for wastewater, and high reproducibility for lead detection [35]. For copper detection, a (4-mercaptobenzoic acid (MBA))-capped AgNPs and fluorine-doped tin oxide decorated with gold nanoislands can be used, on which a (AgNPs-MBA-Cu^{2+}-MBA -AgNPs) carboxylate complex is formed, and the MBA-AgNPs link enhances the signal on the oxidation side of the electrochemical system [36]. Sensors based on those principles exist for most heavy metal detection, inorganic species with nitrogen and derivatives like ammonia (NH_3), phenolic compounds, and many more [34].

Hydrogen gas is currently of significant importance due to the focus on the energy sector and the search for clean and renewable energy sources, but it also requires considerable caution because hydrogen gas is highly explosive. For that reason, especially in places that deal with hydrogen production, the detection of this gas is highly important. A silver nanoparticle-decorated SnO_2 film is known to have the ability to detect H_2 through a redox mechanism on the film surface. The results show a limit of detection of 50 ppm at higher operating temperatures and 600 ppm at ambient temperature, where the presence of AgNPs increases the sensitivity of the sensor by 67% [37].

As part of the advancements in treatments within the healthcare system, it also becomes critical to accurately track important biomolecules in the body to prevent and regulate unhealthy conditions in general. Despite their strategic usage in drug and treatment development, silver nanoparticles can also act as biosensors to provide critical information about dangers related to unbalanced vital molecules necessary for optimal body function. Their application in this regard operates on basically the same principles already cited, many of which involve the plasmonic resonance phenomenon, but with high selectivity toward the target biomolecule and with health-friendly compounds [38, 39]. In addition, recent developments toward their use toward bio-imaging have been showing promising results [39].

3.3 Catalysis applications

Dyes are used on a large scale for commercial and industrial purposes, representing a high demand for the textile industry, for example, but also present themselves as potential environmental hazards due to their easy water contamination potential. Therefore, the more common use for silver nanoparticles regarding catalytic activity involves the acceleration of photo/chemical degradation processes for dyes. The degradation reaction of methylene blue dye by $NaBH_4$ can be catalytically improved by silver nanoparticles. The reduction reaction immediately occurs with the addition of AgNPs, and the reaction time for the sample without the nanoparticles' action is increased by more than an hour [40]. Congo red, which is a toxic dye used regularly in industrial applications for paper and textile production, has its degradation through $NaBH_4$ reduction, which is the most common method for many dyes, kinetically prevented. Silver nanoparticles can perform a catalytic action toward Congo red degradation by providing a pathway through which the electrons can pass and accelerate the reaction [41]. AgNPs can act on the photocatalytic degradation of Rhodamine B, another dye, with sufficient stability to be reusable and potentially applicable for water cleaning processes [42]. AgNPs also showed high photocatalytic activity against Basic Blue 9 and Eriochrome Black T under UV-Light, with no considerable contaminant destruction observed for samples without the silver nanoparticles present [43].

It can also be used for other contaminants like 4-nitrophenol, which is a toxic molecule commonly present in industrial wastewater. Piper chaba-synthesized silver nanoparticles were used as a catalyst for the 4-nitrophenol reduction reaction by $NaBH_4$, during which the presence of AgNPs instantaneously changed the solution color from light yellow to yellowish green (peak absorbance shifted from 316 to 401 nm), indicating the formation of the reduced species and taking 9 minutes for almost full reduction of the reagent, whereas the same shift only occurred 1 hour later for the control solution without the silver nanoparticles [40]. SPR can generate “hot electrons” – energetic electrons that can be injected into nearby catalysts to boost reduction reactions (e.g., CO_2 reduction).

Despite the usage of AgNPs in contaminant detection being a prominent field of research, it is known that they can also act as catalysts in more complex systems to accelerate reactions, such as photocatalytic nitrogen fixation reactions, which are considered an ideal route for ammonia synthesis. In this context, a AgNPs-modified Bi_2S_3/MnOS composite can increase catalytic activity through surface plasmon resonance phenomena and enhanced charge-transfer at the composite interface [44].

3.4 Biomedical applications

Between the vast applications of silver nanoparticles in science, its usage in the medical field has always been one of the leading areas of study on the matter, with great importance in the discovery of new modified materials. Due to the unique properties that emerge when the "nano" aspect is applied, silver nanoparticles differ from other metals in its usage in compounds that interact within biological environments.

Within the medical science field, bacterial resistance to medicine has become a considerable problem in today's health system, due to self-medication being so common nowadays and the general adaptation to pre-existing procedures available. The usage of silver nanoparticles for antibacterial treatment is well established within the research field and leads to new developments in medicines and treatments [45]. The key mechanism behind the great effectiveness of silver nanoparticles against bacteria relies on the ability of the metal cations to bind with proteins in bacterial cellular walls, harming the biological activity of their internal components and leading to the death of the organism [46, 47].

As examples of studies developed regarding the combat against bacteria, a material based on graphene oxide and green-synthesized silver nanoparticles by bacterial biomass was tested and shown to have high antibacterial activity. The graphene oxide's high surface area and porosity offer a great adsorption site for the nanoparticles' fixation. Antibacterial tests using the well-plate method (inhibition zone) for three different human pathogenic bacteria (Escherichia coli, Bacillus subtilis, and Staphylococcus aureus) demonstrated effectiveness for both gram-positive and gram-negative strands, with higher relative activity observed at higher concentrations of the nanocomposite containing silver [48]. In vitro tests conducted with a grafted-chitosan base were also utilized with silver nanoparticles to induce higher activity toward bacterial fronts, revealing activity against Methicillin-resistant Staphylococcus aureus (MRSA) and P. aeruginosa [49]. AgNPs synthesized from Cestrum nocturnum also exhibited activity toward Citrobacter, S. typhi, E. faecalis, P. vulgaris, and V. cholerae [50]. Studies on cellulosic polymers stabilized silver nanoparticles showed the effects of different coating agents on stability and antibacterial activity. EC and HPMC (cellulosic-based polymers) stabilized silver nanoparticles demonstrated better efficiency against Gram-negative Salmonella typhimurium and Serratia marcescens [51]. Biogenic silver nanoparticles synthesized by Bacillus subtilis were tested and showed activity against Staphylococcus epidermidis, Klebsiella pneumoniae, and Candida albicans [52].

Silver nanoparticles are being applied in many textile fields with the objective of taking advantage of the antibacterial character of those particles, and that is especially true for high-humidity skin regions. Socks containing this benefit are already on the market, offering anti-odor and antimicrobial activity.

Along with antibacterial activity, silver nanoparticles also exhibit antifungal, antimicrobial, antiviral, and even anticancer activities that are being explored with great importance [53]. Biosynthesis of AgNPs by B. thuringiensis CFE (cell-free extract) was conducted, where the biostructure formed by the bacteria reduces the Ag ions in silver nitrate and forms the nanoparticles. In vitro antifungal activity was performed against four different types of Aspergillus fungal species, which can cause Aspergillosis (a fungal infection that affects the respiratory system), and resulted in antifungal activity for all four types. Ultrastructure studies showed ruptures in the cellular wall and its separation from the cell's membrane, caused by the presence of silver nanoparticles in the vicinity [54]. Negatively charged citrate-stabilized AgNPs (TCSB-AgNPs) and positively charged cysteamine-capped AgNPs (CHSB-AgNPs) were compared regarding silver ion release in the fungi and its effect on fungicide activity for F. avenaceum and F. equiseti, which showed CHSB-AgNPs having higher activity than TCSB-AgNPs over either shorter or longer treatment times [55].

Despite the danger of fungal/viral infections and bacterial diseases, fortunately, treatments and preventive methods are more accessible and do not present as life-changing processes most of the time. With that knowledge, cancer becomes one of the hardest diseases to get treatment. Cancer has one of the highest mortality rates among any known conditions, with close to 10 million deaths in 2022, according to the International Agency for Research on Cancer [56], and the treatment itself is not only highly expensive but also brings collateral effects to those being treated, bearing a great psychological burden on the affected individuals. Because of this, the anticancer activity of silver nanoparticles becomes increasingly important over the years to elevate the efficiency and minimize the cost of the process itself.

Anticancer activity of AgNPs is explained by its cytotoxic properties, as the presence of these particles leads to an increase in reactive oxygen species (ROS) inside the cell when the silver nanoparticles cause the cellular membrane to rupture. These reactive species provoke oxidative stress within the cell and lead to apoptosis (controlled cell death). Studies in anticancer tests have shown that the increase in ROS in cancerous cells is greater than in normal cells. Therefore, the use of AgNPs toward cancerous cells can be employed to target harmful structures and selectively harm the cancer [57–59].

Besides the direct use of silver nanoparticles to treat health problems, there's also a lot of research going on about using those particles as drug carriers. Drug carriers consist of chemical structures that increase a drug's efficiency by providing a protective barrier around the active pharmaceutical ingredient. That barrier can enhance the drug's efficiency through various mechanisms, such as allowing it to be released at a specific location and rate, altering its solubility within the biological environment, or preventing degradation [60]. The use of nanoparticles plays a key role in the risk-benefit ratio, as their unique properties promote high customization of the medicine toward improved efficiency while also reducing the amount of the "foreign" agent within it [61]. Additionally, nanoparticles can increase control over drug-release timing [62]. There are many routes of administration, such as oral and injectable methods, which are the most common [62], as well as transdermal (through skin absorption), which offers several benefits like easy accessibility and more systematic release into the system [61], intravenous, inhalation, and others.

AgNPs synthesized by Setaria verticillata extract were used with anticancer drugs, including doxorubicin (DOX) and daunorubicin (DNR), showing anticancer activity and potential to reduce chemotherapy side-effects. The study revealed the influence of nanoparticle size and shape on the efficacy of the drug delivery system [63]. Spherical-shaped silver nanoparticles, 5.7 nm in size, were loaded with gefitinib (an anti-cancer drug) and reduced the viability of the MCF-7 cell line (a breast cancer cell population used for testing drugs and treatment development) by more than 50% in comparison to the medicine alone [63]. Carboplatin, a common antineoplastic used for the treatment of various types of cancer such as breast, lung, and brain, had its activity tested with and without the use of silver nanoparticles for comparison, resulting in a higher efficacy level for the AgNPs-loaded drug. The same comparison was made for gemcitabine-loaded AgNPs, which also showed greater activity when combined with the silver nanoparticles [64].

It's important to clarify the distinction between the direct use of silver nanoparticles as anti-cancer and anti-pathogen agents and their use as drug carriers. The direct use of AgNPs exploits their cytotoxicity, attacking cancerous cells or viruses/bacteria on their own. Acting as carriers, the focus lies in improving the path of the acting substance, increasing the drug efficacy toward treatment, meaning that for them to be considered drug carriers, they need to be directly attached to the drug and have a positive relationship with its resulting activity.

Commercial applications of those nanoparticles consist mainly of the modification of common products that are already available. Band-aids containing silver nanoparticles are already used to prevent further infections and promote better cicatrization processes, such as Acticoat™, which is a healing dressing that releases sustainable amounts of AgNPs into the wound, promoting the antibacterial effect described previously, while also keeping the moisture to enable proper healing [65]. Toothpastes with low amounts of silver nanoparticles are used to increase oral microbial hygiene [66]. Bodysplashes/Deodorants and shampoos with AgNPs are designed to increase their cleaning efficiency and bactericidal effect [67, 68]. Creams and serums with silver nanoparticles are formulated to prevent and combat acne [69].

Along with an extensive list of applications already in place, many studies with silver nanoparticles are still in development, seeking to leverage their unique medical properties to address health system problems worldwide. Simultaneously, a new concern arises with the use of these particles in product development, particularly regarding potential safety issues related to exposure to AgNPs in the human body. The toxicity of silver nanoparticles varies, especially with the route and frequency of exposure. Smaller AgNPs could travel through the bloodstream and potentially deposit inside organs, leading to metal accumulation. Currently, studies indicate no considerable damage to organs like the lungs, liver, or nasal cavities regarding 15 to 30 nm particles; however, reports of brain injuries with higher concentrations of AgNPs (up to 2.9 mg/m^3) have already emerged [70–72].

4. Conclusion

The versatility of silver nanoparticles (AgNPs) allows for their application in various scientific and technological fields, especially due to their remarkable properties when confined to quantum size. Soon, it will be possible to obtain AgNPs with effective size and shape control, as well as through environmentally friendly methods and/or techniques, whose application could fit into sustainable practices, providing improvements in materials without causing harm to humans and the environment.

Soon, Surface Engineering studies will focus on modifying the coating of AgNPs to control the rate of ion release, increase stability, and reduce cytotoxicity in healthy cells. The use of AgNPs in wound care, antibacterial coatings for medical devices, and targeted delivery of anticancer agents is promising. Further research into the behavior of AgNPs in vivo is needed to establish safe dosage and exposure limits, ensuring biosafety.

Acknowledgments

L. P. da Costa thanks CNPQ for the PQ scholarship granted, process 311448/2023-2. Duval, V. O. M. thanks CAPES for the scholarship granted.

Conflict of Interest

We declare that the mention of some commercial products in this text is merely to exemplify how AgNPs are already present in our lives, and there is no relationship whatsoever with these products, nor any commercial advertising agreements, nor any connection with the brands. Therefore, there is no conflict of interest with such products.

Author details

Davi L. Rios[1], Vitória O. M. Duval[1] and Luiz P. da Costa[1,2]*

1 Graduate Program in Chemistry, Federal University of Sergipe, São Cristóvão, Brazil

2 Chemistry Department, Federal University of Sergipe, São Cristóvão, Brazil

*Address all correspondence to: lupeco7@hotmail.com

References

[1] Saini I, Himanshi Yadav M. Optoelectronic and sensing applications of plasmonic silver nanoparticles. Special Issue – ICACCG 2020. 2020

[2] Ali W, Shabani V, Linke M, Dinc S, Gebert B, Altinpinar S, Hildebrandt M, Gutmann JS, Mayer-Gall T. Electrical conductivity of silver nanoparticles doped carbon nanofibers measured by CS-AFM. RSC Advances. 2019; **9**(8):4553–4562. DOI: 10.1039/C8RA04594A

[3] Suksomboon M, Kongsawatvoragul K, Duangdangchote S, Sawangphruk M. Reducing the energy band gap of cobalt hydroxide nanosheets with silver atoms and enhancing their electrical conductivity with silver nanoparticles. ACS Omega. 2021;**6**:20804–20811. DOI: 10.1021/acsomega.1c01908

[4] Chi QG, Yang M, Zhang CH, Zhang TD, Feng Y, Chen QG. Nonlinear electrical conductivity and thermal properties of AgNPs/BN/EPDM composites for cable accessory. IEEE Transactions on Dielectrics and Electrical Insulation. 2019;**26**(4):1081–1088. DOI: 10.1109/TDEI.2019.007860

[5] Rac-Rumijowska O, Teterycz H. Electrical conductance Mechanism of Silver–Polyacrylonitrile nanocomposite fibers. Materials. 2023;**16**(8):3085. DOI: 10.3390/ma.16083085

[6] Zayets V, Saito H, Ando K, Yuasa S. Long-distance propagation of a surface plasmon on the surface of a ferromagnetic metal. Optics Express. 2015;**23**(10):12834–12839. DOI: 10.1364/OE.23.012834

[7] Wiley BJ, Im SH, Li Z-Y, McLellan J, Siekkinen A, Xia Y. Maneuvering the surface plasmon resonance of silver nanostructures through shape-controlled synthesis. The Journal of Physical Chemistry B. 2006;**110**(32):15666–15675. DOI: 10.1021/jp06086.28

[8] Huang C-H, Lin H-Y, Lau B-C, Liu C-Y, Chui H-C, Tzeng Y. Plasmon-induced optical switching of electrical conductivity in porous anodic aluminum oxide films encapsulated with silver nanoparticle arrays. Optics Express. 2010;**18**(26):27891–27899. DOI: 10.1364/OE.18.027891

[9] Jamalzadegan S, Zare M, Dickens MJ, Schenk F, Velayati A, Yarema M, Dickey MD, Wei Q. Shape and size-dependent surface plasmonic resonances of liquid metal alloy (EGaIn) nanoparticles. Nanoscale. 2025;**17**:22819–22833. DOI: 10.1039/D5NR025.02H

[10] Kontoleta E, Askes SHC, Garnett EC. Self-optimized catalysts: Hot-electron driven photosynthesis of catalytic photocathodes. ACS Applied Materials & Interfaces. 2019;**11**(39):35713–35719. DOI: 10.1021/acsami.9b10913

[11] Kang M, Jeon B, Park JY. Catalytic boosting by surface-plasmon-driven hot electrons on antenna–reactor Schottky nanodiodes. Nano Letters. 2023;**23**(11):5116–5122. DOI: 10.1021/acs.nanolett.3c01037

[12] Zhang W, Yao Y, Sullivan N, Chen Y. Modeling the primary size effects of citrate-coated silver nanoparticles on their ion release kinetics. Environmental Science &

Technology. 2011;**45**:4422–4428. DOI: 10.1021/es104205a

[13] Rajesh S, Dharanishanthi V, Kanna AV. Antibacterial mechanism of biogenic silver nanoparticles of Lactobacillus acidophilus. Journal of Experimental Nanoscience. 2015;**10** (15):1143–1152. DOI: 10.1080/17458080.2014.985750

[14] Morozova OV. Silver Nanostructures: Limited sensitivity of detection, toxicity and anti-inflammation effects. International Journal of Molecular Sciences. 2021;**22**(18):9928. DOI: 10.3390/ijms2218992.8

[15] Sun X, Shi J, Zou X, Wang C, Yang Y, Zhang H. Silver nanoparticles interact with the cell membrane and increase endothelial permeability by promoting VE-cadherin internalization. Journal of Hazardous Materials. 2016;**317**:570–578. DOI: 10.1016/j.jhazmat

[16] Zhang J, Ahmadi M, Fargas G, Perinka N, Reguera J, Lanceros-Méndez S, Llanes L, Jiménez-Piqué E. Silver nanoparticles for conductive inks: From synthesis and Ink Formulation to their use in printing technologies. Metals. 2022;**12** (2):234. DOI: 10.3390/met12020234

[17] Hong GB, Luo YH, Chuang KJ, Cheng HY, Chang KC, Ma CM. Facile synthesis of silver nanoparticles and Preparation of Conductive Ink. Nanomaterials. 2022;**12**:171. DOI: 10.3390/nano12010171

[18] Ibrahim N, Zubir SA, Manaf AA, Mustapha M. Stability and conductivity of water-based colloidal silver nanoparticles conductive inks for sustainable printed electronics. Journal of the Taiwan Institute of Chemical Engineers. 2023;**153**:105202. DOI: 10.1016/j.jtice.2023.105202

[19] Shahariar H, Kim I, Soewardiman H, Jur JS. Inkjet printing of reactive Silver Ink on textiles. ACS Applied Materials & Interfaces. 2019;**11**(6):6208–6216. DOI: 10.1021/acsami.8b18231

[20] Boumegnane A, Nadi A, Cherkaoui O, Tahiri M. Inkjet printing of silver conductive ink on textiles for wearable electronic applications. In: Materials Today: Proceedings; 2022;**58**:1235–1241. DOI: 10.1016/j.matpr.2022.01.469

[21] Ramadan R, Martín-Palma RJ. Electrical characterization of MIS Schottky barrier diodes based on nanostructured porous silicon and silver nanoparticles with Applications in Solar Cells. Energies. 2020;**13** (9):2165. DOI: 10.3390/en13092165

[22] Moiz SA, Alahmadi ANM, Karimov KS. Improved organic solar cell by incorporating silver nanoparticles embedded polyaniline as buffer layer. Solid-State Electronics. 2020;**163**:107658. DOI: 10.1016/j.sse.2019.107658

[23] Jbira E, Derouiche H, Missaoui K. Enhancing effect of silver nanoparticles (AgNPs) interfacial thin layer on silicon nanowires (SiNWs)/PEDOT: PSS hybrid solar cell. Solar Energy. 2020;**211**:1230–1238. DOI: 10.1016/j.solener.2020.09.090

[24] Prosposito P, Burratti L, Venditti I. Silver Nanoparticles as Colorimetric Sensors for Water Pollutants. Chemosensors. 2020;**8**(2):26. DOI: 10.3390/chemosensors8020026

[25] Roto R, Mellisani B, Kuncaka A, Mudasir M, Suratman A. Colorimetric sensing of Pb^{2+} Ion by Using Ag Nanoparticles in the presence of Dithizone. Chemosensors. 2019;**7**

(23):28. DOI: 10.3390/chemosensors7030028

[26] Shrivas K, Sahu B, Deb MK, Thakur SS, Sahu S, Kurrey R, Kant T, Patle TK, Jangde R. Colorimetric and paper-based detection of lead using PVA capped silver nanoparticles: Experimental and theoretical approach. Microchemical Journal. 2019;**150**:104156. DOI: 10.1016/j.microc.2019.104156

[27] Bhattacharjee Y, Chatterjee D, Chakraborty A. Mercaptobenzoheterocyclic compounds functionalized silver nanoparticle, an ultrasensitive colorimetric probe for Hg (II) detection in water with picomolar precision: A correlation between sensitivity and binding affinity. Sensors and Actuators B: Chemical. 2018;**255**:210–216. DOI: 10.1016/j.snb.20.17.0.8.066

[28] Yoon S-J, Nam Y-S, Lee H-J, Lee Y, Lee K-B. Colorimetric probe for Ni^{2+} based on shape transformation of triangular silver nanoprisms upon H_2O_2 etching. Sensors and Actuators B: Chemical. 2019;**300**:127045. DOI: 10.1016/j.snb.2019.127045

[29] Tashkhourian J, Sheydaei O. Chitosan capped silver nanoparticles as colorimetric sensor for the determination of iron (III). Analytical and Bioanalytical Chemistry Research. 2017;**4**(2):249–260

[30] Bhandari S, Parihar VS, Kellomäki M, Mahato M. Highly selective and flexible silver nanoparticles-based paper sensor for on-site colorimetric detection of paraquat pesticide. RSC Advances. 2024;**14**:2884–28853. DOI: 10.1039/d4ra04557b

[31] Murugesan P, Moses JA. Carbon quantum dots stabilized silver nanoparticles-based colorimetric sensor for detection of pesticide residues. Materials Science and Engineering: B. 2024;**304**:117354. DOI: 10.1016/j.mseb.2024.117354

[32] Sahu B, Kurrey R, Khalkho BR, Deb MK. α-Cyclodextrin functionalized silver nanoparticles as colorimetric sensor for micro extraction and trace level detection of chlorpyrifos pesticide in fruits and vegetables. Colloids and Surfaces A: Physicochemical and Engineering Aspects. 2022;**654**:129947. DOI: 10.1016/j.colsurfa.2022.129947

[33] Abdali K, Al-Bermany E, Abass KH. Impact the silver nanoparticles on properties of new fabricated polyvinyl alcohol-polyacrylamide-polyacrylic acid nanocomposites films for optoelectronics and radiation pollution applications. Journal of Polymer Research. 2023;**30**:138. DOI: 10.1007/s10965-023-03514-y

[34] Ivanišević I. The role of Silver Nanoparticles in electrochemical sensors for Aquatic environmental analysis. Sensors. 2023;**23**:3692. DOI: 10.3390/s23073692

[35] Ganash AA, Alghamdi RA. Fabrication of a novel polyaniline/green-synthesized, silver-nanoparticle-modified carbon paste electrode for electrochemical sensing of lead ions. JCCS. 2021;**68**:2312–2325. DOI: 10.1002/jccs.202100348

[36] Zhou M, Han L, Deng D, Zhang Z, He H, Zhang L, Luo L. 4-mercaptobenzoic acid modified silver nanoparticles-enhanced electrochemical sensor for highly sensitive detection of Cu^{2+}. Sensors and Actuators B: Chemical. 2019;**291**:164–169. DOI: 10.1016/j.snb.2019.04.060

[37] Mohamedkhair AK, A. DQ, Yamani ZH. Silver nanoparticle-decorated tin oxide thin films: Synthesis, characterization, and hydrogen gas sensing. Frontiers in Materials. 2019;**6**:188. DOI: 10.3389/fmats.2019.00188

[38] Beck F, Loessl M, Baeumner AJ. Signaling strategies of silver nanoparticles in optical and electrochemical biosensors: Considering their potential for the point-of-care. Microchimica Acta. 2023;**190**:91. DOI: 10.1007/s00604-023-05666-6

[39] Tan P, Li H, Wang J, B GSC. Silver nanoparticle in biosensor and bioimaging: Clinical perspectives. Biotechnology and Applied Biochemistry. 2020;**38**(6):1236–1242. DOI: 10.1002/bab.2045

[40] Mahiuddin M, Saha P, Ochiai B. Green synthesis and catalytic activity of Silver nanoparticles based on piper chaba stem extracts. Nanomaterials. 2020;**10**(9):1777. DOI: 10.3390/nano10091777

[41] Albeladi SSR, Malik MA, Al-thabaiti SA. Facile biofabrication of silver nanoparticles using Salvia officinalis leaf extract and its catalytic activity towards Congo red dye degradation. Journal of Materials Research and Technology. 2020;**9**(5):10031–10044. DOI: 10.1016/j.jmrt.2020.06.074

[42] Alshehri AA, Malik MA. Phytomediated Photo-induced green synthesis of Silver Nanoparticles using Matricaria chamomilla L. and its catalytic activity against Rhodamine B. Biomolecules. 2020;**10**(12):1604. DOI: 10.3390/biom10121604

[43] Shirzadi-Ahodashti M, Mizwari ZM, Hashemi Z, Rajabalipour S, Ghoreishi SM, Mortazavi-Derazkola S, Ebrahimzadeh MA. Discovery of high antibacterial and catalytic activities of biosynthesized silver nanoparticles using C. fruticosus (CF-AgNPs) against multi-drug-resistant clinical strains and hazardous pollutants. Environmental Technology & Innovation. 2021;**23**:101607. DOI: 10.1016/j.eti.2021.101607

[44] Urgesa MH, Wolde GS, Kuo D-H. Plasmonic silver nanoparticle-deposited n-Bi_2S_3/p-MnOS diode-type catalyst for enhanced photocatalytic nitrogen fixation: Introducing the defective p-MnOS. Chemical Engineering Journal. 2023;**464**:142717. DOI: 10.1016/j.cej.2023.142717

[45] Dawadi S, Katuwal S, Gupta A, Lamichhane U, Thapa R, Jaisi S, Lamichhane G, Bhattarai DP, Parajuli N. Current research on silver nanoparticles: Synthesis, characterization, and applications. Journal of Nanomaterials. 2021;**1**:6687290. DOI: 10.1155/2021/6687290

[46] Das CGA, Kumar VG, Dhas TS, Karthick V, Govindaraju K, Joselin JM, Baalamurugan J. Antibacterial activity of silver nanoparticles (biosynthesis): A short review on recent advances. Biocatalysis and Agricultural Biotechnology. 2020;**27**:101593. DOI: 10.1016/j.bcab.2020.101593

[47] Menichetti A, Mavridi-Printezi A, Mordini D, Montalti M. Effect of size, shape and surface functionalization on the antibacterial activity of Silver Nanoparticles. Journal of Functional Biomaterials. 2023;**14**:244. DOI: 10.3390/jfb14050244

[48] Potbhare AK, Umekar MS, Chouke PB, Bagade MB, Aziz SKT, Abdala AA, Chaudhary RG. Bioinspired

graphene-based silver nanoparticles: Fabrication, characterization and antibacterial activity. Materials Today: Proceedings. 2020;**29**:720–725. DOI: 10.1016/j.matpr.2020.04.212

[49] Dara PK, Mahadevan R, Digita PA, Visnuvinayagam S, Kumar LRG, Mathew S, Ravishankar CN, Anandan R. Synthesis and biochemical characterization of silver nanoparticles grafted chitosan (Chi-Ag-NPs): In vitro studies on antioxidant and antibacterial applications. SN Applied Sciences. 2020;**2**:665. DOI: 10.1007/s42452-020-2261-y

[50] Keshari AK, Srivastava R, Singh P, Yadav VB, Nath G. Antioxidant and antibacterial activity of silver nanoparticles synthesized by Cestrum nocturnum. Journal of Ayurveda and Integrative Medicine. 2020;**11**:37–44. DOI: 10.1016/j.jaim.2017.11.003

[51] Abdellatif AAH, Alturki HNH, Tawfeek HM. Different cellulosic polymers for synthesizing silver nanoparticles with antioxidant and antibacterial activities. Scientific Reports. 2021;**11**:84. DOI: 10.1038/s41598-020-79834-6

[52] Alsamhary KI. Eco-friendly synthesis of silver nanoparticles by Bacillus subtilis and their antibacterial activity. Saudi Journal of Biological Sciences. 2020;**27**:2185–2191. DOI: 10.1016/j.sjbs.2020.04.026

[53] Arshad F, Naikoo GA, Hassan IU, Chava SR, El-Tanani M, Aljabali AA, Tambuwala MM. Bioinspired and green synthesis of Silver Nanoparticles for Medical Applications: A green perspective. Applied Biochemistry and Biotechnology. 2024;**196**:3636–3669. DOI: 10.1007/s12010-023-04719-z

[54] Hashem AH, Saied E, Amin BH, Alotibi FO, Al-Askar AA, Arishi AA, Elkady FM, Elbahnasawy MA. Antifungal Activity of Biosynthesized Silver Nanoparticles (AgNPs) against aspergilli causing Aspergillosis: Ultrastructure study. Journal of Functional Biomaterials. 2022;**13**:242. DOI: 10.3390/jfb13040242

[55] Matras E, Gorczyca A, Przemieniecki SW, Oćwieja M. Surface properties-dependent antifungal activity of silver nanoparticles. Scientific Reports. 2022;**12**:18046. DOI: 10.1038/s41598-022-22659-2

[56] Ferlay J, Ervik M, Lam F, Laversanne M, Colombet M, Mery L, Piñeros M, Znaor A, Soerjomataram I, Bray F. Global Cancer Observatory: Cancer Today. Lyon, France: International Agency for Research on Cancer; 2024. Available from: https://gco.iarc.who.int/media/globocan/factsheets/cancers/39-all-cancers-fact-sheet.pdf., [Accessed: 2026-March-15]

[57] Wypij M, Jędrzejewski T, Trzcińska-Wencel J, Ostrowski M, Rai M, Golińska P. Green Synthesized Silver Nanoparticles: Antibacterial and Anticancer Activities, Biocompatibility, and Analyses of Surface-Attached Proteins. Frontiers in Microbiology. 2021;**12**:632505. DOI: 10.3389/fmicb.2021.632505

[58] Alharbi NS, Alsubhi NS. Green synthesis and anticancer activity of silver nanoparticles prepared using fruit extract of Azadirachta indica. Journal of Radiation Research and Applied Sciences. 2022;**15**:335–345. DOI: 10.1016/j.jrras.2022.08.009

[59] Shaaban MT, Mohamed BS, Zayed M, El-Sabbagh SM. Antibacterial, antibiofilm, and anticancer activity of

silver-nanoparticles synthesized from the cell-filtrate of Streptomyces enissocaesilis. BMC Biotechnology. 2024;**24**:8. DOI: 10.1186/s12896-024-00833-w

[60] Trucillo P. Drug Carriers: Classification, Administration, Release Profiles, and Industrial Approach. Processes. 2021;**9**:470. DOI: 10.3390/pr9030470

[61] Chandrakala V, Aruna V, Angajala G. Review on metal nanoparticles as nanocarriers: Current challenges and perspectives in drug delivery systems. Emergent Materials. 2022;**5**:1593–1615. DOI: 10.1007/s42247-021-00335-x

[62] Gomes HIO, Martins CSM, Prior JAV. Silver Nanoparticles as Carriers of Anticancer Drugs for Efficient Target Treatment of Cancer Cells. Nanomaterials. 2021;**11**:964. DOI: 10.3390/nano11040964

[63] Hussein HA, Abdullah MA. Novel drug delivery systems based on silver nanoparticles, hyaluronic acid, lipid nanoparticles and liposomes for cancer treatment. Applied Nanoscience. 2022;**12**:3071–3096. DOI: 10.1007/s13204-021-02018-9

[64] Tunç T. Synthesis and characterization of silver nanoparticles loaded with carboplatin as a potential antimicrobial and cancer therapy. Cancer Nanotechnology. 2024;**15**:2. DOI: 10.1186/s12645-023-00243-1

[65] Smith & Nephew. ACTICOAT Antimicrobial Barrier Dressings. Available from: https://www.smith-nephew.com/en/health-care-professionals/products/advanced-wound%20management/%20acticoat-global#application [Accessed: 2026-March-15]

[66] NOTINO. WhitePearl Nanocare. Available from: https://www.notino.co.uk/white-pearl/nanocare-toothpaste/ [Accessed: 2026-March-15]

[67] Giovanna Baby. Body Splash Giovanna Baby Silver. Available from: https://www.giovannababy.com.br/body-splash-giovanna-baby-silver-260ml/p?srsltid=AfmBOooNOMtRyF1ljEu8sSVrxAYwPRXyLDbp52EtkgIeXltsTGXH8IEP [Accessed: 2026-March-15]

[68] ONE 'N ONLY. Shiny silver conditioning shampoo. Available from: https://aonebeauty.com/products/one-n-only-shiny-silver-conditioning-shampoo-12oz [Accessed: 2026-March-15]

[69] Colloid. Silver luxury face soap. Available from: https://colloid.at/product/silver-luxury-face-soap/ [Accessed: 2026-March-15]

[70] Bamal D, Singh A, Chaudhary G, Kumar M, Singh M, Rani N, Mundlia P, Sehrawat AR. Silver Nanoparticles biosynthesis, characterization, antimicrobial activities, applications, cytotoxicity and safety issues: An updated review. Nanomaterials. 2021;**11**:2086. DOI: 10.3390/nano11082086

[71] Eustis S, El-Sayed M. Why gold nanoparticles are more precious than pretty gold: Noble metal surface plasmon resonance and its enhancement of the radiative and nonradiative properties of nanocrystals of different shapes. Chemical Society Reviews. 2005;**35**:209–217

[72] Rahman S, Rahman L, Khalil AT, Ali N, Zia D, Ali M, Shinwari ZK. Endophyte-mediated synthesis of silver nanoparticles and their biological applications. Applied Microbiology and Biotechnology. 2019;**103**(6):2551–2569. DOI: 10.1007/s00253-019-09661-x

Section 2

Advanced Functional and Catalytic Properties

Chapter 5

Preparation of Silver Nanoparticles with Crystalline Defects and Application in Electrocatalysis

Yi Feng and Zhe Li

Abstract

Among various metallic materials, silver has excellent electrical conductivity and a low price, making it a promising candidate to replace platinum-group noble metals for various electrocatalytic reactions. However, the electrocatalytic performance of silver-based catalysts is far inferior to that of platinum-group metals. The extremely low catalytic activity of metallic silver originates from its unique electronic structure. According to the d-band center theory, the d-band electrons in silver are completely filled, resulting in a lower d-band center. Additionally, the antibond orbital between the Ag catalyst and the adsorbate is occupied, resulting in the weak adsorption ability of silver to reactants, and weak adsorption directly determines its low catalytic activity. Crystal defects can effectively regulate the electronic structure of silver and affect its adsorption ability, thus improving its catalytic activity. This chapter systematically reviews various preparation methods of silver nanoparticles rich in crystal defects and their applications in various electrocatalytic reactions. At the same time, it explains the impact of crystal defects on the electronic structure of silver nanoparticles and their electrocatalytic mechanism.

Keywords: crystal defects, coordination number, lattice strain, electrocatalysis, electronic structure

1. Introduction

With the rapid development of the world economy, fossil energy continues to be consumed, resulting in a series of energy and environmental problems, such as energy shortages, environmental pollution, and the greenhouse effect [1, 2]. In order to mitigate these crises confronting humanity, researchers from all over the world have been committed to developing efficient clean energy to replace fossil energy over the past few decades. They have made great achievements, mainly focusing on the use of wind energy, hydropower, nuclear energy, solar energy, and other new energy sources to generate clean electricity. The use of new energy has greatly alleviated the energy and environmental crises. However, due to the seasonal and regional unsustainable characteristics of wind energy, water energy, solar energy, etc., and the huge losses associated with the long-distance transmission and storage of electricity, the use and storage of electricity are greatly limited [3, 4]. Therefore, how to use electricity reasonably and efficiently has become the research focus of current scientific researchers.

Electrocatalytic reactions, with their clean and efficient characteristics, have become a promising research direction to solve the current dilemma of inefficient electricity use.

For example, the electrolytic water hydrogen production technology converts electrical energy into clean hydrogen energy through the electrocatalytic hydrogen production reaction (HER) [5, 6]; metal–air batteries use the electrocatalytic oxygen reduction/oxygen production reaction (ORR/OER) to realize the rapid conversion of electrical energy and chemical energy [7, 8]; fuel cells use the electrocatalytic oxygen reduction reaction (ORR) to efficiently convert chemical energy into electrical energy [9, 10]; and artificial photosynthesis uses the carbon dioxide electrical reduction reaction (CO_2RR) to convert electrical energy into organic fuels with higher energy density [11, 12], etc. High-efficiency electrocatalysts are the core of various electrocatalytic reactions. High-efficiency electrocatalysts can effectively reduce energy consumption, accelerate the reaction rate, and improve the energy utilization efficiency of new energy technologies. The efficiency of electrocatalytic energy conversion directly depends on the performance of electrocatalysts, so efficient electrocatalytic reactions require the development of electrocatalysts with excellent performance.

At present, various noble metal catalysts (e.g., platinum, palladium, iridium, and ruthenium) exhibit the highest performance and are the most widely used in various electrocatalytic reactions. Although noble metal catalysts demonstrate excellent catalytic activity, their high price greatly limits their large-scale application [13, 14]. Among various metal materials, silver offers a low price (approximately 1% of the price of platinum metals). At the same time, silver possesses excellent electrical conductivity (it is the most conductive metal), which gives it great potential to replace precious metals in various electrocatalytic reactions [15]. However, the intrinsic electrocatalytic activity of metallic silver is still far inferior to that of noble metal catalysts.

Electrocatalytic processes usually involve a sequence of key steps: the adsorption of the reaction precursors, the adsorption and desorption of the reaction intermediates, and the desorption of the final products. The essence that determines the catalytic performance of the electrocatalyst lies in the electronic structure of the catalyst. Specific electronic interactions occur between the catalyst's unique electronic structures and reaction precursors, intermediates, and final products, thus affecting the adsorption and desorption behavior at the catalytic interface. The most classic and widely used theory describing the relationship between the electronic structure of transition metals and adsorption energy is the d-band center theory. The theory quantitatively characterizes the adsorption strength of metal catalysts on the adsorbent through the position of the d-band center of transition metals relative to the Fermi energy level: the closer the d-band center is to the Fermi energy level (that is, the higher the center of the d-band), the electron interaction between the metal and the adsorbent is significantly enhanced, which leads to an increase in the adsorption capacity. Conversely, the farther the center of the d-band is from the Fermi energy level (that is, the lower the center of the d-band), the weaker the adsorption capacity. The d-band center theory provides important theoretical guidance for explaining and predicting the catalytic activity of metal catalysts

[16, 17]. According to the classical d-band center theory for transition metals, the extremely low catalytic activity of silver metal comes from its unique d^{10} electron structure. The d^{10} electrons in silver are completely filled, resulting in a lower d-band center. The antibonding orbital between the catalyst and the adsorbate is occupied by electrons, resulting in the weak adsorption capacity of silver. This weak adsorption directly determines the low catalytic activity of silver [18]. For example, the adsorption energy of hydrogen atoms on the crystal surface of Ag(111) is approximately 0.5 eV, while the adsorption energy on the crystal surface of Pt(111) is - 0.1 eV, as shown in **Figure 1**. This indicates that the hydrogen adsorption of silver is weak, resulting in its HER catalytic activity being much lower than that of the precious metal platinum. Therefore, it is necessary to regulate the silver structures at the atomic and electronic levels to modify the catalytic activity of silver catalysts.

In recent years, the creation of crystal defects has become an effective strategy for regulating the electronic structure of metal electrocatalysts due to its controllable variables and wide applicability. The creation of crystal defects mainly involves manipulating the electronic structure of the catalyst by altering the coordination number of atoms and inducing lattice strain [19, 20]. Altering the coordination number is achieved by modifying the atomic coordination environment of the catalyst to regulate the electronic structure of specific catalytic sites, thereby influencing the adsorption and desorption behaviors of reactants during the catalytic process and enhancing its catalytic activity. For example, controlling the size and exposed surface of nanoparticles and regulating the morphology of nanomaterials (nanowires, nanosheets, etc.) can create catalytic sites with low coordination numbers on the catalyst surface, thereby enhancing its adsorption capacity. Lattice strain can change the electron distribution of atoms in the crystal structure, thereby achieving the regulation of the electronic structure. Strain can be classified as compressive strain and tensile strain. Generally, tensile strain can increase the atomic spacing and reduce the overlap degree of d-electron orbitals between metal atoms, enhancing the d-band center of the catalyst and its adsorption of reactants, while compressive strain has the opposite effect.

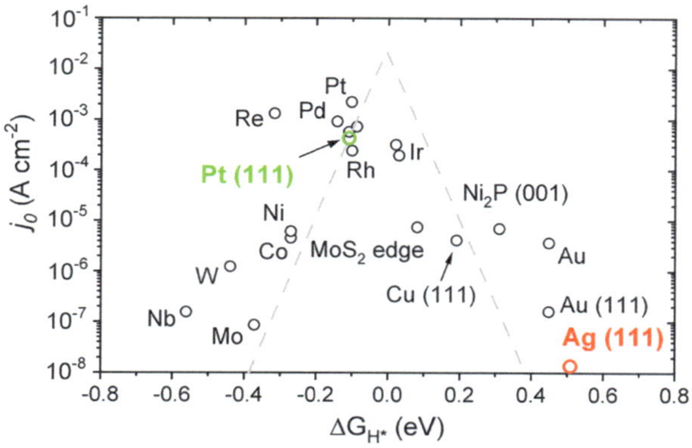

Figure 1.
A volcano plot for HER based on exchange current density (j_0) as a function of the Gibbs free energy of adsorbed hydrogen (ΔG_{H^*}). Catalytic activity for the HER increases in metals that are structurally closer to the apex of a volcano plot.

This chapter focuses on the synthesis strategy of silver nanoparticles with crystal defects, mainly including chemical synthesis methods and physical synthesis methods. At the same time, we discuss the application of silver nanoparticles with crystal defects in HER, ORR, and CO_2RR. In addition, we provide a systematic and in-depth analysis of the mechanistic influence of crystal defects in silver nanoparticles on electrocatalytic activity.

2. Synthesis of silver nanoparticles with crystalline defects

2.1 Chemical method

The chemical method is currently the most widely used and flexible method for synthesizing nanoparticles. The core idea behind creating crystal defects in this method is to influence the normal nucleation and growth of nanoparticles by designing reaction kinetics, thermodynamics, and the synthesis environment. Thus, crystal defects can be created in the material. Usually, a metal salt solution is used as the precursor, and methods such as adding surfactants or using material templates are employed during the chemical preparation process. To date, the main chemical approaches for producing defect-rich nanoparticles include the use of strong reducing agents, sonochemical methods, and photochemical techniques.

Numerous studies have demonstrated that face-centered cubic (FCC) metals tend to form multiply-twinned structures under diverse conditions, including low-pressure argon evaporation and polyol reduction in liquid-phase systems [21]. Yang et al. synthesized five-fold twinned silver nanoparticles by reducing polymeric silver thiolate precursors with sodium borohydride ($NaBH_4$), which exhibited a narrow size distribution of 2–3 nm [22]. Zhang et al. prepared icosahedral silver nanoparticles with abundant nanotwins and crystalline strains via alcohol-mediated reduction of silver ions. Regarding the formation mechanism of icosahedral silver nanoparticles, molecular dynamics simulations suggest that their nucleation may originate from the assembly of magic-number clusters. The thermodynamically stable structure of a cluster is size-dependent: as the cluster grows and the number of atoms increases, sequential structural transformations occur to adopt the most energetically favorable configuration. The transformation rate is regulated by temperature, the type of the dominant capping agent, and particle size, while the final morphology is determined by the competition between structural transformation and layer-by-layer growth [23]. Wiley et al. were the first to achieve selective nucleation and growth of single-(111)-twinned silver nanoparticles, yielding right bipyramids with edge lengths of 75–150 nm. In this synthetic strategy, the key to generating single-twinned seeds was the introduction of sodium bromide (NaBr) into the polyol process, where silver nitrate ($AgNO_3$) is reduced by ethylene glycol in the presence of polyvinylpyrrolidone [24].

Additionally, Tang et al. synthesized nanotwinned silver with a high density of twin boundaries (TBs) via pulsed electrochemical deposition, identifying an optimal twin width of ~ 10 nm for such nanoparticles. To tune the twin-boundary density, the nanoparticles were annealed at varying temperatures, resulting in grains with distinct twin widths and boundary densities [25]. Shi et al. also developed a pulsed electrolysis technique to fabricate silver nanoparticles (Ag NPs) with surface atomic defects. As shown in **Figure 2**, two silver disks are immersed in the electrolyte as the

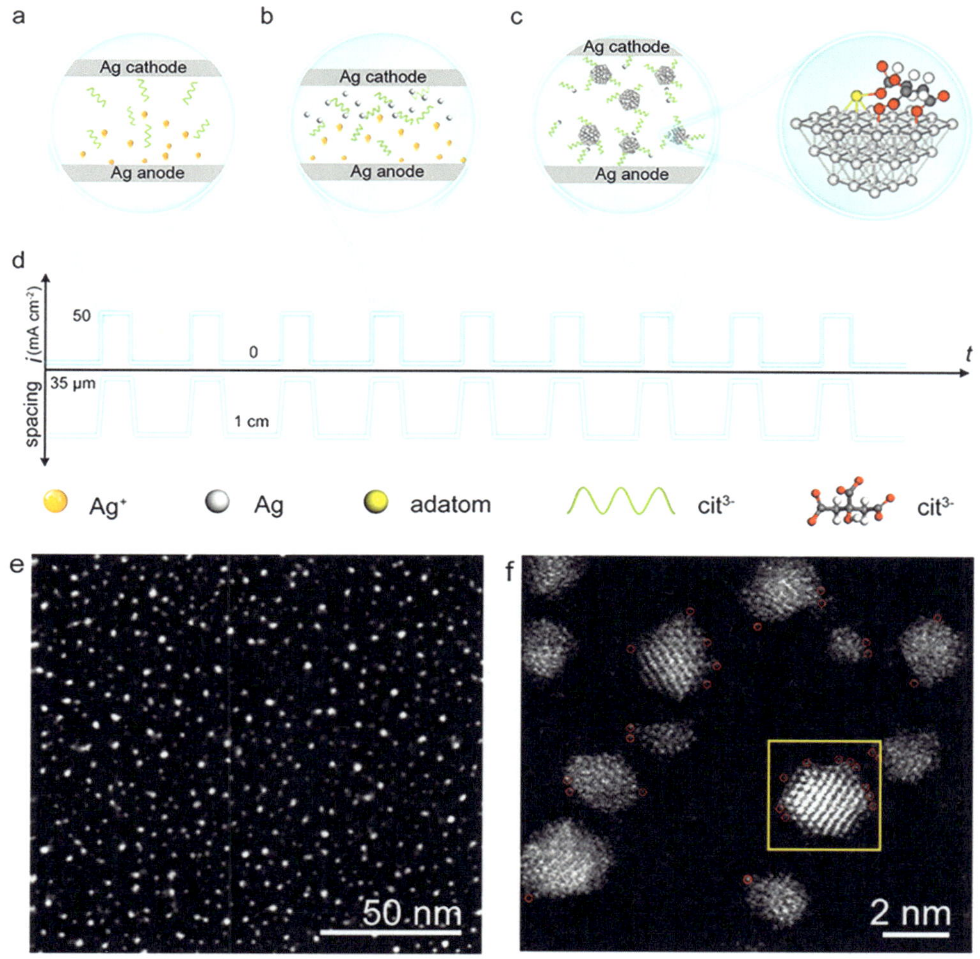

Figure 2.
Schematic illustration of the synthesis of Ag NPs with surface atomic defects by pulse electrolysis. (a)–(c) The evolution process of Ag NPs, where panels respectively show the formation of Ag ions and Ag NPs near the cathode surface. The electrolytic cell consists of a fixed anode and a movable cathode, with the electrolyte. (d) The electrolytic current–time relationship, as well as the electrode spacing–time relationship. (e)–(f) Transmission electron microscope (TEM) images of Ag NPs with surface atomic defects; the atomic defects are marked in red circles. reprinted with permission from Ref[26]. Copyright 2023, American Chemical Society.

cathode and anode, respectively, with the cathode capable of relative movement via servo motor control. During the discharge pulse, the anode undergoes oxidation under a high positive potential and current density, generating a high concentration of silver ions. These ions migrate to the cathode and are reduced by hydrogen radicals to silver monomers. According to the LaMer model, the high-density monomers rapidly aggregate to form a silver nucleus. After the discharge pulse ends, the two electrodes rapidly separate, and a large amount of solvent rushes in, diluting the silver monomers and thus interrupting the further growth of the silver nucleus, resulting in uniformly dispersed silver nanoparticles. At the same time, the citrate ions in the electrolyte can combine with the silver nanoparticles and monomers, and then stably adsorb the monomers onto the nanoparticles as adsorption atoms [26].

2.2 Physical method

The fundamental concept of the physical method is to provide extremely high energy, thereby causing the bulk metallic target material to be ejected in the form of atoms, ions, or clusters. These ejected species subsequently undergo condensation, nucleation, and growth into nanoparticles in an inert gas or liquid environment. The nonequilibrium conditions induced by the rapid quenching of the material during the preparation process play a critical role in the formation of crystalline defects. The synthesis of defective nanoparticles via physical methods relies on high-energy input and rapid nonequilibrium processes that stabilize atoms in metastable states, thereby facilitating the formation of nanostructures enriched with vacancies, dislocations, grain boundaries (GB), and other defects. Currently, the primary physical techniques used for preparing defective nanoparticles include pulsed laser ablation, magnetron sputtering deposition, arc discharge, and ball milling.

Li et al. synthesized silver nanoparticles (Ag NPs) with a high density of stacking faults (SFs) via pulsed laser ablation in liquid (PLAL), a physical approach. In this process, a silver target was immersed in water and irradiated by a high-energy pulsed laser, rapidly heating it to a vapor or plasma state, followed by rapid quenching into the solid state in cold water. This quenching effect induced the aggregation of abundant vacancies on the close-packed {111} planes of silver, which subsequently collapsed to form SFs [27]. Furthermore, as illustrated in **Figure 3**, Kang et al. realized controlled modulation of SF density in Ag NPs by tuning the cooling rate during laser-induced quenching. The silver target immersed in the liquid medium was ablated by a high-power pulsed laser.

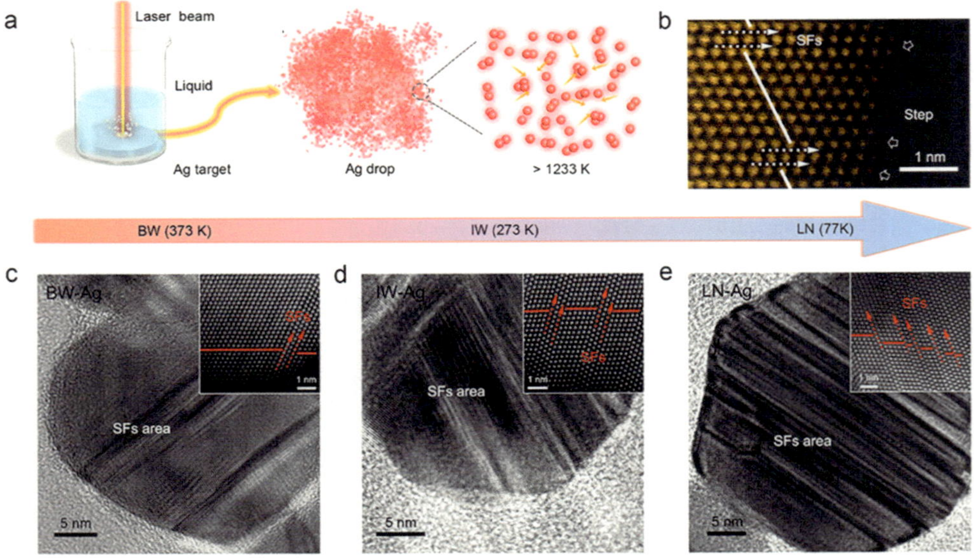

Figure 3.
Preparation and characterizations of Ag NPs with different densities of SFs. (a) Schematic illustrations of the PLAL method. (b) Aberration-corrected TEM image of Ag NPs. (c–e) TEM images of BW–Ag (boiling water, 373 K), IW–Ag (ice water, 273 K), and LN–Ag (liquid nitrogen, 77 K). The inset images are the corresponding aberration-corrected TEM images. Reprinted with permission from [28]. Copyright 2023, American Chemical Society.

The laser instantly heated the surface of the silver target to an extremely high temperature, forming silver droplets. Due to the significant temperature difference between the silver droplets and the liquid medium, the droplets rapidly cooled and formed solid silver nanoparticles. The silver nanoparticles nucleated and grew under the rapid quenching condition of nonequilibrium and then collapsed to form stacking dislocations. In this study, boiling water (373 K), ice water (273 K), and liquid nitrogen (77 K) were selected as the ablation media. Due to the different quenching speeds, the silver nanoparticles prepared in the three liquid media had different densities of SFs [28].

Liang et al. proposed a conceptually new method using the microwave plasma process to reshape the silver film into uniformly distributed submicrometer-scale silver nanoparticles, thereby preparing silver nanoparticles with special crystal plane exposure. Microwave plasma treatment reconfigured the random crystallographic orientations of silver into low-index facets (e.g., {100} and {111}), which are characterized by lower surface energy [29]. Lu et al. developed a novel method for producing defective Ag NPs via spark ablation in gas (SAG). The synthesis process of Ag NPs by SAG is illustrated in **Figure 4**. In this setup, nitrogen gas acts as the carrier gas, and bulk silver rods serve as electrodes. Under high voltage, initial electrons moving toward the anode collide with and ionize background gas atoms, triggering an electric spark. During spark discharge, the silver electrode is partially vaporized by the spark's high temperature; subsequently, Ag NPs form and deposit on the substrate via quenching by the flowing carrier gas. By adjusting the carrier gas flow rate, Ag NPs with different particle sizes can be synthesized [30].

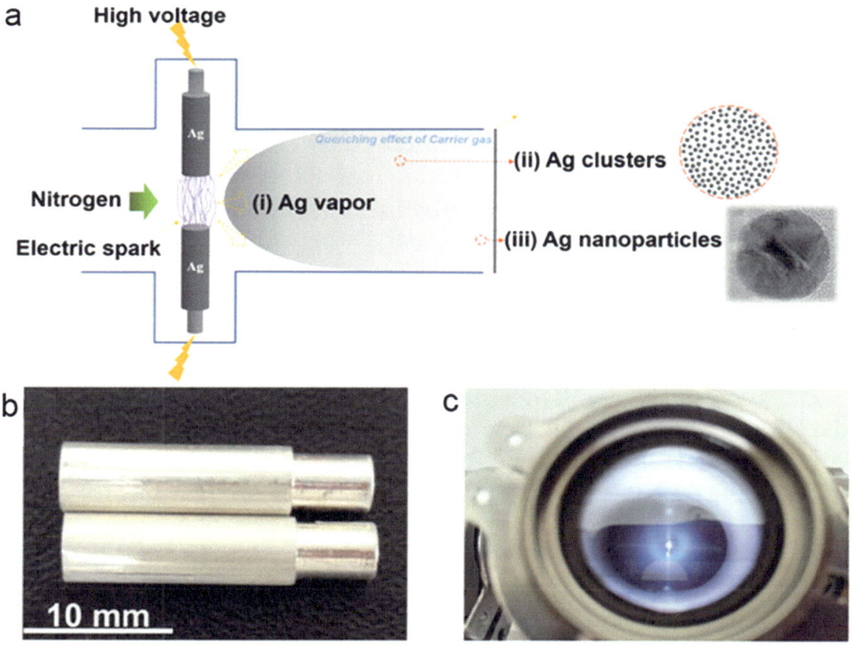

Figure 4.
(a) Scheme illustrating the synthesis of Ag NPs. (b) Photograph of bulk Ag electrodes. (c) Photograph depicting the spark ablation process. Reprinted with permission from [30]. Copyright 2020, Royal Society of Chemistry.

3. Application of silver nanoparticles in electrocatalysis

3.1 Hydrogen evolution reaction (HER)

Hydrogen is expected to play a critical role in future energy systems, owing to its high combustion calorific value (282 kJ mol^{-1}) and the environmental benignity of its combustion byproducts, making it an ideal energy carrier. According to the prediction of the International Energy Agency (IEA), by 2050, the proportion of hydrogen in global energy consumption will exceed 10%. Electrolysis of water is an ideal method for hydrogen production. The reaction can be carried out under mild conditions, the hydrogen production purity is high, and no polluting by-products are generated. It is a "zero-carbon emission" green hydrogen production method [31]. The HER serves as the core electrocatalytic process in water electrolysis, and silver nanoparticles have been widely employed as catalysts in this reaction.

HER is a two-electron transfer process that occurs on the surface of an electrocatalyst and involves multiple reaction steps, with the mechanism being highly dependent on the electrolyte environment. In acidic electrolytes, the HER proceeds via H^+ ions as the primary reactant, whereas in alkaline electrolytes, H_2O molecules serve as the hydrogen source. Compared to acidic media, alkaline HER requires an additional initial step: the adsorption and subsequent dissociation of H_2O molecules on the catalyst surface to form adsorbed hydrogen intermediates (H*) and OH^- ions. This extra kinetic barrier results in the catalytic activity for HER in alkaline electrolytes being typically two or three orders of magnitude lower than that in acidic electrolytes [14, 32, 33].

The research conducted by Li et al. demonstrated the application of SF-rich Ag NPs in acidic HER catalysis. As shown in **Figure 5**, compared to single-crystal and twinned Ag NPs synthesized via conventional chemical methods, the SF-rich Ag NPs exhibited a substantial improvement in catalytic activity for acidic HER, even surpassing the performance of commercial Pt/C catalysts. A series of comparative experiments and theoretical calculations revealed that the synergistic effects of low coordination numbers and tensile strain induced by SFs effectively modulate the hydrogen adsorption free energy on Ag NPs. This optimization shifts the position of SF-rich Ag NPs from the lower right corner of the HER volcano diagram to the top of the diagram, explaining their superior catalytic performance. Moreover, a series of comparative experiments and theoretical calculations have also been conducted to distinguish the individual effects of particle size and crystal defects on catalytic activity, and proved that the high catalytic activity originated from the crystal defects [27]. Kang et al. prepared Ag NPs with surface defects of average diameters of 3 nm, 6 nm, and 9 nm by regulating the pulse laser ablation energy, and applied them to acidic HER catalysis. The electrochemical test results showed that as the particle size decreased, the HER catalytic activity of the Ag NPs significantly increased. Among them, the Ag NPs with an average diameter of 3 nm showed an overpotential of 96 mV at 10 mA cm^{-2}, which was much lower than that of commercial Ag nanopowder (406 mV). The theoretical calculation results indicated that reducing the coordination number could decrease the charge density between adjacent Ag atoms, shift the d-band center of Ag upward, enhance the binding strength between Ag and hydrogen intermediates, and improve the intrinsic catalytic activity of Ag for HER [34]. Feng et al. employed laser ablation in liquid to fabricate Ag NPs containing abundant oxide-induced bulk defects using silver-based alloy targets for alkaline

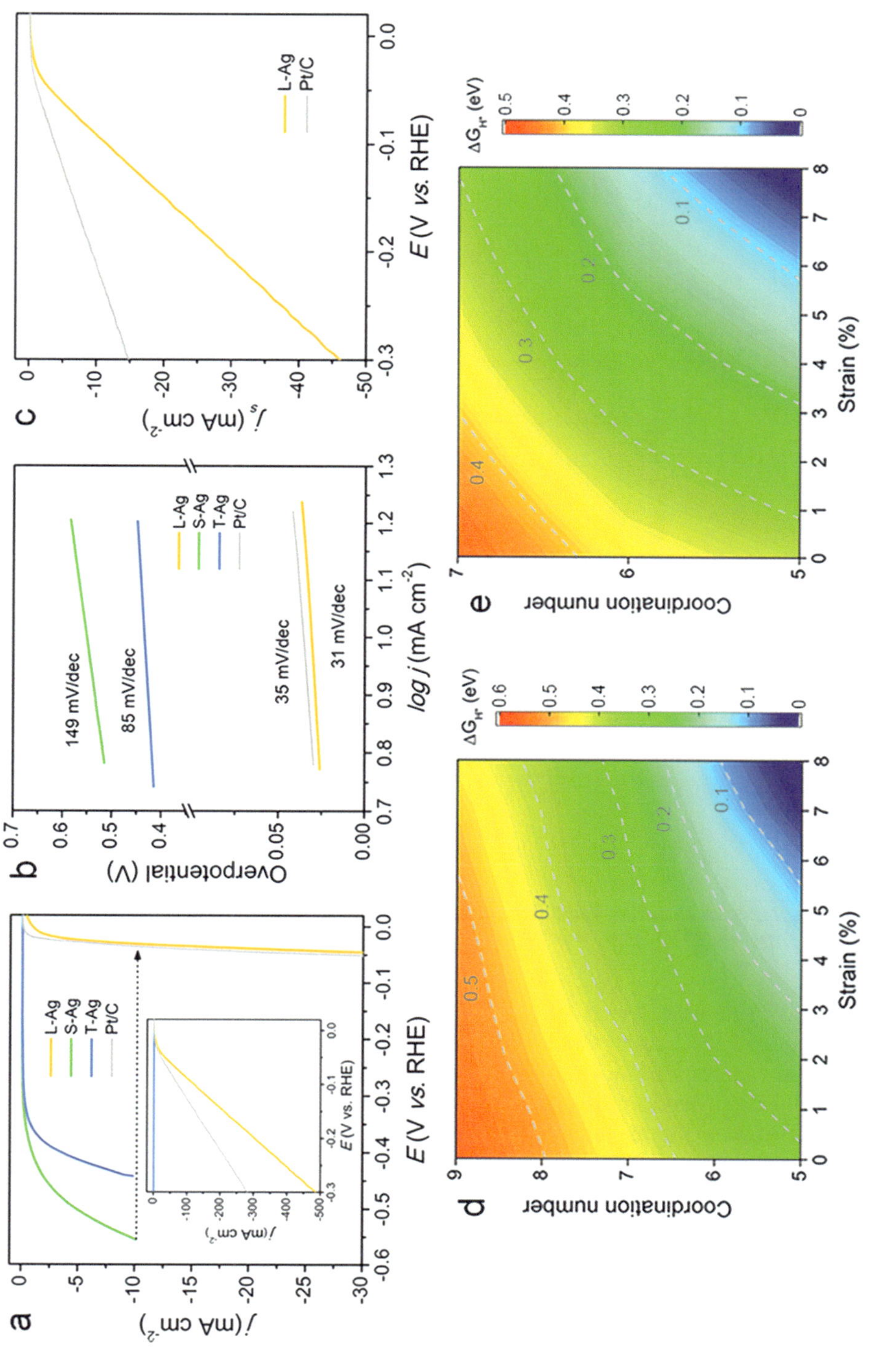

Figure 5.
HER activity of Ag NPs with abundant SFs. (a) Polarization curves; the inset shows the corresponding polarization curves under high applied potential. (b) Corresponding Tafel plots. (c) Specific activity. (d) and (e) Gibbs free energy (ΔG_{H^*}) as a function of tensile strain and coordination number on Ag (111) and Ag (110) surfaces, respectively. Reprinted with permission from Ref. 27. Copyright 2019, Springer Nature.

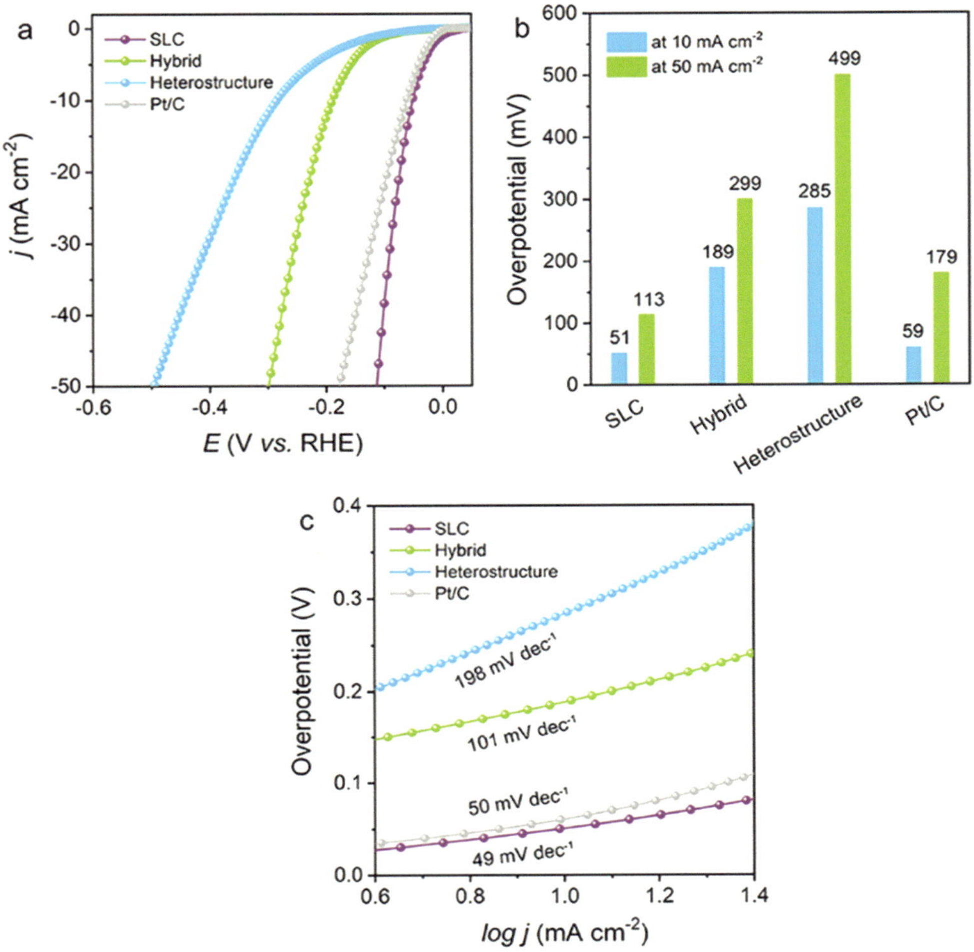

Figure 6.
HER activity of Ag NPs with abundant bulk defects. (a) Polarization curves. (b) The comparison of overpotentials. (c) Corresponding Tafel plots. Reprinted with permission from [36]. Copyright 2021, Elsevier.

HER catalysis. These defect-rich Ag NPs exhibited catalytic performance superior to that of commercial Pt/C catalysts, as illustrated in **Figure 6**. Comprehensive characterization and computational studies indicated that the enhanced HER activity originates from the introduction of tensile strain and a reduction in surface coordination number due to the bulk defects. The combined effects of tensile strain and lower coordination number significantly strengthen the adsorption capability of Ag toward hydrogen reaction intermediates [35].

3.2 Oxygen reduction reaction (ORR)

Emerging energy conversion technologies, such as hydrogen fuel cells, ethanol fuel cells, and metal-air batteries, enable the highly efficient conversion of chemical energy into electrical energy [36]. These systems boast significantly higher theoretical energy densities than lithium-ion batteries and operate without emitting pollutants, making them promising candidates for environmentally sustainable and clean energy solutions that are increasingly being integrated into practical applications [37]. The core catalytic reaction of this new type of battery technology is the ORR, and silver nanoparticles are widely used in the ORR.

The ideal ORR is a four-electron reaction that occurs on the surface of an electrocatalyst and consists of multiple reaction steps [38, 39]. In acidic electrolyte, the final product of the four-electron ORR reaction is H_2O, while in alkaline electrolyte, it is OH^-. Alkaline ORR is currently widely studied by researchers due to its mild reaction conditions. However, when the ORR electrocatalytic reaction is not complete, the entire reaction is a two-electron reaction, and the final reaction product is H_2O_2. Generally, the four-electron reaction is considered to be the favorable ORR reaction pathway because the two-electron reaction significantly reduces the energy conversion efficiency and accelerates the degradation of proton-conducting polymers [14, 40].

Kang et al. prepared silver nanoparticles with different stacking layer defect densities by changing the temperature of the liquid cooling medium to regulate the cooling rate of the metal plasma. These nanoparticles were then applied to the ORR catalytic reaction. The electrochemical test results showed that as the SF density increased, the catalytic activity of the silver nanoparticles gradually improved. The ORR onset potential and half-wave potential of the Ag NPs with optimal performance were superior to those of the commercial Pt/C catalyst (**Figure** 7). At the same time, the kinetic current density was significantly higher than that of the commercial Pt/C catalyst. A series of advanced characterizations indicated that the high-density SFs could generate high tensile strain and a large number of unsaturated coordination atoms in the Ag NPs. The theoretical calculation results further showed that the tensile strain and low coordination atoms could significantly reduce the charge density between adjacent Ag atoms, causing the d-band center of Ag to shift upward, enhancing the adsorption strength of Ag with ORR reactants and intermediates, and improving the intrinsic catalytic activity of Ag for ORR. This research work indicates that the high tensile strain and low coordination atoms brought by high-density stacking layer defects can significantly enhance the ORR activity of Ag catalysts, making Ag potentially replace the noble metal Pt and become a commercial ORR catalyst [28].

In a complementary study, Shi et al. utilized pulse-electrodeposited Ag NPs featuring surface adatom defects for alkaline ORR catalysis, achieving exceptional catalytic performance surpassing that of commercial Pt/C catalysts. The mass activity of these defect-rich Ag NPs was orders of magnitude higher than that of conventionally synthesized silver nanoparticles. Density functional theory (DFT) calculations attributed the enhanced activity to the low coordination number of surface adatoms, which reinforces the adsorption of key reaction intermediates. Furthermore, it was applied to metal-air battery devices, where it demonstrated superior battery performance compared to platinum catalysts [26]. Zhou et al. prepared a nanoporous silver catalyst with abundant internal defects by acid-etching AgAl alloys and applied it to ORR catalysis. The porous silver exhibited excellent catalytic performance. The catalytic results indicated that the nanoporous silver catalyst could achieve comparable or even better ORR performance than state-of-the-art Pt/C catalysts at a low overpotential. This is extremely ideal for improving energy efficiency in electrochemical energy applications. This

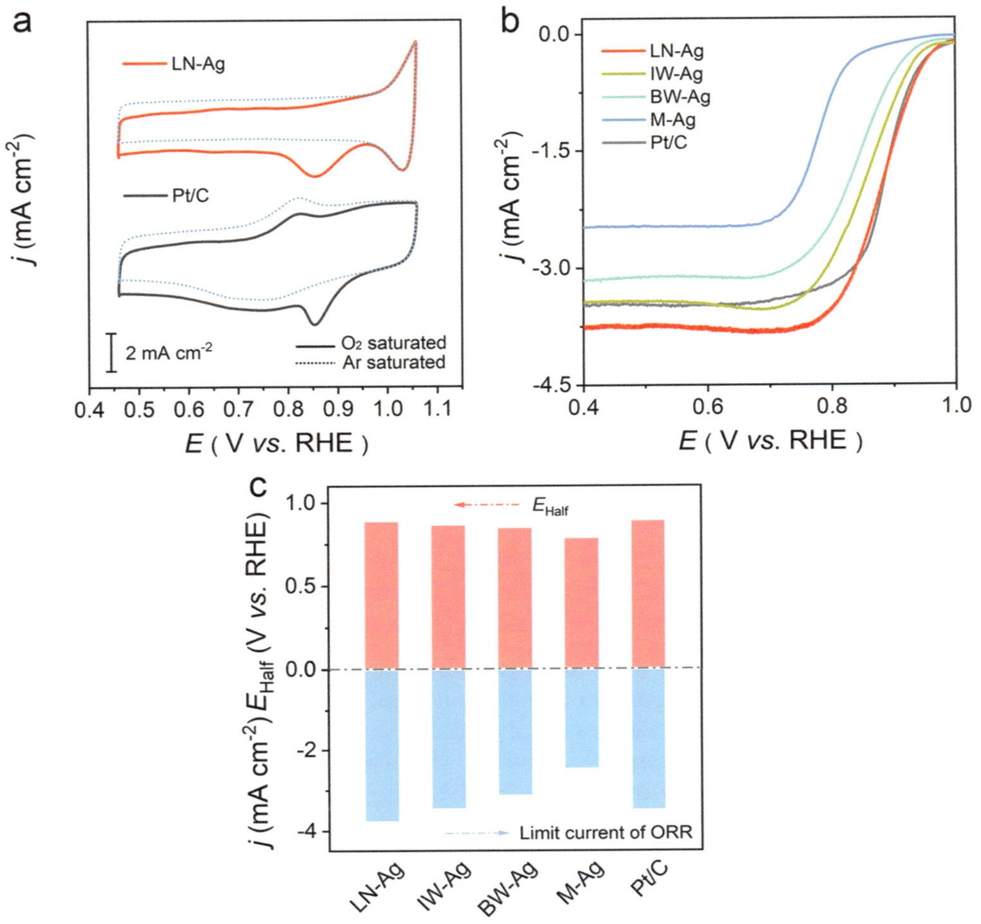

Figure 7.
ORR activity of Ag NPs with different densities of SFs. (a) CV curves. (b) Polarization curves. (c) The comparison of half-wave potential (red bar) and limiting current density (blue bar). Reprinted with permission from [28]. Copyright 2023, American Chemical Society.

significant enhancement in activity is attributed to a geometric effect caused by the nanoporous structure [41].

3.3 Carbon dioxide reduction reaction (CO_2RR)

In recent years, the overexploitation and use of fossil fuels have led to a significant increase in the concentration of carbon dioxide in the atmosphere, from about 280 parts per million (ppm) before the Industrial Revolution (1750) to 415 ppm by 2021. This continuous rise has triggered a series of consequences associated with the greenhouse effect, including an increase in global average temperature and an accelerated rise in sea levels [42]. Therefore, reducing carbon dioxide emissions and promoting carbon dioxide conversion technology has become a core strategy to meet the urgent challenges posed by the accumulation of greenhouse gases. Using renewable electricity to convert carbon dioxide into fuel and chemical raw materials can not only effectively reduce emissions, but also produce chemical products with high added value, thus realizing "artificial photosynthesis." Electrochemical reduction of carbon dioxide is a key catalytic process in artificial photosynthesis systems [43–45]. Among the various catalysts, silver nanoparticles are highly selective in reducing carbon dioxide to carbon monoxide (a fuel and chemical intermediate with important industrial value), making it a key material in CO_2RR applications.

Electrochemical reduction of CO_2 to CO involves a two-electron transfer pathway [46]. In the first step, a CO_2 molecule is adsorbed on the surface of the catalyst and receives an electron and a proton simultaneously to form a *COOH intermediate. Then, the intermediate is further reduced to generate *CO. If the active site has weak adsorption affinity, *CO will be easily desorbed, releasing CO as the final product. However, due to the d^{10} electron configuration of silver, its interaction with the key *COOH intermediate is inherently weak, resulting in high energy barriers during its formation process [27, 47, 48]. In addition, CO_2RR contains multiple competitive reaction pathways and potential products, among which there is significant kinetic competition in HER. Therefore, a large amount of research focuses on adjusting the microstructure of silver catalysts through defect engineering to improve Faraday efficiency (FE) and the current density of carbon monoxide generation simultaneously [49, 50].

GBs and TBs are a class of planar defects in crystal solids. Through the electrochemical reduction of Ag_2O, silver nanocatalysts can form rich defect structures, including GBs and TBs [51]. These defective silver crystals enhance the adsorption of *COOH intermediates, thus achieving ~100% CO FE at an overpotential of 0.7 V in H-type batteries, with an overpotential of 0.7 V, and a current density of 180 mA cm^{-2} at – 1.0 V *vs.* RHE in a flow cell (**Figure 8**). Tang et al. used pulsed electrochemical deposition technology to synthesize silver

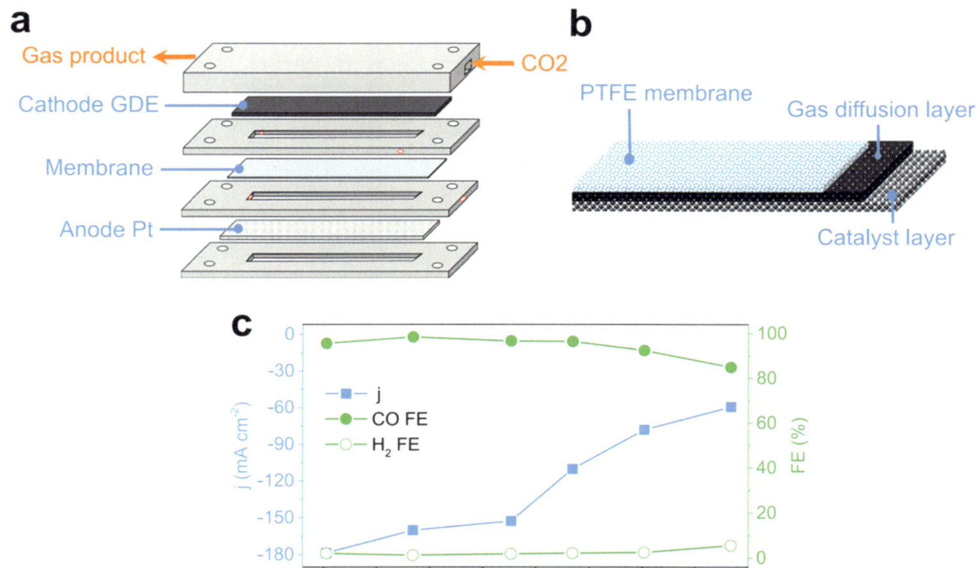

Figure 8.
CO_2RR activities of defect-rich Ag nanocrystals. (a) Schematic illustration of the flow cell. (b) Cathode gas diffusion electrode (GDE). (c) Total current density and FEs of CO and H_2. Reprinted with permission from [52]. Copyright 2021, Springer Nature.

nanocatalysts with high twin grain boundary density [25]. At a pulse current density of 1 A cm^{-2}, the width of the twin is 10 nm, and the twin crystal density reaches 1.5 × 10^4 cm^{-1}. These lengthy and winding TBs enable the silver catalyst to achieve a turnover frequency of 92% carbon monoxide FE and 127 s^{-1} at - 1.0 V *vs.* RHE. Xu et al. prepared a porous coral-like silver catalyst with high grain boundary density, which maintains carbon monoxide selectivity (CO FE above 90%) over a wide potential range from - 0.6 to - 1.0 V vs. RHE [52]. Like the grain boundary and the twin crystal boundary, SFs are also effective active sites to enhance the catalytic activity of silver. Research shows that in silver nanorods with dense accumulation defects, these defects can be coupled together to form a unique structure composed of opposing atoms [53]. Rietveld refinement of XRD data combined with aberration-corrected HAADF-STEM analysis revealed significant tensile strain, averaging up to 0.7%. Geometric phase analysis indicated that up to 59% of the lattice area was subjected to tensile strain. This cooperative effect of coupled SFs facilitates the adsorption and activation of CO_2 molecules, thereby improving the performance of the CO_2 electroreduction reaction.

Point defects represent another significant type of crystalline defect capable of enhancing the CO_2RR performance of silver catalysts. Electrochemical reconstruction of silver halides enables the formation of homogeneously distributed vacancies on both the surface and in the bulk of silver catalysts [54]. These vacancies induce a low coordination number and lattice strain in silver, resulting in an upshifted d-band center. This electronic modulation strengthens the hybridization between the Ag_d orbitals and the C_p orbitals of the *COOH intermediate, thereby promoting *COOH adsorption and facilitating the conversion of CO_2 to CO (**Figure 9**). Furthermore, doping-induced defects can effectively tune the catalytic activity of silver catalysts. For instance, Ma et al. employed electrochemical treatment to reconstruct the catalyst surface, yielding single-atom copper species supported on defective silver nanowires (Cu/Ag(S)) [55]. The synergistic interaction between the copper single atoms and the adjacent defective silver introduces microstrain and lowers the d-band center of Cu/Ag(S), thus reducing the energy barrier for *CO formation.

4. Conclusion(s)

This chapter focuses on the synthesis methodologies of silver nanoparticles with enriched crystalline defects, encompassing both chemical and physical approaches. The chemical method mainly affects the nucleation and growth of nanoparticles through chemical additives, while the physical method primarily enables the nanoparticles to achieve nonequilibrium nucleation and growth through the rapid quenching effect. Moreover, Ag nanoparticles with abundant defects are widely applied in HER, ORR, and carbon dioxide reduction reaction, and they exhibit excellent catalytic performance. These excellent electrocatalytic properties are attributed to the changes in surface coordination numbers and lattice strains caused by the crystal defects, which, in turn, regulate the electronic structure of the silver metal and optimize the adsorption of silver on the catalytic reaction intermediates. The statement in this chapter indicates that introducing crystal defects can effectively control the electrocatalytic activity of silver nanoparticles, providing a reference for the efficient application of silver nanoparticles in electrocatalytic reactions.

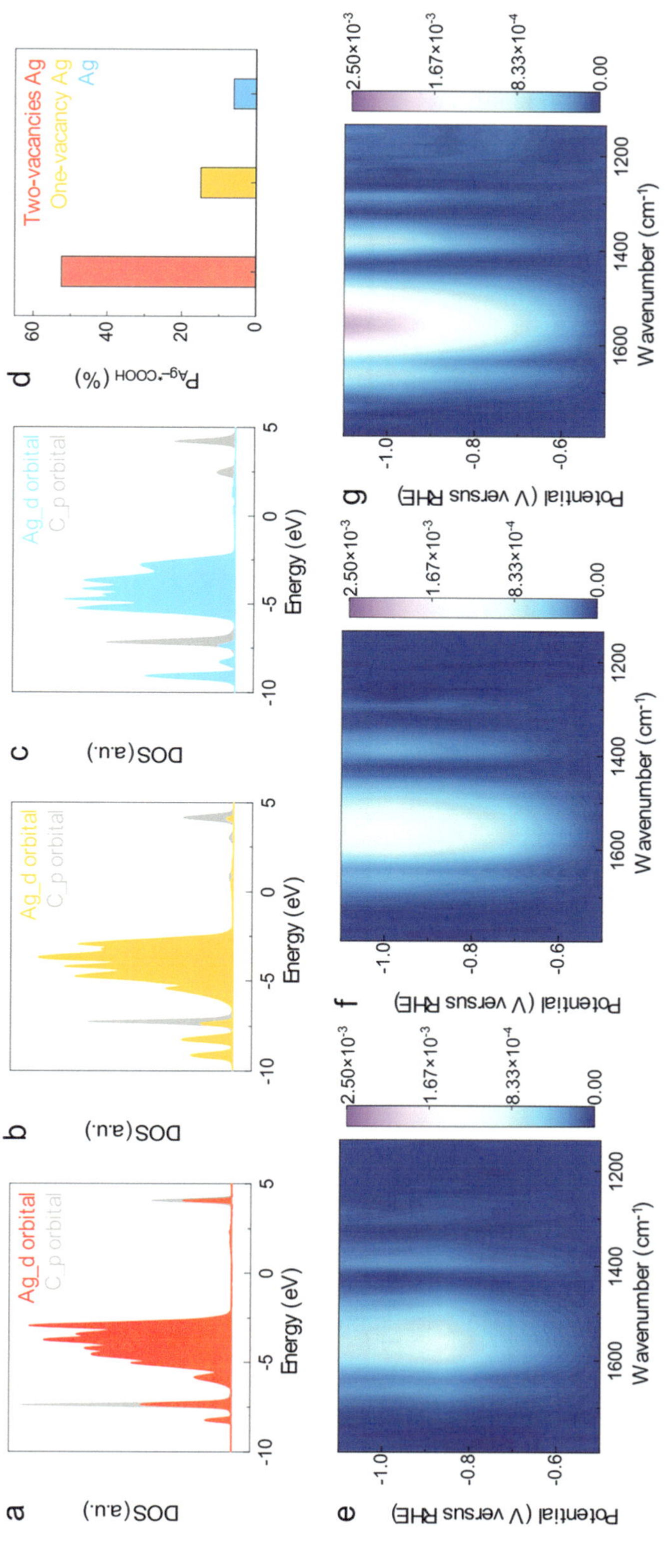

Figure 9.
*Origin of the enhanced CO_2RR performance imparted by homogeneous Ag vacancies. (a–c) Density of states (DOS) of the c_p orbital in *COOH and the Ag_d orbital of two-vacancies Ag (a), one-vacancy Ag (b), and Ag (c). (d) The degree of orbital hybridization for two-vacancies Ag, one-vacancy Ag, and Ag. (e–g) In situ Fourier transform infrared spectroscopy (FTIR) patterns of Ag with inappreciable vacancies (e), Ag with only surficial vacancies (f), and Ag with homogeneous vacancies (g). Reprinted with permission from [55]. Copyright 2023, American Chemical Society.*

Acknowledgments

This work was supported by the National Natural Science Foundation of China (No. 52401292), the Beijing-Tianjin-Hebei Fundamental Research Cooperation Project (No. B2024203051), the Science and Technology Research Program of Chongqing Municipal Education Commission (No. KJQN202400531), and the Startup Foundation of Chongqing Normal University (No. 24XLB021). The author acknowledges the use of Youdao for language polishing of the manuscript.

Conflict of Interest

The authors declare no conflict of interest.

Author details

Yi Feng[2] and Zhe Li[1*]

1 School of Chemical Engineering and Technology, Institute of Molecular Plus, Tianjin University, Tianjin, China

2 College of Chemistry and Materials Science, Chongqing Normal University, Chongqing, China

*Address all correspondence to: zhli@tju.edu.cn

References

[1] Kondratenko EV, Mul G, Baltrusaitis J, Larrazábalc GO, Pérez-Ramírez J. Status and perspectives of CO_2 conversion into fuels and chemicals by catalytic, photocatalytic and electrocatalytic processes. Energy & Environmental Science. 2013;**6**:3112–3135. DOI: 10.1039/C3EE41272E

[2] Geng DS, Ding N, Andy Hor TS, Liu Z, Sun X, Zong Y. Potential of metal-free "graphene alloy" as electrocatalysts for oxygen reduction reaction. Journal of Materials Chemistry A. 2015;**3**:1795–1810. DOI: 10.1039/C4TA06008C

[3] Turner JA. Sustainable hydrogen production. Science. 2004;**305**:972–974. DOI: 10.1126/science.1103197

[4] Chu S, Majumdar A. Opportunities and challenges for a sustainable energy future. Nature. 2012;**488**:294–303. DOI: 10.1038/nature11475

[5] Yin J, Fan QH, Li YX, Cheng FY, Zhou PP, Xi PX, Sun SH. Ni-C-N nanosheets as catalyst for hydrogen evolution reaction. Journal of the American Chemical Society. 2016;**138**:14546–14549. DOI: 10.1021/jacs.6b09351

[6] He Y, Ji SF, Wang RY, Lv HF, Lopes PP, Paulikas AP, Li HY, Mao SX, Wang CM, Markovic NM, Li J, Stamenkovic VR, Li YD. High-performance Rh_2P electrocatalyst for efficient water splitting. Journal of the American Chemical Society. 2017;**139**:5494–5502. DOI: 10.1021/jacs.7b01376

[7] Liang JY, Wang YZ, Wang CC, Lu SY. In situ formation of NiO on Ni foam prepared with a novel leaven dough method as an outstanding electrocatalyst for oxygen evolution reactions. Journal of Materials Chemistry A. 2016;**4**:9797–9806. DOI: 10.1039/C6TA03729A

[8] Feng YY, Zhang HJ, Fang L, Mu YP, Wang Y. Uniquely monodispersing NiFe alloyed nanoparticles in three-dimensional strongly linked sandwiched graphitized carbon sheets for high-efficiency oxygen evolution reaction. ACS Catalysis. 2016;**6**:4477–4485. DOI: 10.1021/acscatal.6b00481

[9] Varnell JA, Tse ECM, Schulz CE, Fister TT, Haasch RT, Timoshenko J, Frenkel AI, Gewirth AA. Identification of carbon-encapsulated iron nanoparticles as active species in non-precious metal oxygen reduction catalysts. Nature Communications. 2016;**7**:12582. DOI: 10.1038/ncomms12582

[10] Liu Q, Wang YB, Dai LM, Yao JN. Scalable fabrication of nanoporous carbon fiber films as bifunctional catalytic electrodes for flexible Zn-air batteries. Advanced Materials. 2016;**28**:3000–3006. DOI: 10.1002/adma.201506112

[11] Zheng Y, Vasileff A, Zhou XL, Jiao Y, Jaroniec M, Qiao SZ. Understanding the roadmap for electrochemical reduction of CO_2 to multi-carbon oxygenates and hydrocarbons on copper-based catalysts. Journal of the American Chemical Society. 2019;**141**:7646–7659. DOI: 10.1021/jacs.9b02124

[12] Liu S, Yang HB, Hung SF, Ding J, Cai WZ, Liu LH, Gao JJ, Li XN, Ren XY,

Kuang ZC, Huang YQ, Zhang T, Liu B. Electrifying model single-Atom catalyst for elucidating the CO_2 reduction reaction. Angewandte Chemie International Edition. 2020;**59**:798–803. DOI: 10.1002/anie.201911995

[13] Zhang L, Doyle-Davis K, Sun XL. Pt-Based electrocatalysts with high atom utilization efficiency: From nanostructures to single atoms. Energy & Environmental Science. 2019;**12**: 492–517. DOI: 10.1039/C8EE02939C

[14] Jiao Y, Zheng Y, Jaroniec M, Qiao SZ. Design of electrocatalysts for oxygen- and hydrogen-involving energy conversion reactions. Chemical Society Reviews. 2015;**44**:2060–2086. DOI: 10.1039/C4CS00470A

[15] Holewinski A, Idrobo JC, Linic S. High-performance Ag-Co alloy catalysts for electrochemical oxygen reduction. Nature Chemistry. 2014;**6**:828–834. DOI: 10.1038/nchem.2032

[16] Hammer B, Nørskov JK. Electronic factors determining the reactivity of metal surfaces. Surface Science. 1995;**243**:211–220. DOI: 10.1016/0039-6028(96)00588-2

[17] Kitchin JR, Nørskov JK, Barteau MA, Chen JG. Modification of the surface electronic and chemical properties of Pt(111) by subsurface 3d transition metals. The Journal of Chemical Physics. 2004;**120**: 10240–10246. DOI: 10.1063/1.1737365

[18] Mo JY, Stefanov BI, Lau THM, Chen TY, Wu S, Wang ZQ, Gong XQ, Wilkinson I, Schmid G, Tsang SCE. Superior performance of Ag over Pt for hydrogen evolution reaction in water electrolysis under high overpotentials. ACS Applied Energy Materials. 2019;**2**:1221–1228. DOI: 10.1021/acsaem.8b01777

[19] Falicov L, Somorjai G. Correlation between catalytic activity and bonding and coordination number of atoms and molecules on transition metal surfaces: Theory and experimental evidence. Proceedings of the National Academy of Sciences. 1985;**82**:2207–2211. DOI: 10.1073/pnas.82.8.2207

[20] Tedsree K, Li T, Jones S, Chan CWA, Yu KMK, Bagot PAJ, Marquis EA, Smith GDW, Tsang SCE. Hydrogen production from formic acid decomposition at room temperature using an Ag-Pd core-shell nanocatalyst. Nature Nanotechnology. 2011;**6**:302–307. DOI: 10.1038/nnano.2011.42

[21] Gao Y, Song L, Jiang P, Liu LF, Yan XQ, Zhou ZP, Liu DF, Wang JX, Yuan HJ, Zhang ZX, Zhao XW, Dou XY, Zhou WY, Wang G, Xie SS, Chen HY, Li JQ. Silver nanowires with five-fold symmetric cross-section. Journal of Crystal Growth. 2005;**276**:606–612. DOI: 10.1016/j.jcrysgro.2004.11.396

[22] Yang HY, Wang Y, Chen X, Zhao XJ, Gu L, Huang HQ, Yan JZ, Xu CF, Li G, Wu JC, Edwards AJ, Dittrich B, Tang ZC, Wang DD, Lehtovaara L, Hakkinen H, Zheng NF. Plasmonic twinned silver nanoparticles with molecular precision. Nature Communications. 2016;**7**:12809. DOI: 10.1038/ncomms12809

[23] Zhang QB, Xie JP, Yang JH, Lee JY. Monodisperse icosahedral Ag, Au, and Pd nanoparticles: Size control strategy and superlattice formation. ACS Nano. 2009;**3**:139–148. DOI: 10.1021/nn800531q

[24] Wiley BJ, Xiong YJ, Li ZY, Yin YD, Xia YN. Right Bipyramids of Silver:

A new shape derived from single twinned seeds. Nano Letters. 2006;**6**:765–768. DOI: 10.1021/nl060069q

[25] Tang C, Gong P, Xiao TS, Sun ZZ. Direct electrosynthesis of 52% concentrated CO on silver's twin boundary. Nature Communications. 2021;**12**:2139. DOI: 10.1038/s41467-021-22428-1

[26] Shi ZZ, Li WP, Kang WJ, Feng Y, Li Z, Zhong XY, Yang J, Liu H, Dong CK, Yin PF, Chen FR, Du XW. Silver clusters with adatoms as a catalyst for the oxygen reduction reaction. ACS Catalysis. 2023;**13**:9181–9189. DOI: 10.1021/acscatal.3c01400

[27] Li Z, Fu JY, Feng Y, Dong CK, Liu H, Du XW. A silver catalyst activated by stacking faults for the hydrogen evolution reaction. Nature Catalysis. 2019;**2**:1107–1114. DOI: 10.1038/s41929-019-0365-9

[28] Kang WJ, Feng Y, Li Z, Chen ZN, Dong CK, Yang J, Yin PF, Liu H, Du XW. Engineering dense stacking faults in silver nanoparticles for boosting the oxygen reduction reaction. ACS Energy Letters. 2023;**8**:3512–3519. DOI: 10.1021/acsenergylett.3c01173

[29] Liang FL, Zhou W, Li J, Zhu ZH. Microwave-plasma induced reconstruction of silver catalysts for highly efficient oxygen reduction. Journal of Materials Chemistry A. 2013;**1**:13746–13749. DOI: 10.1039/c3ta13656f

[30] Lu JD, Guo J, Song SH, Yu GF, Liu H, Yang XJ, Lu ZM. Preparation of Ag nanoparticles by spark ablation in gas as catalysts for electrocatalytic hydrogen production. RSC Advances. 2020;**10**:38583–38587. DOI: 10.1039/d0ra06682f

[31] Shih AJ, Monteiro MCO, Dattila F, Pavesi D, Philips M, Silva AHM, Vos RE, Ojha K, Park S, Heijden OVD, Marcandalli G, Goyal A, Villalba M, Chen XT, Gunasooriya KK, McCrrum I, Mom R, Lopez N, Koper MTM. Water electrolysis. Nature Reviews Methods Primers. 2022;**2**:84. DOI: 10.1038/s43586-022-00164-0

[32] Yang JU, Mohmad AR, Wang Y, Fullon R, Song XJ, Zhao F, Bozkurt I, Augustin M, Santos EJG, Shin HS, Zhang WJ, Voiry D, Jeong HY, Chhowalla M. Ultrahigh-current-density niobium disulfide catalysts for hydrogen evolution. Nature Materials. 2019;**18**:1309–1314. DOI: 10.1038/s41563-019-0463-8

[33] Lu XF, Yu L, Lou XW. Highly crystalline Ni-doped FeP/carbon hollow nanorods as all-pH efficient and durable hydrogen evolving electrocatalysts. Science Advances. 2019;**5**:eaav6009. DOI: 10.1126/sciadv.aav6009

[34] Kang WJ, Cheng CQ, Li Z, Feng Y, Shen GR, Du XW. Ultrafine Ag nanoparticles as active catalyst for electrocatalytic hydrogen production. ChemCatChem. 2019;**11**:5976–5981. DOI: 10.1002/cctc.201901364

[35] Feng Y, Li Z, Cheng CQ, Kang WJ, Mao J, Shen GR, Yang J, Dong CK, Liu H, Du XW. Strawberry-like Co_3O_4-Ag bifunctional catalyst for overall water splitting. Applied Catalysis B-Environmental. 2021;**299**:120658. DOI: 10.1016/j.apcatb.2021.120658

[36] Lee JS, Kim ST, Cao R, Choi NS, Liu ML, Lee KT, Cho J. Metal-Air batteries with high energy density: Li-Air versus Zn-Air. Advanced Energy Materials. 2011;**1**:34–50. DOI: 10.1002/aenm.201000010

[37] Cheng FY, Chen J. Metal-air batteries: From oxygen reduction electrochemistry to cathode catalysts. Chemical Society Reviews. 2012;**41**:2172–2192. DOI: 10.1039/C1CS15228A

[38] Norskov JK, Rossmeisl J, Logadottir A, Lindqvist L, Kitchin JR, Bligaard T, Jonsson H. Origin of the overpotential for oxygen reduction at a fuel-cell cathode. Journal of Physical Chemistry B. 2004;**108**:17886–17892. DOI: 10.1021/jp047349j

[39] Ge XM, Sumboja A, Wuu D, An T, Li B, Goh FW, Hor TS, Zong Y, Liu ZL. Oxygen reduction in alkaline media: From mechanisms to recent advances of catalysts. ACS Catalysis. 2015;**5**:4643–4667. DOI: 10.1021/acscatal.5b00524

[40] Wang XQ, Li ZJ, Qu YT, Yuan TW, Wang WY, Wu Y, Li YD. Review of metal catalysts for oxygen reduction reaction: From nanoscale engineering to atomic design. Chemistry. 2019;**5**: 1486–1511. DOI: 10.1016/j.chempr.2019.03.002

[41] Zhou Y, Lu Q, Zhuang ZB, Hutchings GS, Kattel S, Yan YS, Chen JG, Xiao JQ, Jiao F. Oxygen reduction at very low overpotential on nanoporous Ag catalysts. Advanced Energy Materials. 2015;**5**:1500149. DOI: 10.1002/aenm.201500149

[42] Jin S, Hao ZM, Zhang K, Yan ZH, Chen J. Advances and challenges for the electrochemical reduction of CO_2 to CO: From fundamentals to industrialization. Angewandte Chemie International Edition. 2021;**60**:20627–20648. DOI: 10.1002/anie.202101818

[43] Popović S, Smiljanić M, Jovanovič P, Vavra J, Buonsanti R, Hodnik N. Stability and degradation mechanisms of copper-based catalysts for electrochemical CO_2 reduction. Angewandte Chemie International Edition. 2020;**59**:14736–14746. DOI: 10.1002/anie.202000617

[44] Huang JNE, Li FW, Ozden A, Rasouli AS, Pelayo García de Arquer F, Liu SJ, Zhang SZ, Luo MC, Wang X, Lum YW, Xu Y, Bertens K, Miao RK, Dinh CT, Sinton D, Sargent EH. CO_2 electrolysis to multicarbon products in strong acid. Science. 2021;**372**:1074–1078. DOI: 10.1126/science.abg6582

[45] Huang JF, Hörmann N, Oveisi E, Loiudice A, Gregorio GLD, Andreussi O, Marzari N, Buonsanti R. Potential-induced nanoclustering of metallic catalysts during electrochemical CO_2 reduction. Nature Communications. 2018;**9**:3117. DOI: 10.1038/s41467-018-05544-3

[46] Zhao R, Zhu ZY, Ouyang T, Liu ZQ. Selective CO-to-syngas conversion enabled by bimetallic gold/zinc sites in partially reductived gold/zinc oxide arrays. Angewandte Chemie International Edition. 2024;**63**: e202313597. DOI: 10.1002/anie.202313597

[47] Chen Q, Liu K, Zhou YJ, Wang XQ, Wu KZ, Li HM, Pensa E, Fu JW, Miyauchi M, Cortes E, Liu M. Ordered Ag nanoneedle arrays with enhanced electrocatalytic CO reduction via structure-induced inhibition of hydrogen evolution. Nano Letters. 2022;**22**:6276–6284. DOI: 10.1021/acs.nanolett.2c01853

[48] Pan FP, Yang Y. Designing CO reduction electrode materials by morphology and interface engineering. Energy & Environmental Science. 2020;**13**:2275–2309. DOI: 10.1039/D0EE00900H

[49] Ross MB, De Luna P, Li YF, Dinh CT, Kim DY, Yang PD, Sargent EH. Designing materials for electrochemical carbon dioxide recycling. Nature Catalysis. 2019;**2**:648–658. DOI: 10.1038/s41929-019-0306-7

[50] Birdja YY, Pérez-Gallent E, Figueiredo MC, Göttle AJ, Calle-Vallejo F, Koper MTM. Advances and challenges in understanding the electrocatalytic conversion of carbon dioxide to fuels. Nature Energy. 2019;**4**:732–745. DOI: 10.1038/s41560-019-0450-y

[51] Wu X, Guo Y, Sun Z, Xie F, Guan D, Dai J, Yu F, Hu Z, Huang YC, Pao CW, Chen JL, Zhou W, Shao Z. Fast operando spectroscopy tracking in situ generation of rich defects in silver nanocrystals for highly selective electrochemical CO_2 reduction. Nature Communications. 2021;**12**:660. DOI: 10.1038/s41467-021-20960-8

[52] Xu X, Yang S, Wang Y, Chen Y, Devi AAS, Hu F. Grain boundary engineering in 3D porous silver electrocatalysts for enhanced CO_2-to-CO conversion. Molecules. 2025;**30**:3475. DOI: 10.3390/molecules30173475

[53] Kang WJ, Li Z, Feng Y, Shi ZZ, Hu XZ, Dong CK, Yang J, Liu H, Yin PF, Zhang R, Du XW. Coupled stacking faults in silver nanorods for CO_2 electroreduction. Nano Letters. 2025;**25**:63–70. DOI: 10.1021/acs.nanolett.4c04204

[54] Qin Y, Zhan G, Tang C, Yang D, Wang X, Yang J, Mao C, Hao Z, Wang S, Qin Y, Li H, Chen K, Liu M, Li J. Homogeneous vacancies-enhanced orbital hybridization for selective and efficient CO_2-to-CO electrocatalysis. Nano Letters. 2023;**23**:9227–9234. DOI: 10.1021/acs.nanolett.3c01905

[55] Ma Z, Wan T, Zhang D, Yuwono JA, Tsounis C, Jiang J, Chou YH, Lu X, Kumar PV, Ng YH, Chu D, Toe CY, Han Z, Amal R. Atomically dispersed Cu catalysts on sulfide-derived defective Ag nanowires for electrochemical CO_2 reduction. ACS Nano. 2023;**17**:2387–2398. DOI: 10.1021/acsnano.2c09473

Section 3

Biomedical and Therapeutic Applications

Chapter 6

Green-Synthesized Silver Nanoparticles in Topical Drug Delivery

Olutayo Ademola Adeleye, Bernard Opatimidi Patani and Adepero Olubukola Awolesi

Abstract

This chapter provides an in-depth overview of the development and application of silver nanoparticles (AgNPs) in topical drug delivery. Topical drug delivery systems utilizing green-synthesized AgNPs offer a sustainable and multifunctional approach to overcome the skin's stratum corneum barrier, enhancing localized therapeutic efficacy for dermatological conditions such as infected wounds, burns, inflammatory skin disorders, and chronic ulcers. By employing plant-derived phytochemicals as reducing and capping agents, these AgNPs exhibit improved biocompatibility, stability, and reduced toxicity compared to chemically synthesized counterparts, while retaining intrinsic broad-spectrum antimicrobial, anti-inflammatory, antioxidant, and wound-healing properties. The chapter explores skin anatomy, barrier function, and penetration pathways, highlighting how nanoscale AgNPs exploit intercellular, transcellular, and appendageal routes for better drug retention in epidermal and dermal layers with minimal systemic exposure. Various topical formulations – including ointments, creams, gels, hydrogels, transdermal patches, film-forming systems, and wound dressings – are discussed for optimal incorporation of green-synthesized AgNPs, ensuring controlled release and synergy with excipients. Mechanistic insights reveal multifaceted actions involving silver ion release, membrane disruption, reactive oxygen species generation, cytokine modulation, angiogenesis promotion, and tissue regeneration. Characterization techniques, biological evaluations, clinical applications, and future perspectives emphasize the potential of these systems in addressing antibiotic resistance and improving patient outcomes, while underscoring the need for further toxicological studies and regulatory standardization to facilitate clinical translation.

Keywords: green synthesis, silver nanoparticles, topical drug delivery, wound healing, antimicrobial activity

IntechOpen

1. Introduction

Topical drug delivery involves the direct application of therapeutic agents to the skin or mucosal surfaces. It is widely employed in the management of dermatological and localized disorders due to its ability to deliver therapeutic agents directly to the site of action while minimizing systemic exposure and associated adverse effects. This route of administration offers several advantages, including improved patient compliance, ease of application, avoidance of first-pass hepatic metabolism, and the potential for sustained drug action at the target site [1, 2]. Consequently, topical formulations such as creams, ointments, gels, and wound dressings remain central to the treatment of skin infections, inflammation, burns, and chronic wounds.

Despite these advantages, the effectiveness of topical drug delivery is often limited by the unique structure and protective function of the skin. The stratum corneum, the outermost layer of the epidermis, is a highly organized and lipophilic barrier designed to prevent excessive water loss and protect against environmental stressors. Its dense "brick-and-mortar" architecture, composed of corneocytes embedded in a lipid matrix, severely restricts the penetration of many therapeutic agents, particularly hydrophilic molecules and high-molecular-weight compounds [3, 4]. As a result, conventional topical formulations frequently exhibit poor skin penetration, low bioavailability, and limited therapeutic efficacy.

To overcome these limitations, advanced formulation strategies have been developed to enhance drug transport across the skin barrier. Among these strategies, nanotechnology-based drug delivery systems have gained considerable attention due to their ability to improve drug solubility, control release profiles, and enhance interaction with skin components [5]. Nanoscale carriers can exploit intercellular and appendageal pathways, thereby improving penetration through the stratum corneum and prolonging drug residence time within the skin [6].

Within the broad range of nanotechnology-enabled systems, silver nanoparticles (AgNPs) have attracted particular interest because of their unique physicochemical and biological properties. In addition to serving as nanoscale delivery enhancers, AgNPs possess intrinsic broad-spectrum antimicrobial, anti-inflammatory, and wound-healing activities, which are highly advantageous for topical and dermatological applications [7–10]. Their small particle size, ranging from 1 to 100 nm, large surface-to-volume ratio, and surface plasmon resonance contribute to enhanced biological interactions, making them especially effective in the management of infected wounds, burns, and inflammatory skin disorders. Historically, silver has been utilized for its healing properties since ancient times, but nanotechnology has revolutionized its efficacy by enhancing surface area, reactivity, and bioavailability.

The biomedical performance of AgNPs, however, is strongly influenced by their method of synthesis. Conventional chemical and physical synthesis techniques, although effective, often involve toxic reducing agents and harsh reaction conditions, which raise concerns regarding environmental safety and biocompatibility [11–13]. These challenges have driven growing interest in green synthesis approaches, which employ plant-derived phytochemicals as natural reducing and stabilizing agents. Phytoconstituents such as phenolics, flavonoids, terpenoids, and reducing sugars facilitate nanoparticle formation while simultaneously improving stability and biological compatibility [14].

Medicinal plants, including *Aloe vera*, *Azadirachta indica*, and *Curcuma longa*, have been successfully utilized for the biosynthesis of AgNPs with enhanced antimicrobial and wound-healing activities [15–18]. Similarly, *Ehretia cymosa*, a medicinal shrub widely distributed in Africa and traditionally used in the treatment of inflammation and wounds, has demonstrated significant potential as a biogenic source for AgNP synthesis. Studies have shown that *E. cymosa*–mediated AgNPs exhibit favorable physicochemical characteristics and superior antibacterial and anti-inflammatory performance when incorporated into topical formulations [7–9].

To fully appreciate how nanotechnology-based delivery systems, particularly green-synthesized AgNPs, enhance topical drug delivery, a clear understanding of skin structure and barrier function is essential. Accordingly, the following section examines the anatomy of the skin and its relevance to topical and transdermal drug delivery.

2. Skin structure and its relevance to topical drug delivery

The skin is the largest organ of the human body, accounting for approximately 15%–20% of total body weight, and serves as the primary interface between the body and the external environment. It functions as a dynamic protective barrier against physical, chemical, and biological threats, while also playing essential roles in thermoregulation, sensory perception, immune defense, and metabolic processes [19]. In the context of topical and transdermal drug delivery, the skin acts both as the "target site for therapy" and as a "selective barrier" that regulates the penetration of exogenous substances. Therefore, an understanding of skin structure is fundamental to the design of effective topical drug delivery systems (see **Figure 1**).

2.1 Epidermis

The epidermis is the outermost layer of the skin and is primarily composed of keratinocytes arranged in distinct strata, namely the stratum basale, stratum spinosum, stratum granulosum, stratum lucidum (in thick skin), and stratum corneum. Among these layers, the stratum corneum plays the most important role in regulating drug penetration. It consists of terminally differentiated, non-viable corneocytes embedded in a highly ordered lipid matrix composed mainly of ceramides, cholesterol, and free fatty acids [20, 21].

This unique "brick-and-mortar" organization provides remarkable resistance to the penetration of foreign molecules, making the stratum corneum the principal rate-limiting barrier in topical drug delivery [22, 23]. Only molecules with suitable physicochemical properties – such as low molecular weight, optimal lipophilicity, and sufficient thermodynamic activity – can effectively traverse this layer. As a result, many conventional topical formulations fail to achieve therapeutically effective drug concentrations within deeper skin layers.

2.2 Dermis

The dermis lies beneath the epidermis and is composed of dense connective tissue rich in collagen and elastin fibers, which confer mechanical strength and elasticity to the skin. It contains an extensive network of blood vessels, lymphatics, and nerve

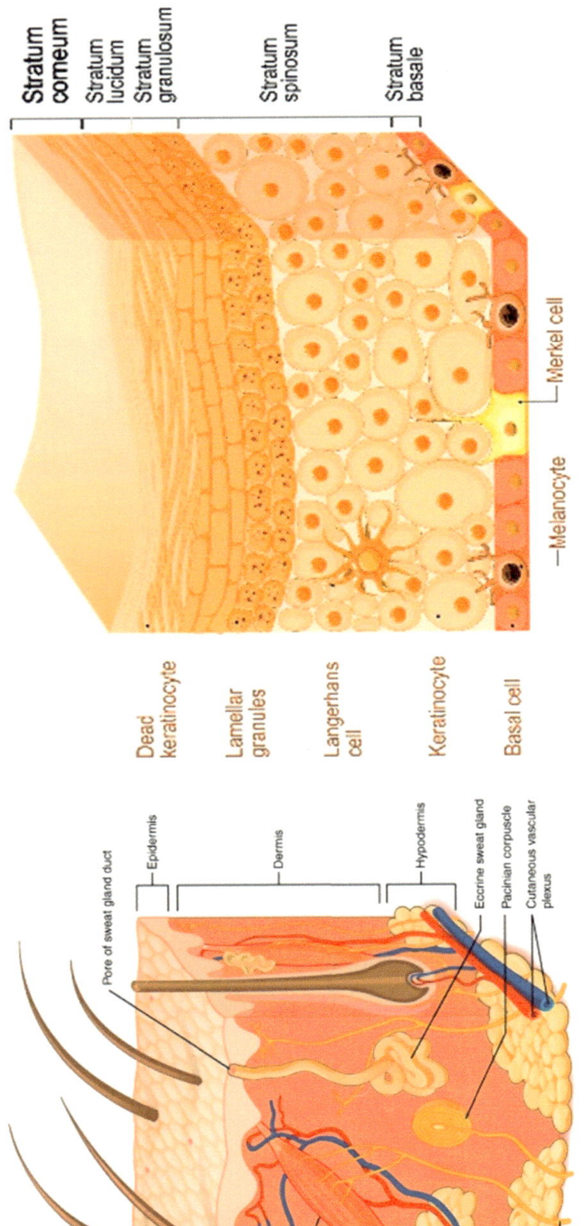

Figure 1.
Cross-sectional diagram of human skin anatomy showing the main layers (epidermis, dermis, and hypodermis), including the stratum corneum, hair follicles, sweat glands, and blood vessels.

endings, as well as immune cells such as macrophages and mast cells. From a drug delivery perspective, the dermis is a key therapeutic target for the treatment of inflammatory skin disorders, microbial infections, and wound healing.

Once a drug penetrates the stratum corneum and viable epidermis, it can readily diffuse within the dermal matrix and interact with local tissues or be absorbed into the systemic circulation. Therefore, successful delivery to the dermis is often associated with enhanced therapeutic efficacy for topical formulations. However, excessive dermal absorption may also increase the risk of systemic exposure, highlighting the need for controlled and targeted delivery strategies.

2.3 Hypodermis

The hypodermis, also referred to as the subcutaneous layer, is composed predominantly of adipose tissue interspersed with connective tissue and blood vessels. This layer provides thermal insulation, mechanical cushioning, and energy storage. Although the hypodermis does not constitute a significant barrier to drug permeation, it influences drug distribution, depot formation, and systemic absorption, particularly for lipophilic compounds.

In topical drug delivery, penetration into the hypodermis is generally undesirable unless systemic action is intended. Therefore, formulation strategies are often designed to retain the drug within the epidermal and dermal layers to achieve localized therapeutic effects while minimizing systemic exposure.

2.4 Relevance of skin structure to topical drug delivery

Each layer of the skin contributes uniquely to its protective and physiological functions; however, the stratum corneum remains the principal determinant of drug permeation. Its structural complexity and lipid-rich composition pose significant challenges to effective topical drug delivery. Consequently, modern formulation approaches aim to either transiently disrupt the stratum corneum barrier or exploit alternative penetration pathways, such as hair follicles and sweat glands.

Nanotechnology-based systems, including AgNPs, have demonstrated the ability to interact with skin lipids, enhance follicular penetration, and improve localized drug retention within the epidermal and dermal layers [24]. Understanding the structural organization of the skin is, therefore, essential for the rational development of advanced topical delivery systems capable of overcoming biological barriers while maintaining skin integrity and safety.

This structural perspective provides the foundation for the subsequent discussion on skin barrier function and drug penetration pathways, which further elucidates the challenges and opportunities in topical drug delivery.

3. Skin barrier function and drug penetration

The primary function of the skin barrier is to protect the body from excessive water loss and the ingress of potentially harmful substances. While this barrier function is essential for physiological homeostasis, it presents a major challenge for topical and transdermal drug delivery. Drug permeation across the skin is governed by the structural organization of the stratum corneum and the physicochemical properties of both the drug and the formulation [25, 26]. Understanding the pathways and factors influencing skin penetration is, therefore, essential for the rational design of effective topical delivery systems.

3.1 Pathways of drug penetration through the skin

Drug molecules can permeate the skin through three principal pathways: intercellular, transcellular, and appendageal routes.

3.1.1 Intercellular pathway

The intercellular route is the most dominant pathway for percutaneous absorption. In this route, drug molecules diffuse through the lipid matrix surrounding the corneocytes in the stratum corneum. Because this pathway is lipid-rich and highly tortuous, it favors the penetration of lipophilic drugs with appropriate partitioning characteristics. The organization and composition of stratum corneum lipids strongly influence permeability, making this pathway a key target for penetration enhancement strategies.

3.1.2 Transcellular pathway

In the transcellular route, drug molecules pass directly through the corneocytes and the surrounding lipid domains. This pathway requires repeated partitioning between hydrophilic and lipophilic environments, making it energetically unfavorable for most compounds. Consequently, transcellular transport contributes minimally to overall drug permeation under normal conditions.

3.1.3 Appendageal pathway

The appendageal route involves penetration through skin appendages such as hair follicles, sebaceous glands, and sweat ducts. Although these structures occupy a relatively small fraction of the total skin surface area, they provide low-resistance pathways that are particularly relevant for nanoparticles and particulate delivery systems. This route has gained increasing attention in nanotechnology-based topical drug delivery due to its role in follicular targeting and reservoir formation.

3.2 Factors influencing skin permeation

Several physiological and pathological factors influence drug penetration across the skin. Skin hydration significantly affects permeability, as increased hydration disrupts lipid organization within the stratum corneum, enhancing drug diffusion [27]. Lipid composition and thickness of the stratum corneum vary with age, anatomical site, and environmental conditions, further influencing barrier properties.

Disease states such as eczema, psoriasis, burns, and wounds can alter skin integrity, leading to increased permeability and unpredictable drug absorption [28, 29]. While this may enhance therapeutic efficacy in some cases, it also raises concerns regarding systemic exposure and toxicity. Therefore, formulation design must account for both healthy and diseased skin conditions.

3.3 Influence of formulation properties

The physicochemical properties of the drug – such as molecular weight, lipophilicity, ionization state, and solubility – play a central role in determining skin permeation [30]. In general, drugs with low molecular weight (<500 Da), moderate lipophilicity, and high thermodynamic activity exhibit better skin penetration.

Formulation factors, including dosage form, excipient selection, pH, viscosity, and the presence of penetration enhancers, significantly influence drug transport across the skin. Occlusive formulations, for example, increase skin hydration and thereby enhance permeability, whereas highly viscous systems may reduce drug diffusion rates. These considerations highlight the limitations of conventional topical formulations in achieving controlled and efficient skin delivery.

3.4 Implications for advanced drug delivery systems

The inherent resistance of the skin barrier underscores the need for advanced delivery strategies capable of enhancing drug penetration while preserving skin integrity and safety. Approaches such as chemical penetration enhancers, physical methods (e.g., microneedles and iontophoresis), and nanotechnology-based systems have been explored to overcome these challenges.

Nanocarriers, including metallic nanoparticles such as AgNPs, offer unique advantages by interacting with skin lipids, exploiting follicular pathways, and providing localized drug retention within the epidermal and dermal layers [24]. These properties make nanotechnology-based systems particularly attractive for topical drug delivery, setting the stage for the subsequent discussion on the role of AgNPs in enhancing skin penetration and therapeutic efficacy.

4. Relevance of nanotechnology and AgNPs in topical drug delivery

Nanotechnology has introduced innovative strategies for overcoming the limitations associated with conventional topical drug delivery systems. By reducing materials to the nanoscale, drug carriers can exhibit altered physicochemical properties, enhanced surface reactivity, and improved interactions with biological membranes. In topical applications, nanotechnology-based delivery systems have demonstrated the ability to enhance drug solubility, control release kinetics, increase skin retention, and improve penetration through the stratum corneum while minimizing systemic exposure [31, 32].

Nanocarriers such as liposomes, niosomes, ethosomes, nanoemulsions, polymeric nanoparticles, and metallic nanoparticles have been extensively investigated for skin-targeted drug delivery. These systems can interact with stratum corneum lipids, exploit appendageal pathways, and form drug reservoirs within hair follicles and skin layers, thereby enhancing localized therapeutic efficacy. Among these approaches, metallic nanoparticles – particularly AgNPs – have gained significant attention due to their multifunctional properties.

4.1 Unique properties of AgNPs

AgNPs possess distinct physicochemical characteristics that make them highly suitable for topical drug delivery. Their nanoscale size (typically 1–100 nm) provides a large surface-area-to-volume ratio, enabling enhanced interaction with skin components and microbial membranes. In addition, the surface plasmon resonance of AgNPs contributes to increased reactivity and biological activity.

4.2 Advantages of AgNPs over conventional topical systems

AgNPs present distinct and multifunctional advantages over conventional topical drug delivery systems such as creams, gels, and liposomes. Unlike traditional formulations that primarily function as passive carriers, AgNPs play a dual role as nanoscale delivery enhancers and intrinsic therapeutic agents, exhibiting broad-spectrum antimicrobial, anti-inflammatory, antioxidant, antiviral, anticancer, and wound-healing activities. Their nanoscale dimensions and high surface-area-to-volume ratio facilitate enhanced drug loading, controlled and sustained release, and superior skin penetration through intercellular, transcellular, and appendageal pathways – particularly through hair follicles and sweat glands – resulting in improved localized retention within the epidermal and dermal layers.

In addition to these inherent advantages, surface functionalization represents a key strength of AgNP-based delivery systems. Functionalization of AgNPs with polymers, ligands, or biomolecules enhances nanoparticle stability, prevents aggregation, and improves biocompatibility, while simultaneously modulating drug-loading efficiency and biological interactions with skin tissues. Such functional coatings enable controlled and targeted interactions with cutaneous components, thereby improving therapeutic efficacy while minimizing local and systemic toxicity.

Furthermore, AgNPs enable the gradual release of bioactive silver ions (Ag^{+}), providing prolonged therapeutic action compared to the burst release commonly observed with conventional topical formulations. The physicochemical properties of AgNPs, including size, shape, and surface chemistry, can be precisely tailored to optimize cellular uptake, tissue interaction, and biocompatibility. Remarkably, green-synthesized AgNPs, produced using plant-derived phytochemicals as reducing and capping agents, demonstrate enhanced stability and safety profiles, with surface-bound phytochemicals contributing additional antioxidant and anti-inflammatory effects. This synergistic functionality enables effective activity against multidrug-resistant pathogens and biofilms, a major limitation of many conventional antibiotics. Consequently, these attributes position AgNPs as a versatile and powerful platform for topical and localized drug delivery, capable of addressing key challenges such as skin barrier penetration, microbial resistance, and the need for sustained, site-specific therapeutic action [33, 34].

4.3 Antimicrobial activity

Beyond their role as passive carriers, AgNPs exhibit **intrinsic** broad-spectrum antimicrobial activity against Gram-positive and Gram-negative bacteria, fungi, and some viruses. This activity is primarily attributed to the release of silver ions, disruption of microbial cell membranes, interference with enzymatic systems, and induction of oxidative stress [35, 36]. These properties are particularly advantageous in topical formulations intended for infected wounds, burns, and chronic ulcers.

4.4 Anti-inflammatory and wound-healing effects

AgNPs have demonstrated significant anti-inflammatory and wound-healing properties. Studies have shown that AgNPs can modulate inflammatory cytokine production, reduce edema, and promote angiogenesis and re-epithelialization during the wound-healing process [37, 38]. These effects make AgNPs valuable for the treatment of inflammatory skin disorders and tissue injuries, where both infection control and inflammation suppression are required.

When incorporated into topical formulations, AgNPs can act synergistically with formulation excipients and bioactive compounds, enhancing therapeutic outcomes. Their ability to localize within the epidermal and dermal layers further supports sustained and targeted skin therapy with minimal systemic absorption.

4.5 Green-synthesized AgNPs for skin applications

The method of AgNP synthesis plays an essential role in determining their biological performance and safety. Conventional chemical synthesis techniques often involve toxic reducing agents and stabilizers that may compromise biocompatibility. In contrast, green synthesis approaches utilize plant-derived phytochemicals as natural reducing and capping agents, resulting in nanoparticles with improved stability, reduced toxicity, and enhanced biological activity.

Plant-mediated AgNPs synthesized using extracts from medicinal plants such as *Aloe vera*, *Azadirachta indica*, *Curcuma longa*, and *Ehretia cymosa* have demonstrated superior antimicrobial, anti-inflammatory, and wound-healing activities when incorporated into topical formulations [7–9, 15–18]. The presence of phytochemical capping agents on the nanoparticle surface may further enhance skin compatibility and provide additional antioxidant and therapeutic benefits.

4.6 Implications for topical drug delivery systems

The ability of AgNPs to enhance skin penetration, provide localized retention, and exert intrinsic therapeutic effects underscores their relevance in modern topical drug delivery. Incorporating AgNPs into suitable topical systems – such as creams, ointments, gels, hydrogels, and wound dressings – represents a rational and multifunctional approach to localized skin therapy. This strategy not only improves drug delivery efficiency but also addresses common challenges such as microbial infections, inflammation, and delayed wound healing.

The growing body of evidence supporting AgNP-based topical systems provides a strong foundation for their continued development and clinical translation. The subsequent sections of this chapter, therefore, focus on the formulation of AgNP-based topical delivery systems and the evaluation of their biological performance.

5. Formulation of AgNP-based topical drug delivery systems

The successful application of AgNPs in topical therapy depends not only on their physicochemical and biological properties but also on their incorporation into suitable dosage forms. An ideal topical formulation should ensure uniform nanoparticle distribution, physicochemical stability, drug-excipient compatibility, skin compatibility, controlled drug release, and adequate residence time at the site of application. Various topical delivery systems have, therefore, been explored to optimize the therapeutic performance of AgNPs for dermatological applications.

5.1 General considerations in formulating AGNP-based topical systems

Formulation of AgNP-containing topical systems requires careful consideration of several factors, including particle size, surface charge, concentration, and compatibility with formulation excipients. Nanoparticles must remain stable within the formulation without aggregation or excessive release of silver ions, which could cause skin irritation or toxicity. The pH of topical formulations is particularly important, as skin-compatible pH values (typically 4.5–6.0) help maintain barrier integrity and reduce irritation.

Excipients such as emulsifiers, gelling agents, humectants, and preservatives should not interfere with the stability or biological activity of AgNPs. In green-synthesized systems, phytochemical capping agents often enhance nanoparticle stability and biocompatibility, making them especially suitable for topical formulations.

5.2 Ointments

Ointments are semisolid, hydrophobic formulations that provide occlusive effects, thereby increasing skin hydration and enhancing drug penetration. AgNP-loaded ointments are particularly useful for wound healing and inflammatory skin conditions due to their prolonged residence time and sustained-release characteristics.

Studies have shown that AgNP-based ointments formulated using bases such as white soft paraffin, wool fat, and hard paraffin exhibit good homogeneity, acceptable spreadability, and enhanced antimicrobial and anti-inflammatory activity compared with conventional formulations [9, 39, 40]. The occlusive nature of ointments also supports moisture retention, which is beneficial for wound repair and epithelial regeneration.

5.3 Creams

Creams are emulsion-based systems, typically oil-in-water (O/W) or water-in-oil (W/O), that offer improved cosmetic acceptability and ease of application compared with ointments. AgNP-loaded creams have gained popularity for dermatological use due to their non-greasy nature and enhanced patient compliance.

Incorporation of AgNPs into O/W creams facilitates better diffusion of nanoparticles and active agents within the aqueous phase, leading to improved antimicrobial efficacy and skin penetration. Green-synthesized AgNP creams derived from medicinal plants have demonstrated favorable physicochemical properties, including appropriate pH, viscosity, spreadability, and stability, along with superior antibacterial and anti-inflammatory performance [9, 41, 42].

5.4 Gels and hydrogels

Gels and hydrogels are highly hydrated polymeric networks that provide cooling effects, ease of application, and controlled drug release. They are particularly suitable for inflamed, burned, or sensitive skin. AgNP-loaded gels, prepared using polymers such as carbopol, chitosan, polyvinyl alcohol (PVA), and alginate, have shown excellent antimicrobial activity and wound-healing potential [43, 44].

Hydrogels containing AgNPs offer additional advantages, including moisture retention, oxygen permeability, and sustained silver ion release, which are essential for effective wound management. Several studies have reported accelerated wound closure, reduced microbial load, and enhanced re-epithelialization with AgNP-based hydrogel systems [45, 46].

5.5 Lotions and sprays

Lotions are low-viscosity emulsions suitable for application over large skin surfaces, making them ideal for conditions such as dermatitis, acne, and superficial infections. AgNP-based lotions provide uniform coverage and rapid absorption while maintaining antimicrobial activity [47].

AgNP-containing sprays and aerosol formulations offer advantages like easy, contact-free application for rapid and uniform coverage on wounds, in situ film formation for moisture management, and sustained antimicrobial release, making them promising for infected or burn wounds. Spray-based systems are emerging but less common. They are particularly useful in burn care and emergency settings, where minimizing physical contact with damaged tissue is essential.

5.6 Transdermal patches and film-forming systems

Transdermal patches and film-forming systems incorporating AgNPs represent advanced topical platforms that combine controlled drug release with antimicrobial protection. Polymeric films and patches fabricated using materials such as chitosan, gelatin, and synthetic polymers have demonstrated sustained release of silver ions and enhanced wound healing outcomes [48, 49].

These systems form protective barriers over the skin, reducing microbial contamination while allowing gas exchange. Electrospun nanofiber mats containing AgNPs have also been investigated for chronic wound management and diabetic ulcers, showing promising results in both experimental and clinical studies [48–50].

5.7 Wound dressings

AgNP-based wound dressings are among the most successful clinical applications of silver nanotechnology. These include hydrofiber dressings, foams, films, and nanofiber scaffolds designed to provide antimicrobial protection, absorb exudates, and promote tissue regeneration [50]. Commercial silver-containing dressings have been widely used in burn units and chronic wound care due to their broad-spectrum antimicrobial activity and effectiveness against antibiotic-resistant pathogens [51].

5.8 Summary and relevance to topical therapy

Each topical dosage form offers distinct advantages in delivering AgNPs to the skin. The choice of formulation depends on the intended therapeutic application, site of action, and patient-related factors. Overall, the incorporation of AgNPs into suitable topical delivery systems represents a rational and versatile strategy for enhancing antimicrobial efficacy, controlling inflammation, and promoting wound healing.

Following formulation development, it is essential to evaluate the biological performance and safety of AgNP-based topical systems. Accordingly, the next section focuses on the biological activities of AgNP formulations, including their antibacterial, anti-inflammatory, and wound-healing effects.

6. Biological activities of AgNP-based topical formulations

The therapeutic relevance of AgNP-based topical formulations extends beyond their role as drug delivery systems. AgNPs possess intrinsic biological activities that contribute directly to their clinical effectiveness in dermatological applications. These activities include broad-spectrum antimicrobial action, modulation of inflammatory responses, antioxidant effects, and promotion of wound healing. When incorporated into suitable topical formulations, these properties act synergistically to enhance therapeutic outcomes.

6.1 Antimicrobial activity

The antimicrobial activity of AgNPs is one of their most extensively studied and clinically relevant properties. AgNP-based topical formulations have demonstrated effectiveness against a wide range of pathogenic microorganisms, including Gram-positive bacteria (*Staphylococcus aureus*), Gram-negative bacteria (*Escherichia coli*, *Pseudomonas aeruginosa*), fungi (*Candida albicans*), and multidrug-resistant strains [35, 36].

The antimicrobial mechanism of AgNPs is multifactorial. AgNPs adhere to microbial cell membranes, causing structural damage and increased membrane permeability. They also release silver ions (Ag^+), which interact with thiol groups in proteins and enzymes, leading to enzyme inactivation and disruption of cellular respiration [52, 53]. Additionally, AgNPs induce the generation of reactive oxygen species (ROS), resulting in oxidative stress, DNA damage, and eventual microbial cell death [35, 36].

Topical formulations containing AgNPs provide sustained antimicrobial activity at the site of application, reducing microbial load and minimizing the risk of infection. This is particularly advantageous in wound management, burn care, and chronic ulcer treatment, where microbial colonization significantly delays healing.

6.2 Anti-inflammatory activity

Inflammation is a common pathological feature of many skin disorders, including dermatitis, psoriasis, burns, and chronic wounds. AgNP-based topical formulations have demonstrated significant anti-inflammatory effects through the modulation of inflammatory mediators and immune responses [37, 38].

Studies have shown that AgNPs can downregulate pro-inflammatory cytokines, such as tumor necrosis factor-alpha (TNF-α), interleukin-6 (IL-6), and IL-1β, while promoting the expression of anti-inflammatory cytokines [37, 38, 54]. AgNPs also inhibit excessive neutrophil infiltration and reduce edema and erythema at inflamed sites.

Green-synthesized AgNPs exhibit enhanced anti-inflammatory activity due to the presence of phytochemical capping agents, such as flavonoids and phenolic compounds, which possess inherent anti-inflammatory and antioxidant properties [14]. When incorporated into topical gels, creams, or ointments, these nanoparticles provide localized inflammation control with minimal systemic side effects.

6.3 Antioxidant effects

Oxidative stress plays a critical role in skin aging, inflammation, and delayed wound healing. AgNP-based topical formulations have been shown to exhibit antioxidant activity by scavenging free radicals and reducing oxidative damage in skin tissues [10, 42, 55–57].

In plant-mediated AgNP systems, antioxidant effects are often enhanced due to surface-bound phytochemicals, which contribute additional radical-scavenging activity. This antioxidant potential supports tissue repair processes and protects skin cells from oxidative injury, particularly in inflamed or wounded skin.

6.4 Wound-healing activity

One of the most promising applications of AgNP-based topical formulations is in wound healing. Wound repair is a complex biological process involving hemostasis, inflammation, proliferation, and remodeling. AgNPs positively influence multiple stages of this process.

AgNP-containing formulations have been reported to accelerate wound closure, enhance fibroblast proliferation, stimulate collagen deposition, and promote angiogenesis [10, 17, 38–41, 43–49]. These effects result in faster re-epithelialization and improved tissue regeneration. In addition, the antimicrobial and anti-inflammatory properties of AgNPs help create a favorable wound environment by preventing infection and reducing excessive inflammation.

Animal studies and clinical evaluations have demonstrated improved healing outcomes with AgNP-loaded hydrogels, ointments, creams, and dressings compared with conventional formulations [7–9, 41–46]. These systems are particularly beneficial in the management of chronic wounds, diabetic ulcers, and burn injuries.

6.5 Safety and biocompatibility considerations

Despite their therapeutic benefits, the safety and biocompatibility of AgNP-based topical formulations remain important considerations. Factors such as particle size, concentration, surface coating, and duration of exposure influence the biological response of skin tissues to AgNPs.

Topical application generally limits systemic absorption, reducing the risk of systemic toxicity. Green-synthesized AgNPs are often associated with improved biocompatibility due to the absence of toxic chemical residues and the presence of natural capping agents. However, excessive silver accumulation may lead to local irritation or cytotoxic effects, emphasizing the importance of dose optimization and comprehensive safety evaluation [58].

6.6 Summary and clinical implications

The intrinsic biological activities of AgNPs significantly enhance the therapeutic value of AgNP-based topical formulations. Their combined antimicrobial, anti-inflammatory, antioxidant, and wound-healing properties make them highly suitable for the treatment of a wide range of dermatological conditions. When formulated appropriately, AgNP-based systems offer effective, localized therapy with improved safety and patient outcomes.

The following section focuses on the evaluation methods and characterization techniques used to assess the physicochemical properties, biological performance, and safety of AgNP-based topical formulations.

7. Characterization and evaluation of AgNP-based topical formulations

Comprehensive characterization and evaluation are essential to ensure the quality, safety, stability, and therapeutic efficacy of AgNP-based topical formulations. These assessments encompass physicochemical characterization of nanoparticles, evaluation of formulation properties, *in vitro* and *ex vivo* performance testing, and preliminary safety studies. Such systematic evaluation provides the scientific basis for formulation optimization and supports regulatory and clinical translation.

7.1 Physicochemical characterization of AgNPs

The physicochemical properties of AgNPs strongly influence their biological activity and performance in topical formulations.

7.1.1 Particle size and size distribution

Particle size is typically determined using dynamic light scattering (DLS), transmission electron microscopy (TEM), or scanning electron microscopy (SEM). Nanoparticles intended for topical delivery usually range from 10 to 100 nm, as this size range favors enhanced skin interaction and follicular penetration while minimizing toxicity.

7.1.2 Surface charge (zeta potential)

Zeta potential provides insight into nanoparticle stability and surface interactions. AgNPs with higher absolute zeta potential values (±30 mV) exhibit better colloidal stability and reduced aggregation within formulations.

7.1.3 Morphology and crystallinity

TEM and SEM are used to examine nanoparticle shape and morphology, while X-ray diffraction (XRD) confirms crystalline structure and phase purity. The face-centered cubic structure of silver is commonly observed in well-synthesized AgNPs.

7.1.4 Surface chemistry

Fourier transform infrared spectroscopy (FTIR) is used to identify functional groups associated with capping agents, particularly in green-synthesized AgNPs. These surface-bound phytochemicals contribute to nanoparticle stability and biological activity.

7.2 Evaluation of topical formulation properties

After nanoparticle incorporation, the final topical formulation must be evaluated to ensure consistency, stability, and patient acceptability.

7.2.1 pH determination

The pH of topical formulations is measured to ensure compatibility with the skin's natural pH (4.5–6.0). Maintaining an appropriate pH helps preserve skin barrier integrity and minimizes irritation.

7.2.2 Viscosity and rheological behavior

Viscosity and flow properties are evaluated using viscometers or rheometers to assess spreadability, extrudability, and application behavior. Pseudoplastic or shear-thinning behavior is generally desirable for topical formulations.

7.2.3 Spreadability and homogeneity

Spreadability tests evaluate the ease of application, while visual and microscopic examinations confirm uniform nanoparticle distribution throughout the formulation.

7.2.4 Stability studies

Physical and chemical stability studies are conducted under various storage conditions to assess changes in appearance, pH, viscosity, and nanoparticle integrity over time. Stability is significant for ensuring consistent therapeutic performance.

7.3 *In vitro* drug release and permeation studies

In vitro release studies provide insight into the release kinetics of AgNPs and associated bioactive compounds from topical formulations. These studies are commonly performed using Franz diffusion cells with synthetic membranes or dialysis membranes.

Ex vivo skin permeation studies using excised human or animal skin (e.g., rat or porcine skin) further evaluate nanoparticle penetration and retention within different skin layers. These studies help predict *in vivo* performance and assess the potential for systemic absorption.

7.4 Evaluation of biological performance and safety

The biological performance and safety of AgNP-based topical formulations must be systematically evaluated to establish their therapeutic relevance and biocompatibility. These evaluations are essential for confirming that the incorporated AgNPs retain their intended antimicrobial and anti-inflammatory functions while promoting wound repair without inducing adverse skin reactions. The antimicrobial effectiveness, wound-healing potential, and skin compatibility are essential parameters for topical systems, as they directly influence clinical outcomes, patient tolerance, and regulatory acceptance.

7.4.1 Antimicrobial activity

The antimicrobial efficacy of AgNP-based topical formulations is evaluated using standard microbiological assays, such as agar well diffusion, minimum inhibitory concentration (MIC), and time-kill studies against clinically relevant pathogens.

7.4.2 Anti-inflammatory and wound-healing assays

In vitro cell culture models and *in vivo* animal models are used to assess anti-inflammatory effects and wound-healing potential. Parameters such as wound contraction rate, epithelialization time, collagen deposition, and histopathological changes are commonly evaluated.

7.5 Skin irritation and safety studies

Skin irritation and safety assessments are essential for topical products. Primary skin irritation tests are conducted using animal models or reconstructed human epidermis models to evaluate erythema, edema, and histological changes following topical application.

Cytotoxicity studies using skin-related cell lines (e.g., keratinocytes and fibroblasts) provide additional information on formulation safety. Green-synthesized AgNPs generally demonstrate lower cytotoxicity due to their biocompatible surface chemistry.

7.6 Challenges in correlating *in vitro* activities with *in vivo* and clinical outcomes

Although AgNPs demonstrate promising skin permeation and biological activity *in vitro*, establishing a reliable correlation with *in vivo* and clinical outcomes remains challenging due to fundamental biological and methodological differences between experimental models and living systems. *In vitro* diffusion platforms lack active physiological processes such as blood flow, immune surveillance, lymphatic clearance, and dynamic metabolism, all of which play critical roles in regulating nanoparticle penetration, retention, transformation, and clearance *in vivo*. While *ex vivo* studies frequently report that AgNPs are largely retained within the stratum corneum with limited dermal penetration under controlled conditions, *in vivo* skin environments characterized by inflammation, vascular perfusion, tissue remodeling, and biochemical heterogeneity can substantially alter nanoparticle distribution and persistence in ways not captured by static models [59]. In addition, the physicochemical variability of AgNPs – including size, surface chemistry, coating, and dissolution behavior – and their dynamic interactions with biological components such as proteins and enzymes lead to changes in bioavailability, cellular interactions, and biodistribution *in vivo* that are not adequately reflected in simplified *in vitro* systems. Discrepancies between nominal exposure conditions used *in vitro*, which often involve constant and elevated concentrations, and the transient, regulated exposure profiles encountered *in vivo* further complicate extrapolation. Collectively, these factors – biological complexity, nanoparticle transformation under physiological conditions, and inherent limitations of current experimental models – limit the predictive power of *in vitro* permeation data alone and underscore the need for integrated translational approaches and standardized *in vitro–in vivo* correlation frameworks when assessing the dermal behavior and clinical relevance of nanoparticle-based systems.

7.7 Summary

Systematic characterization and evaluation of AgNP-based topical formulations ensure their quality, safety, and therapeutic effectiveness. These studies form a significant bridge between formulation development and clinical application, supporting regulatory approval and translational research. The next section addresses the mechanism of action of AgNPs.

8. Mechanistic basis of action of AgNPs in topical therapy

The therapeutic effectiveness of AgNP-based topical formulations is underpinned by well-defined molecular and cellular mechanisms. The mechanistic process is not attributable to a singular mechanism but arises from a complex, multifaceted interplay of physicochemical and biological interactions that explain their broad-spectrum antimicrobial activity, anti-inflammatory effects, and ability to promote tissue repair. Their nanoscale dimensions (typically 1–100 nm) confer a high surface area-to-volume ratio, which is fundamental to their enhanced reactivity and sustained release of bioactive silver ions (Ag^+). A clear understanding of these pathways is essential for optimizing formulation design, ensuring safety, and expanding the clinical applications of AgNPs in dermatological therapy. This section provides the primary and synergistic mechanisms through which AgNPs exert their antimicrobial, anti-inflammatory, and pro-healing effects, forming the scientific foundation for their clinical use in wound dressings.

8.1 Antimicrobial mechanism

The broad-spectrum antimicrobial activity of AgNPs against bacteria, fungi, and some viruses is the cornerstone of their topical application, particularly for infected wounds and burns. This activity operates through a sequential and often concurrent series of events (see **Figure 2**).

8.1.1 Sustained release of silver ions (Ag^+)

The principal antimicrobial action of AgNPs involves the slow, sustained oxidation and release of Ag^+ ions from the nanoparticle at the application site. The release of the ions is influenced by nanoparticle size, shape, surface coating, and local pH. Smaller particles release ions more rapidly due to their larger relative surface

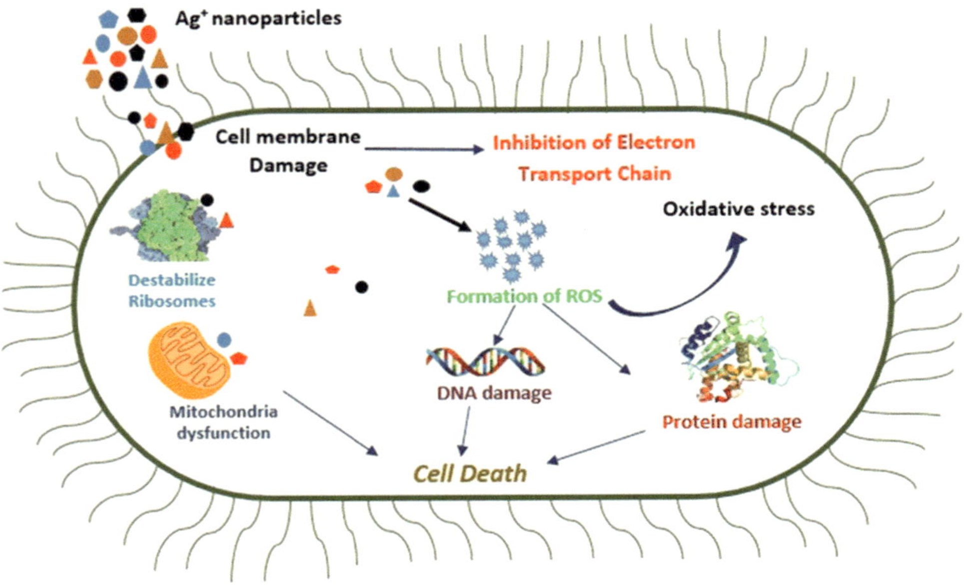

Figure 2.
Antimicrobial mechanism of action of AgNP [69].

area. The released Ag^+ ions exhibit a high affinity for thiol (–SH) groups in proteins and phosphate in DNA [60, 61]. This interaction leads to:

- Enzyme and protein denaturation: Ag^+ binds to enzymes in the bacterial cell membrane and cytoplasm, disrupting the electron transport chain and inhibiting ATP production, which leads to metabolic arrest.

- DNA damage: Ag^+ ions can intercalate with bacterial DNA, disrupting replication and transcription processes.

- Membrane protein disruption: Binding to membrane proteins alters their structure and function, compromising membrane integrity.

8.1.2 Direct nanoparticle interaction and membrane disruption

AgNPs can physically interact with microbial cell walls and membranes in distinctly different ways, which are [62]:

- Adhesion: Positively charged AgNPs are electrostatically attracted to the negatively charged components of bacterial cell walls, which physically and chemically compromise the cell.

- Intracellular penetration and damage: Upon adhesion, AgNPs penetrate microbial cells through membrane pores or endocytosis-like processes. Inside, they denature ribosomes, halting protein synthesis; bind to DNA, preventing replication and transcription; and inactivate enzymes involved in metabolism. This metabolic disruption leads to cell death, effectively targeting biofilms where traditional antibiotics fail [63].

- Membrane pit formation and permeation: Adhered nanoparticles can anchor to the membrane, causing physical pits (localized structural collapse), local dissipation of the proton motive force, and increased permeability. This leads to uncontrolled leakage of ions (K^+, Mg^{2+}), metabolites, and ultimately, cell lysis [64].

8.1.3 ROS generation

AgNPs catalyze the generation of ROS within microbial cells through several pathways [65].

- Catalytic activity: The high surface-area-to-volume ratio of AgNPs makes them highly reactive. Oxygen molecules in the surrounding medium can react with the surface of the silver, leading to the formation of free radicals.

- Oxidative stress cascade: This ROS surge overwhelms the microbial antioxidant defense system, leading to oxidative damage of lipids (membrane peroxidation), proteins, and DNA, resulting in irreversible cellular damage and death.

8.1.4 Modulation of cellular signaling

AgNPs disrupt signal transduction by dephosphorylating tyrosine residues, interfering with cell cycle progression and metabolism. They oxidize thiol groups in proteins, forming disulfide bonds that inhibit growth [66].

8.1.5 Synergistic and secondary effects

Beyond the primary physical and chemical attacks exhibited, AgNPs exhibit a series of synergistic and secondary effects that make them particularly effective against multidrug-resistant "superbugs."

- Biofilms are slimy, protective extracellular polymeric substances that bacteria build to hide from the immune system and drugs. AgNPs have the unique secondary ability to penetrate these layers, killing the "dormant" bacteria inside and preventing the biofilm from reforming [67].

- Potentiation of antibiotics: Some antibiotics can form chemical complexes with AgNPs. The nanoparticle acts as a delivery vehicle by damaging the bacterial membrane, thereby increasing the concentration of the drug directly into the bacterial cell [68].

8.2 Anti-inflammatory mechanisms

Beyond their antimicrobial effect, AgNPs contribute to the modulation of the inflammatory response, a key factor in resolving acute and chronic wounds. They exert significant anti-inflammatory effects that are highly relevant in topical therapy.

Inflammatory skin conditions are often characterized by excessive production of pro-inflammatory cytokines and prolonged immune activation. AgNPs have been shown to modulate inflammatory signaling pathways as explained below.

8.2.1 Cytokine modulation

AgNPs downregulate the expression of pro-inflammatory cytokines such as TNF-α, IL-1β, and IL-6 from macrophages and other immune cells. Concurrently, they may promote the release of anti-inflammatory cytokines like IL-10 [70].

8.2.2 Neutrophil and macrophage regulation

AgNPs can reduce excessive neutrophil infiltration and promote the transition of macrophages from the pro-inflammatory phenotype to the pro-healing phenotype, which is essential for tissue repair and remodeling [71].

8.2.3 NF-κB pathway inhibition

A key molecular mechanism involves the suppression of the nuclear factor kappa-light-chain-enhancer of activated B cells (NF-κB) signaling pathway, a master regulator of inflammation [37].

8.2.4 Phytochemical capping

Green-synthesized AgNPs often exhibit enhanced anti-inflammatory activity due to phytochemical capping agents, such as flavonoids and phenolic compounds. These natural compounds provide additional antioxidant and cytokine-modulating effects, further improving the therapeutic profile of AgNP-based topical formulations [72].

8.3 Role in angiogenesis and wound healing

AgNPs positively influence the wound healing cascade through their combined antimicrobial and anti-inflammatory actions, as well as through several supportive mechanisms with effects on tissue regeneration. AgNPs stimulate "angiogenesis" and foster an environment conducive to fibroblast proliferation, keratinocyte migration, and extracellular matrix deposition, leading to accelerated wound closure.

8.3.1 Angiogenesis promotion

Studies indicate that low, therapeutic concentrations of AgNPs can stimulate the proliferation and migration of endothelial cells and upregulate angiogenic factors like vascular endothelial growth factor (VEGF), enhancing the formation of new blood vessels (neovascularization), which is crucial for delivering oxygen and nutrients [73, 74].

8.3.2 Fibroblast proliferation and collagen synthesis

AgNPs have been shown to promote the activity of fibroblasts, the cells responsible for producing extracellular matrix components like collagen and fibronectin [46, 74–77]. This leads to improved granulation tissue formation and tensile strength of the healed wound.

8.3.3 Keratinocyte migration (re-epithelialization)

AgNPs can accelerate the migration and proliferation of keratinocytes across the wound bed, facilitating faster wound closure and restoration of the epidermal barrier [75,78].

By reducing microbial burden and controlling inflammation, AgNPs create a favorable microenvironment for tissue regeneration and remodeling [8]. Importantly, the controlled release of silver ions from nanoparticle systems ensures sustained biological activity without excessive cytotoxicity, making AgNP-based formulations suitable for long-term topical use in chronic wounds and burns.

Clinical and experimental models have demonstrated accelerated wound closure, improved granulation tissue formation, and enhanced re-epithelialization following topical AgNP application. These processes are further augmented when AgNPs are incorporated into advanced formulations (hydrogels, nanogels, or dressing composites) that sustain local exposure and augment tissue regeneration.

8.4 Summary and clinical implications

The mechanistic basis of AgNPs in topical therapy is a dynamic convergence of chemistry, physics, and cell biology that acts "simultaneously and synergistically." Their primary antimicrobial action – driven by sustained Ag^+ release, direct membrane damage, and ROS generation – is effectively complemented by sophisticated immunomodulatory and pro-regenerative effects on host tissue. This multi-target approach, attacking pathogens while actively resolving inflammation and stimulating repair, makes AgNPs uniquely effective in managing complex wounds. Understanding the mechanistic basis of AgNPs' action provides a scientific foundation for expanding their clinical applications in dermatology. These mechanisms justify the use of AgNP-based systems in infected wounds, inflammatory skin disorders, burns, and chronic ulcers. Moreover, mechanistic insights guide rational formulation design, dose optimization, and safety evaluation, facilitating regulatory approval and clinical translation.

9. Applications and future perspectives of AgNP-based topical systems

AgNP-based topical delivery systems have attracted considerable attention due to their multifunctional therapeutic properties and versatility in formulation design. Their combined antimicrobial, anti-inflammatory, antioxidant, and wound-healing activities make them suitable for a wide range of dermatological applications. In particular, green-synthesized AgNPs have emerged as promising candidates for safe and sustainable skin-targeted therapy.

9.1 Emerging and advanced applications

Beyond conventional dermatological uses, AgNP-based topical systems are being investigated for advanced applications such as "transdermal drug delivery," "cosmeceuticals," and "antimicrobial coatings" for medical devices. In cosmeceuticals, AgNPs are incorporated into formulations aimed at controlling skin microbiota, reducing inflammation, and preventing oxidative skin damage [79].

Additionally, AgNP-loaded nanofiber mats and film-forming systems are being explored for smart wound dressings capable of controlled drug release and responsive behavior [79–81]. These innovations highlight the expanding role of nanotechnology in personalized and advanced skin therapy.

9.2 Future perspectives and research directions

Despite their promising potential, several challenges must be addressed to enable the widespread clinical translation of AgNP-based topical systems. "Long-term safety and toxicity" remain critical concerns, particularly with prolonged or repeated exposure. Comprehensive toxicological studies are needed to evaluate nanoparticle accumulation, systemic absorption, and potential environmental impact [82–86].

"Stability and scalability" of green-synthesized AgNPs also require further investigation to ensure batch-to-batch consistency and reproducibility during large-scale production. Moreover, regulatory frameworks governing nanoparticle-based topical products are still evolving, necessitating clear guidelines for quality control, safety evaluation, and clinical testing [84, 87, 88].

Future research should focus on optimizing formulation strategies, standardizing green synthesis protocols, and conducting well-designed clinical trials to establish efficacy and safety. These efforts will be essential for translating laboratory-scale innovations into clinically approved and commercially viable topical products.

9.3 Outlook

The expanding range of applications and ongoing technological advancements underscore the growing role of AgNPs in topical drug delivery. Green-synthesized AgNPs, in particular, represent a sustainable and biocompatible approach that aligns with current trends in green chemistry and nanomedicine. The integration of AgNP-based dermatological therapy into clinical practice could address antibiotic resistance and improve outcomes for chronic wounds. Continued interdisciplinary efforts are essential to harness their full potential while ensuring safety. With continued research and regulatory clarity, AgNP-based topical systems are poised to play an increasingly important role in the future of dermatological therapy.

10. Conclusion

Green-synthesized AgNPs represent a sustainable, multifunctional, and highly promising platform for topical drug delivery. By integrating the principles of green chemistry with nanotechnology, plant-mediated AgNPs offer improved biocompatibility, reduced environmental impact, and enhanced therapeutic performance compared with conventionally synthesized counterparts. Their nanoscale size, large surface area, and phytochemical capping agents collectively contribute to improved skin interaction, localized retention, and controlled biological activity.

Throughout this chapter, the structure and barrier function of the skin were highlighted as major challenges in topical drug delivery. AgNP-based systems effectively address these challenges by enhancing penetration through appendageal pathways, improving drug retention within the epidermal and dermal layers, and simultaneously providing intrinsic antimicrobial, anti-inflammatory, antioxidant, and wound-healing effects. These combined properties make AgNPs particularly suitable for the management of infected wounds, burns, acne, inflammatory skin disorders, and chronic ulcers.

The formulation of AgNPs into diverse topical dosage forms – including ointments, creams, gels, hydrogels, and wound dressings – demonstrates their versatility and adaptability to different clinical needs. Systematic physicochemical characterization and biological evaluation are essential to ensure formulation stability, safety, and therapeutic efficacy. Mechanistic insights into silver ion release, membrane disruption, ROS generation, cytokine modulation, and angiogenesis further strengthen the scientific basis for their clinical use.

Despite the significant progress achieved, challenges related to long-term safety, large-scale production, stability, and regulatory standardization remain. Addressing these issues through rigorous toxicological studies, standardized green synthesis protocols, and well-designed clinical trials will be critical for successful clinical translation. Overall, green-synthesized AgNPs represent a forward-looking and sustainable strategy for next-generation dermatological therapies, with strong potential to bridge the gap between nanotechnology research and real-world clinical applications.

Acknowledgments

The authors acknowledge the use of ChatGPT (OpenAI) as an assistive tool for language polishing in the chapter.

Author details

Olutayo Ademola Adeleye[1]*, Bernard Opatimidi Patani[2] and Adepero Olubukola Awolesi[3]

1 Department of Pharmaceutics and Pharmaceutical Technology, Federal University Oye-Ekiti, Oye-Ekiti, Nigeria

2 Medical Department, Oando Energy Resources Ltd, Port Harcourt, Rivers, Nigeria

3 Department of Pharmacognosy and Herbal Medicine, Federal University Oye-Ekiti, Oye-Ekiti, Nigeria

*Address all correspondence to: Olutayo.adeleye@fuoye.edu.ng

References

[1] Gupta N, Gupta GD, Singh D. Localized topical drug delivery systems for skin cancer: Current approaches and future prospects. Frontiers in Nanotechnology. 2022;**4**:1006628. DOI: 10.3389/fnano.2022.1006628

[2] Zhao L, Chen J, Bai B, Song G, Zhang J, Yu H, et al. Topical drug delivery strategies for enhancing drug effectiveness by skin barriers, drug delivery systems and individualized dosing. Frontiers in Pharmacology. 2024;**14**:1333986

[3] Zamani ZA, Jepps OG, Gould T, Anissimov YG. Permeable cornified envelope layer regulates solute transport in human stratum corneum. Journal of Pharmaceutical Sciences. 2023;**112**(7):1939–1946. DOI: 10.1016/j.xphs.2023.03.002

[4] Sjövall P, Gregoire S, Wargniez W, Skedung L, Detroyer A, Gustavo SL. Spatial distribution of active compounds in stratum corneum—partitioning between corneocytes and lipid matrix. Scientific Reports. 2024;**14**:18681. DOI: 10.1038/s41598-024-66418-x

[5] Alnaim AS. Nanocrystals in dermal drug delivery: A breakthrough for enhanced skin penetration and targeted skin disorder treatments. Pharmaceutics. 2024;**16**(12):1561. DOI: 10.3390/pharmaceutics16121561

[6] Liu L, Zhao W, Ma Q, Gao Y, Wang W, Zhang X, et al. Functional nano-systems for transdermal drug delivery and skin therapy. Nanoscale Advances. 2023;**5**(6):1527–1558. DOI: 10.1039/d2na00530a

[7] Adeleye OA, Badru AO, Oyinloye OE, et al. Green synthesized silver nanoparticles for cream formulation: Its anti-inflammatory and healing activities. IOP Conference Series: Materials Science and Engineering. 2020;**805**:012016

[8] Adeleye OA, Bamiro OA, Bakre LG, Badru AO, Balogun-Agbaje OA, Olusola A, et al. In vivo anti-inflammatory assessment of a topical formulation containing Ehretia cymosa extract mediated silver nanoparticles. Nigerian Journal of Pharmaceutical Research. 2022;**17**(2):179–188. DOI: 10.4314/njpr.v17i2.4

[9] Adeleye OA, Aremu OK, Iqbal H, Adedokun MO, Bamiro OA, Okunye OL, et al. Green synthesis of silver nanoparticles using Ehretia cymosa extracts and evaluation of their antibacterial activity in cream and ointment drug delivery systems. Journal of Nanotechnology. 2023;**2023**:2808015

[10] Lakkim V, Reddy MC, Lekkala VVV, Lebaka VR, Korivi M, Lomada D. Antioxidant efficacy of green-synthesized silver nanoparticles promotes wound healing in mice. Pharmaceutics. 2023;**15**(5):1517. DOI: 10.3390/pharmaceutics15051517

[11] Sondi I, Salopek-Sondi B. Silver nanoparticles as antimicrobial agent: A case study on E. coli as a model for Gram-negative bacteria. Journal of Colloid and Interface Science. 2004;**275**(1):177–182. DOI: 10.1016/j.jcis.2004.02.012

[12] Amooaghaie R, Saeri MR, Azizi M. Synthesis, characterization and

biocompatibility of silver nanoparticles synthesized from Nigella sativa leaf extract in comparison with chemical silver nanoparticles. Ecotoxicology and Environmental Safety. 2015;**120**:400–408. DOI: 10.1016/j.ecoenv.2015.06.025

[13] Simon S, Sibuyi NRS, Fadaka AO, Meyer S, Josephs J, Onani MO, et al. Biomedical applications of plant extract-synthesized silver nanoparticles. Biomedicines. 2022;**10**(11):2792. DOI: 10.3390/biomedicines10112792

[14] Lakhani KG, Hamid R, Motamedi E, Marviya GV. A review on plant metabolite-mediated nanoparticle synthesis: Sustainable applications in horticultural crops. Frontiers in Nanotechnology. 2025;7:1545413. DOI: 10.3389/fnano.2025.1545413

[15] Begum Q, Kalam M, Kamal M, Mahboob T. Biosynthesis, characterization, and antibacterial activity of silver nanoparticles derived from Aloe barbadensis Miller leaf extract. Iranian Journal of Biotechnology. 2020;**18**(2):e2383. DOI: 10.30498/IJB.2020.145075.2383

[16] Liknaw T, Belay Y, Ramesh R, et al. Aloe vera leaf extract as a sustainable route for silver nanoparticle synthesis with enhanced antimicrobial activity. Scientific Reports. 2025;**15**:22481

[17] Chinnasamy G, Chandrasekharan S, Koh TW, Bhatnagar S. Synthesis, characterization, antibacterial and wound healing efficacy of silver nanoparticles from Azadirachta indica. Frontiers in Microbiology. 2021;**12**:611560. DOI: 10.3389/fmicb.2021.611560

[18] Singh SR, Akshatha SJ, Abbigeri MB, Padti AC, Bhavi SM, Kulkarni SR, et al. Eco-synthesized silver nanoparticles from Curcuma longa leaves: Phytochemical and biomedical applications. Next Nanotechnology. 2025;**8**:100249

[19] Proksch E, Brandner JM, Jensen JM. The skin: An indispensable barrier. Experimental Dermatology. 2008;**17**(12):1063–1072. DOI: 10.1111/j.1600-0625.2008.00786.x

[20] Bouwstra JA, Ponec M. Stratum corneum lipid composition and organization. In: Elias PM, Feingold KR, editors. Skin Barrier. New York: Taylor & Francis; 2006. p. 37–56

[21] Lefèvre-Utile A, Braun C, Haftek M, Aubin F. Five functional aspects of the epidermal barrier. International Journal of Molecular Sciences. 2021;**22**(21):11676. DOI: 10.3390/ijms222111676

[22] Vitorino C, Sousa J, Pais A. Overcoming the skin permeation barrier: Challenges and opportunities. Current Pharmaceutical Design. 2015;**21**(20):2698–2712. DOI: 10.2174/1381612821666150428124053

[23] Walicka A, Iwanowska-Chomiak B. Drug diffusion transport through human skin. International Journal of Applied Mechanics and Engineering. 2018;**23**(4):977–988. DOI: 10.2478/ijame-2018-0055

[24] Erdoğar N, Gür B, Örgül D. Recent developments of novel nanotechnology-based drug delivery systems for dermal and transdermal applications. European Journal of Pharmaceutical Sciences. 2025;**186**:107413. DOI: 10.1016/j.ejps.2025.107413

[25] Neubert RHH. Mechanisms of penetration and diffusion of drugs and cosmetic actives across the human stratum corneum. European Journal of Pharmaceutics and Biopharmaceutics. 2024;**202**:114394. DOI: 10.1016/j.ejpb.2024.114394

[26] Bouwer F, Brits M, Viljoen JM. Cracking the skin barrier: Models and methods driving dermal drug delivery. Pharmaceutics. 2025;**17**(12):1586. DOI: 10.3390/pharmaceutics17121586

[27] Kim B, Cho HE, Moon SH, Ahn HJ, Bae S, Cho HD, et al. Transdermal delivery systems in cosmetics. Biomedical Dermatology. 2020;**4**:10

[28] Chiang A, Tudela E, Maibach HI. Percutaneous absorption in diseased skin: An overview. Journal of Applied Toxicology. 2012;**32**(8):537–563. DOI: 10.1002/jat.1773

[29] Kocsis D, Horváth S, Kemény Á, Varga-Medveczky Z, Pongor C, Molnar R, et al. Drug delivery through the psoriatic epidermal barrier: A skin-on-a-chip permeability study and ex vivo optical imaging. International Journal of Molecular Sciences. 2022;**23**(8):4237. DOI: 10.3390/ijms23084237

[30] Souto EB, Fangueiro JF, Fernandes AR, Cano A, Sanchez-Lopez E, Garcia ML, et al. Physicochemical and biopharmaceutical aspects influencing skin permeation and role of SLN and NLC for skin drug delivery. Heliyon. 2022;**8**(2):e08938. DOI: 10.1016/j.heliyon.2022.e08938

[31] Akombaetwa N, Ilangala AB, Thom L, Memvanga PB, Witika BA, Buya AB. Current advances in lipid nanosystems intended for topical and transdermal drug delivery applications. Pharmaceutics. 2023;**15**(2):656. DOI: 10.3390/pharmaceutics15020656

[32] Ko J, Kim JU, Choi S, Kim YS, Park SB, Kim JY, et al. Recent advances in nanotechnology-mediated noninvasive transdermal and topical delivery of proteins. Small Science. 2024;**4**(10):2400175. DOI: 10.1002/smsc.202400175

[33] Ammar MM, Ali R, Abd Elaziz NA, Habib H, Abbas FM, Yassin MT, et al. Nanotechnology in oncology: Advances in biosynthesis, drug delivery, and theranostics. Discover Oncology. 2025;**16**:1172

[34] Hussein HA, Abdullah MA. Novel drug delivery systems based on silver nanoparticles, hyaluronic acid, lipid nanoparticles and liposomes for cancer treatment. Applied Nanoscience. 2022;**12**:3071–3096. DOI: 10.1007/s13204-021-02018-9

[35] Khalifa HO, Oreiby A, Mohammed T, Abdelhamid MAA, Sholkamy EN, Hashem H, et al. Silver nanoparticles as next-generation antimicrobial agents: Mechanisms, challenges, and innovations against multidrug-resistant bacteria. Frontiers in Cellular and Infection Microbiology. 2025;**15**:1599113

[36] Rodrigues AS, Batista JGS, Rodrigues MÁV, Thipe VC, Minarini LAR, Lopes PS, et al. Advances in silver nanoparticles: A comprehensive review on their potential as antimicrobial agents and their mechanisms of action elucidated by proteomics. Frontiers in Microbiology. 2024;**15**:1440065

[37] Carvalho-Silva JM, Reis ACD. Anti-inflammatory action of silver nanoparticles in vivo: Systematic review

and meta-analysis. Heliyon. 2024;**10**(14):e34564. DOI: 10.1016/j.heliyon.2024.e34564

[38] Guzzatti MFM, de Moura AB, Venturini LM, de Roch Casagrande L, Lima IR, da Costa C, et al. Biosynthesized silver nanoparticles modulate inflammation in a palatine wound model. Clinical and Experimental Dental Research. 2025;**11**(5):e70213. DOI: 10.1002/cre2.70213

[39] Cheng L, Zhang S, Zhang Q, Gao W, Mu S, Wang B. Wound healing potential of silver nanoparticles from Hybanthus enneaspermus on rats. Heliyon. 2024;**10**(17):e36118. DOI: 10.1016/j.heliyon.2024.e36118

[40] Bold BE, Urnukhsaikhan E, Mishig-Ochir T. Biosynthesis of silver nanoparticles with antibacterial, antioxidant, anti-inflammatory properties and their burn wound healing efficacy. Frontiers in Chemistry. 2022;**10**:972534. DOI: 10.3389/fchem.2022.972534

[41] Marslin G, Selvakesavan RK, Franklin G, Sarmento B, Dias AC. Antimicrobial activity of cream incorporated with silver nanoparticles biosynthesized from Withania somnifera. International Journal of Nanomedicine. 2015;**10**:5955–5963. DOI: 10.2147/IJN.S81271

[42] Odeniyi MA, Okumah VC, Adebayo-Tayo BC, Odeniyi OA. Green synthesis and cream formulations of silver nanoparticles of Nauclea latifolia fruit extracts and evaluation of antimicrobial and antioxidant activities. Sustainable Chemistry and Pharmacy. 2020;**17**:100298. DOI: 10.1016/j.scp.2020.100298

[43] Masood N, Ahmed R, Tariq M, Ahmed Z, Masoud MS, Ali I, et al. Silver nanoparticle impregnated chitosan-PEG hydrogel enhances wound healing in diabetes induced rabbits. International Journal of Pharmaceutics. 2019;**559**:23–36

[44] Aldakheel FM, Mohsen D, El Sayed MM, Alawam KA, Binshaya AS, Alduraywish SA. Silver nanoparticles loaded on chitosan-g-PVA hydrogel for wound-healing applications. Molecules. 2023;**28**(7):3241. DOI: 10.3390/molecules28073241

[45] Haidari H, Bright R, Garg S, Vasilev K, Cowin AJ, Kopecki Z. Eradication of mature bacterial biofilms with concurrent improvement in chronic wound healing using silver nanoparticle hydrogel treatment. Biomedicines. 2021;**9**(9):1182. DOI: 10.3390/biomedicines9091182

[46] Bhubhanil S, Talodthaisong C, Khongkow M, Namdee K, Wongchitrat P, Yingmema W, et al. Enhanced wound healing properties of guar gum/curcumin-stabilized silver nanoparticle hydrogels. Scientific Reports. 2021;**11**:21836

[47] Nqakala ZB, Sibuyi NRS, Fadaka AO, Meyer M, Onani MO, Madiehe AM. Advances in nanotechnology towards development of silver nanoparticle-based wound-healing agents. International Journal of Molecular Sciences. 2021;**22**(20):11272. DOI: 10.3390/ijms222011272

[48] He Y, Chen J, Xu Z, Nie J, Wang F, Ma C, et al. Silver functionalized chitosan composite hydrogel with sustained silver release and enhanced antibacterial properties promotes healing of infected wounds.

International Journal of Biological Macromolecules. 2025;**285**:138290

[49] Paul S, Sasikumar CS, Cherian SM. Biomedical evaluation of chitosan-gelatin transdermal patch embedded with bio-silver nanoparticles as a wound dressing material: An in vitro study. International Journal of ChemTech Research. 2015;**8**(6):107–115

[50] Lv H, Cui S, Yang Q, Song X, Wang D, Hu J, et al. AgNPs-incorporated nanofiber mats: Relationship between AgNP size/content, silver release, cytotoxicity, and antibacterial activity. Materials Science and Engineering: C. 2021;**118**:111331

[51] Shrestha S, Wang B, Dutta PK. Commercial silver-based dressings: In vitro and clinical studies in treatment of chronic and burn wounds. Antibiotics (Basel). 2024;**13**(9):910. DOI: 10.3390/antibiotics13090910

[52] Mikhailova EO. Silver nanoparticles: Mechanism of action and probable bio-application. Journal of Functional Biomaterials. 2020;**11**(4):84. DOI: 10.3390/jfb11040084

[53] Burdușel AC, Gherasim O, Grumezescu AM, Mogoantă L, Ficai A, Andronescu E. Biomedical applications of silver nanoparticles: An up-to-date overview. Nanomaterials (Basel). 2018;**8**(9):681. DOI: 10.3390/nano8090681

[54] Mishra NK, Yadav K, Mohanty SR, Parmar AS, Yadav SK, Haldor C. Tinospora cordifolia silver nanoparticles attenuated lipopolysaccharide-induced testicular inflammation in golden hamster. Regenerative Engineering and Translational Medicine. 2025;**11**:434–448. DOI: 10.1007/s40883-024-00363-z

[55] Sharma S, Bose A, Biswas S, Sen S, Roy I. Cyperus rotundus mediated green synthesis of silver nanoparticles for antibacterial wound dressing applications. Scientific Reports. 2025;**15**:18394. DOI: 10.1038/s41598-025-03555-x

[56] Bedlovičová Z, Strapáč I, Baláž M, Salayová A. A brief overview on antioxidant activity determination of silver nanoparticles. Molecules. 2020;**25**(14):3191. DOI: 10.3390/molecules25143191

[57] Qubtia M, Ghumman SA, Noreen S, Hameed H, Noureen S, Kausar R, et al. Evaluation of plant-based silver nanoparticles for antioxidant activity and promising wound-healing applications. ACS Omega. 2024;**9**(10):12146–12157. DOI: 10.1021/acsomega.3c10489

[58] Samberg ME, Oldenburg SJ, Monteiro-Riviere NA. Evaluation of silver nanoparticle toxicity in skin in vivo and keratinocytes in vitro. Environmental Health Perspectives. 2010;**118**(3):407–413. DOI: 10.1289/ehp.0901398

[59] Kraeling MEK, Topping VD, Keltner ZM, Belgrave KR, Bailey KD, Gao X, et al. In vitro percutaneous penetration of silver nanoparticles in pig and human skin. Regulatory Toxicology and Pharmacology. 2018;**95**:314–322

[60] Leung BO, Jalilehvand F, Mah V, Parvez M, Wu Q. Silver(I) complex formation with cysteine, penicillamine, and glutathione. Inorganic Chemistry. 2013;**52**(8):4593–4602. DOI: 10.1021/ic400192c

[61] Nemčeková K, Dudoňová P, Holka T, Balážová S, Hornychová M, Szebellaiová V, et al. Silver

nanoparticles for biosensing and drug delivery: A mechanical study on DNA interaction. Biosensors (Basel). 2025;**15**(5):331. DOI: 10.3390/bios15050331

[62] Qing Y, Cheng L, Li R, Liu G, Zhang Y, Tang X, et al. Potential antibacterial mechanism of silver nanoparticles and optimization of orthopedic implants by advanced modification technologies. International Journal of Nanomedicine. 2018;**13**: 3311–3327

[63] Dakal TC, Kumar A, Majumdar RS, Yadav V. Mechanistic basis of antimicrobial actions of silver nanoparticles. Frontiers in Microbiology. 2016;**7**:1831. DOI: 10.3389/fmicb.2016.01831

[64] Yin IX, Zhang J, Zhao IS, Mei ML, Li Q, Chu CH. The antibacterial mechanism of silver nanoparticles and its application in dentistry. International Journal of Nanomedicine. 2020;**15**:2555–2562. DOI: 10.2147/IJN.S246764

[65] Ameh T, Gibb M, Stevens D, Pradhan SH, Braswell E, Sayes CM. Silver and copper nanoparticles induce oxidative stress in bacteria and mammalian cells. Nanomaterials (Basel). 2022;**12**(14):2402. DOI: 10.3390/nano12142402

[66] Roy A, Bulut O, Some S, Mandal AK, Yilmaz MD. Green synthesis of silver nanoparticles: Biomolecule–nanoparticle organizations targeting antimicrobial activity. RSC Advances. 2019;**9**(5):2673–2702. DOI: 10.1039/c8ra08982e

[67] Casals E, Gusta MF, Bastus N, Rello J, Puntes V. Silver nanoparticles and antibiotics: A promising synergistic approach to multidrug-resistant infections. Microorganisms. 2025;**13**(4):952. DOI: 10.3390/microorganisms13040952

[68] Deng H, McShan D, Zhang Y, Sinha SS, Arslan Z, Ray PC, et al. Mechanistic study of the synergistic antibacterial activity of combined silver nanoparticles and common antibiotics. Environmental Science & Technology. 2016;**50**(16):8840–8848. DOI: 10.1021/acs.est.6b00998

[69] Mudhafar M, Zainol I, Jaafar A, Desa S. Review synthesis methods of Ag nanoparticles: Antibacterial and cytotoxicity. International Journal of Drug Delivery Technology. 2021;**11**(2):635–648

[70] Zhang XF, Shen W, Gurunathan S. Silver nanoparticle-mediated cellular responses in various cell lines: An in vitro model. International Journal of Molecular Sciences. 2016;**17**(10):1603. DOI: 10.3390/ijms17101603

[71] Haidari H, Bright R, Strudwick XL, Garg S, Vasilev K, Cowin AJ, et al. Multifunctional ultrasmall AgNP hydrogel accelerates healing of Staphylococcus aureus-infected wounds. Acta Biomaterialia. 2021;**128**:420–434

[72] Althumairy DA. Green-synthesized silver nanoparticles from garlic peel target NF-κB and redox imbalance: A novel therapeutic strategy against pyrogallol-induced hepatotoxicity in rats. Nanomaterials (Basel). 2025;**15**(21):1610. DOI: 10.3390/nano15211610

[73] Kalishwaralal K, Banumathi E, Pandian SRK, Deepak V, Muniyandi J, Eom SH, et al. Silver nanoparticles inhibit VEGF-induced cell proliferation and migration in bovine retinal

endothelial cells. Colloids and Surfaces B: Biointerfaces. 2009;**73**(1):51–57. DOI: 10.1016/j.colsurfb.2009.04.025

[74] Palanisamy CP, Poompradub S, Sansanaphongpricha K, Jayaraman S, Subramani K, Sonsudin F. Increased expression levels of PDGF and VEGF magnify the wound healing potential facilitated by biogenic synthesis of silver nanoparticles. Nano-Structures & Nano-Objects. 2024;**39**:101236. DOI: 10.1016/j.nanoso.2024.101236

[75] You C, Li Q, Wang X, Wu P, Ho JK, Jin R, et al. Silver nanoparticle-loaded collagen/chitosan scaffolds promote wound healing via regulating fibroblast migration and macrophage activation. Scientific Reports. 2017;**7**(1):10489. DOI: 10.1038/s41598-017-10481-0

[76] Stefan LM, Iosageanu A, Ilie D, Stanciuc AM, Matel C, Berger D, et al. Extracellular matrix biomimetic polymeric membranes enriched with silver nanoparticles for wound healing. Biomedical Materials. 2021;**16**(3):035010. DOI: 10.1088/1748-605X/abe55d

[77] Kaya M, Akdaşçi E, Eker F, Bechelany M, Karav S. Recent advances of silver nanoparticles in wound healing: Evaluation of in vivo and in vitro studies. International Journal of Molecular Sciences. 2025;**26**(20):9889. DOI: 10.3390/ijms26209889

[78] Majhi RK, Mohanty S, Khan MI, Brauner A. Ag@ZnO nanoparticles induce antimicrobial peptides and promote migration and antibacterial activity of keratinocytes. ACS Infectious Diseases. 2021;7(8):2068–2072. DOI: 10.1021/acsinfecdis.0c00903

[79] Ong WTJ, Nyam KL. Evaluation of silver nanoparticles in cosmeceutical and potential biosafety complications. Saudi Journal of Biological Sciences. 2022;**29**(4):2085–2094. DOI: 10.1016/j.sjbs.2022.01.035

[80] Pansara C, Mishra R, Mehta T, Parikh A, Garg S. Formulation of chitosan-stabilized silver nanoparticle-containing wound healing film: In vitro and in vivo characterization. Journal of Pharmaceutical Sciences. 2020;**109**(7):2196–2205. DOI: 10.1016/j.xphs.2020.03.028

[81] Sabarees G, Velmurugan V, Tamilarasi GP, Alagarsamy V, Raja Solomon V. Recent advances in silver nanoparticles containing nanofibers for chronic wound management. Polymers (Basel). 2022;**14**(19):3994. DOI: 10.3390/polym14193994

[82] Alfalluji AL, Kadhim QS, Mahdi AA. Electrospun poly(ε-caprolactone)/silver nanoparticle nanofibrous scaffolds with antibacterial activity for wound-dressing applications. RSC Advances. 2025;**15**(45):37899–37907. DOI: 10.1039/d5ra06746d

[83] Krishnan PD, Banas D, Durai RD, Kabanov D, Hosnedlova B, Kepinska M, et al. Silver nanomaterials for wound dressing applications. Pharmaceutics. 2020;**12**(9):821. DOI: 10.3390/pharmaceutics12090821

[84] Ma X, Tian Y, Yang R, Wang H, Allahou LW, Chang J, et al. Nanotechnology in healthcare, and its safety and environmental risks. Journal of Nanobiotechnology. 2024;**22**:715

[85] Sati A, Ranade TN, Mali SN, Yasin HKA, Pratap A. Silver nanoparticles (AgNPs): Comprehensive insights into bio/synthesis, key

influencing factors, multifaceted applications, and toxicity – A 2024 update. ACS Omega. 2025;**10**(8):7549–7582. DOI: 10.1021/acsomega.4c11045

[86] Hosny S, Gaber GA, Ragab MS, Ragheb MA, Anter M, Mohamed L, et al. A comprehensive review of silver nanoparticles (AgNPs): Synthesis strategies, toxicity concerns, biomedical applications, AI-driven advancements, challenges, and future perspectives. Arabian Journal for Science and Engineering. 2025. DOI: 10.1007/s13369-025-10612-0

[87] Balaure PC, Niculescu AG, Anghel D, Grumezescu AM, Alberts A. Role of silver nanoparticles in wound healing: Mechanisms, efficacy, and clinical applications. Inorganics. 2025;**13**(12):401. DOI: 10.3390/inorganics13120401

[88] Muslim MRF, Chabib L, Suryaningsih BE, Hayati F, Annisa V. Green-synthesized silver nanoparticle hydrogels for biofilm-infected wounds: Bridging sustainability and clinical translation. Frontiers in Pharmacology. 2025;**16**:1694144. DOI: 10.3389/fphar.2025.1694144

Chapter 7

Neurotoxicity of Silver Nanoparticles: Histopathological and Genetic Insights, Microbiological Interactions and Implications for Nanomedicine

Abubaker M. Hamad, Mohammed Hashim Albashir, Abrar Khalid Aloufi and Sara E. Ibrahim

Abstract

Silver nanoparticles (AgNPs) are among the most widely produced and used nanomaterials in both biomedicine and other industries, due to their antimicrobial capabilities and flexible physical and chemical characteristics. However, nervous system might be at risk due to AgNP exposure. Central nervous system is highly sensitive owing to its high specificity in metabolic demand, low regenerative ability, and dependence on tightly regulated protective barriers such as the blood–brain barrier (BBB). This chapter provides an integrative and systems-level analysis of AgNP-induced neurotoxicity, with a focus on genetic and transcriptomic alterations, cellular and histopathological neural responses, microbiological interactions, and implications for nanomedicine. Across different experimental biological models, AgNP exposure consistently activates oxidative stress pathways, disrupts mitochondrial function, induces inflammatory signaling, and triggers genotoxic responses. Transcriptomic analyses reveal differential vulnerability among neural cell types, with neurons exhibiting sensitivity in energy metabolism and synaptic pathways, while glial cells predominantly mediate inflammatory and antioxidant responses. In parallel, protein corona formation emerges as a critical determinant of nanoparticle bioidentity, influencing cellular uptake, BBB interactions, and neural exposure. Additionally, perturbation of the gut microbiota by AgNPs introduces an indirect yet potentially amplifying pathway of neurotoxicity via the gut–brain axis. This chapter also addresses neurotoxic effect in context of roles of particle size, surface coating, dose, and exposure duration. Therefore, a systems toxicology framework is proposed to link nanoparticle physicochemical origin, molecular pathways, and neural outcomes. Such multi-omics integration advances a mechanistic understanding of AgNP-induced neurotoxicity and outlines risk-aware strategies for safer nanomaterial design and practices.

IntechOpen

Keywords: neurotoxicity, nanomedicine, silver nanoparticles, genetic alteration, microbiology, Histopathological change

1. Introduction

Silver nanoparticles (AgNPs) are among the most widely utilized engineered nanomaterials in contemporary biomedical and industrial applications. Their unique physicochemical properties – most notably their high surface-area-to-volume ratio and potent antimicrobial activity – have enabled extensive use in wound dressings, medical devices, coatings, drug delivery systems, and diagnostic platforms [1–3]. Despite these advantages, the rapid expansion of AgNP-based technologies has raised increasing concerns regarding their potential adverse biological effects, particularly on the nervous system.

The central nervous system (CNS) represents a uniquely vulnerable target for nanomaterial-induced toxicity due to its high metabolic demand, complex cellular organization, limited regenerative capacity, and reliance on tightly regulated protective barriers, such as the blood–brain barrier (BBB) [4, 5]. Emerging evidence indicates that AgNPs, unlike bulk silver or ionic silver alone, can interact with neural tissues through multiple convergent mechanisms, including oxidative stress, mitochondrial dysfunction, neuroinflammatory signaling, genotoxicity, and disruption of neurovascular integrity [6–8].

At the molecular level, transcriptomic and genomic studies have begun to reveal conserved stress-response signatures induced by AgNP exposure across diverse biological systems. These include the upregulation of redox-sensitive genes (e.g., HMOX1, NQO1), inflammatory mediators, metallothioneins, and DNA damage response (DDR) pathways [9–11]. Importantly, many of these pathways are highly relevant to neuronal and glial physiology, suggesting that AgNP-induced molecular perturbations may translate into functional neurotoxicity rather than transient cellular stress.

Beyond intrinsic cellular effects, the biological identity of AgNPs *in vivo* is profoundly shaped by the formation of a protein corona upon contact with biological fluids. This dynamic corona determines cellular recognition, uptake mechanisms, intracellular trafficking, and ultimately tissue distribution [12, 13]. In the context of neurotoxicity, protein corona composition has been shown to influence interactions with BBB endothelial cells, transcytosis efficiency, and the propensity of AgNPs to access neural parenchyma [14]. Thus, neurotoxic outcomes cannot be interpreted solely on the basis of core particle properties but must be understood as emergent phenomena arising from nano–bio interface interactions.

Experimental evidence from *in vivo* vertebrate models further supports the neurotoxic potential of AgNPs. Developmental studies in zebrafish embryos have demonstrated consistent transcriptomic disruptions in mitochondrial energy metabolism, protein synthesis, and stress-response pathways following AgNP exposure [15–17]. Complementary studies in rodent models suggest that AgNPs can induce neuroinflammation, oxidative damage, and synaptic alterations in mammalian brains, reinforcing the translational relevance of these findings [18, 19].

In parallel, the antimicrobial nature of AgNPs introduces an additional and often overlooked dimension of neurotoxicity: indirect effects mediated through perturbation of the gut microbiota. Alterations in microbial composition and function may influence the gut–brain axis by modulating immune signaling, microbial metabolites, and systemic oxidative stress, thereby amplifying neuroinflammatory processes [20–22]. Integrating microbiological interactions into neurotoxicity frameworks is therefore essential for a comprehensive understanding of AgNP-induced effects.

While the majority of studies emphasize toxic outcomes, it is important to acknowledge that AgNP-induced neurotoxicity is highly context-dependent. Factors such as particle size, surface coating, aggregation state, dose, exposure duration, and biological milieu critically shape observed effects [23–25]. Several studies report minimal or no detectable neurotoxic effects under conditions of low-dose exposure, specific surface functionalization, or controlled release of silver ions, highlighting the need for balanced interpretation and risk-aware design [26].

Against this backdrop, this chapter provides an integrative, systems-level analysis of AgNP-induced neurotoxicity, with a particular emphasis on genetic and transcriptomic alterations, cell-type-specific vulnerability (neuronal and glial), microbiological interactions, and implications for nanomedicine. By synthesizing evidence from open-access omics datasets, *in vitro* neural models, and *in vivo* vertebrate studies, the chapter aims to (i) identify conserved molecular pathways underlying neurotoxicity, (ii) clarify critical determinants of neural susceptibility, and (iii) inform safer-by-design strategies for future biomedical applications of silver nanomaterials.

2. Physicochemical properties of AgNPs and their biological identity

The biological behavior, biodistribution, and toxicity of AgNPs are fundamentally governed by their physicochemical properties. Key determinants include particle size, shape, surface charge, coating chemistry, aggregation state, and dissolution kinetics [27–29]. These parameters do not act independently; rather, they collectively define how AgNPs interact with biological systems at the nano–bio interface.

2.1 Particle size, surface area, and silver ion release

Among all physicochemical features, particle size is one of the most critical determinants of AgNP toxicity. Smaller nanoparticles (<20 nm) exhibit a disproportionately large surface-area-to-volume ratio, enhancing surface reactivity and accelerating the release of bioavailable silver ions (Ag^+) [30, 31]. Transcriptomic and toxicological studies consistently demonstrate stronger oxidative stress responses, mitochondrial impairment, and inflammatory signaling following exposure to smaller AgNPs compared to larger counterparts at equivalent mass concentrations [32].

The release of Ag^+ plays a central role in mediating toxicity, particularly through interactions with thiol-containing proteins, mitochondrial enzymes, and antioxidant systems. However, increasing evidence suggests that nanoparticle-specific effects persist even when ionic silver contributions are controlled, indicating that particle-mediated mechanisms contribute independently to biological outcomes [33].

2.2 Surface coating, charge, and cellular uptake

Surface functionalization significantly modulates AgNP interactions with cells. Common stabilizing agents, such as citrate, polyvinylpyrrolidone (PVP), polyethylene glycol (PEG), and protein-based coatings, alter surface charge, hydrophobicity, and colloidal stability [34, 35]. These properties influence cellular uptake pathways, including clathrin-mediated endocytosis, caveolae-dependent internalization, and macropinocytosis.

Positively charged or weakly stabilized AgNPs generally exhibit enhanced cellular uptake but also increased cytotoxicity, particularly in neural and endothelial cells [36, 37]. In contrast, polymer-coated or PEGylated AgNPs often display reduced nonspecific uptake and attenuated acute toxicity, highlighting the importance of surface chemistry in risk mitigation strategies.

2.3 Protein corona formation: Defining the biological identity of AgNPs

Upon exposure to biological fluids such as blood plasma, cerebrospinal fluid, or interstitial media, AgNPs rapidly adsorb proteins and other biomolecules onto their surface, forming a dynamic protein corona [37, 38]. This corona effectively replaces the synthetic surface of the nanoparticle with a biological interface, thereby defining its "biological identity."

The composition of the protein corona depends on nanoparticle size, surface charge, coating, and the surrounding biological milieu. Abundant plasma proteins, such as albumin, fibrinogen, apolipoproteins, and immunoglobulins, are frequently enriched, while lower-abundance proteins with high surface affinity may dominate the tightly bound "hard corona" [39].

Crucially, the protein corona determines how AgNPs are recognized by cells, influencing receptor engagement, uptake efficiency, intracellular trafficking, and clearance mechanisms. From a neurotoxicity perspective, this corona-mediated identity is a decisive factor in governing interactions with the BBB [40].

2.4 Interaction with the BBB

The BBB is composed primarily of specialized endothelial cells connected by tight junctions, supported by astrocytic end-feet and pericytes. Its integrity is essential for maintaining CNS homeostasis. Protein corona-modified AgNPs may interact with BBB endothelial cells via receptor-mediated processes, including scavenger receptors and lipoprotein receptors, particularly when apolipoproteins are present in the corona [39, 41].

Experimental studies indicate that certain corona compositions facilitate transcytosis across endothelial layers, increasing the likelihood of AgNP penetration into the brain parenchyma [42]. Concurrently, AgNP exposure has been shown to induce endothelial oxidative stress, inflammatory signaling, and tight junction disruption, collectively compromising BBB integrity and amplifying neurotoxic risk [43].

Thus, AgNP-induced neurotoxicity may arise not only from direct neuronal exposure but also from indirect effects mediated by BBB dysfunction, allowing secondary toxicants or inflammatory mediators to access neural tissue.

2.5 Intracellular fate and subcellular localization

Following cellular internalization, AgNPs are frequently sequestered within endo-lysosomal compartments. The acidic lysosomal environment promotes nanoparticle dissolution, resulting in localized Ag^+ release and elevated intracellular silver concentrations [35]. This process triggers lysosomal stress, mitochondrial dysfunction, and activation of oxidative and inflammatory signaling cascades.

In neural cells, such subcellular perturbations are particularly consequential due to the high energy demands of neurons and the sensitivity of synaptic function to mitochondrial integrity. Accumulation of AgNPs or released Ag^+ within mitochondria has been linked to impaired oxidative phosphorylation, ATP depletion, and the initiation of apoptotic pathways [44].

3. Oxidative stress and mitochondrial dysfunction in neural contexts

Oxidative stress represents one of the most consistently reported and mechanistically central pathways in AgNP-induced neurotoxicity. Neural tissues are particularly susceptible to redox imbalance due to their high oxygen consumption, abundance of polyunsaturated fatty acids, limited antioxidant reserves, and strong dependence on mitochondrial energy metabolism [45, 46]. Consequently, even modest disruptions in redox homeostasis can precipitate disproportionate neuronal and glial dysfunction (see **Figure 1**).

Figure 1 illustrates the sequential pathway of neurotoxicity induced by AgNPs. Following cellular uptake via endocytosis, AgNPs undergo lysosomal dissolution,

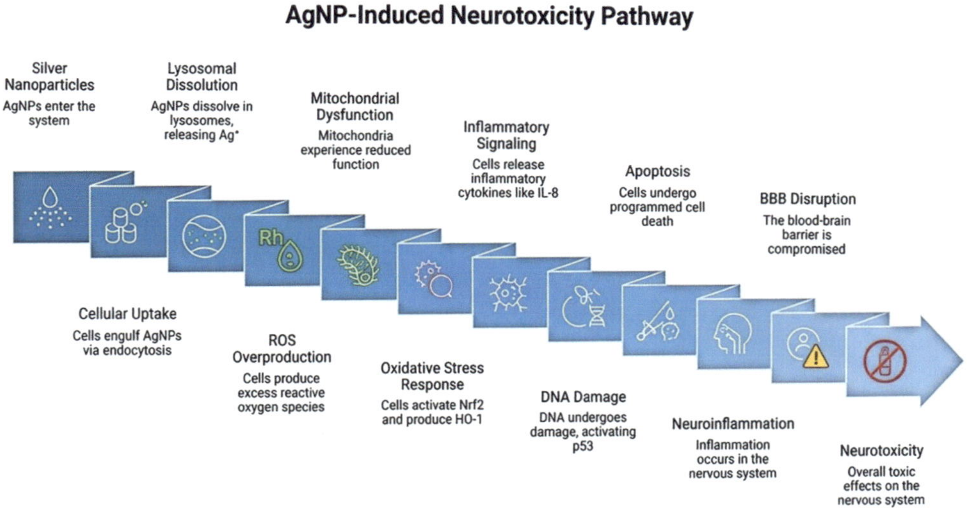

Figure 1.
AgNP-induced neurotoxicity pathway.

leading to the release of bioavailable silver ions. This process promotes excessive reactive oxygen species (ROS) generation, activation of oxidative stress responses, mitochondrial dysfunction, DNA damage, inflammatory signaling, apoptosis, and disruption of the BBB, ultimately resulting in adverse neurological outcomes.

3.1 ROS generation and redox imbalance

Exposure to AgNPs has been repeatedly shown to induce excessive generation of ROS, including superoxide anions, hydrogen peroxide, and hydroxyl radicals [47]. This ROS overproduction arises from multiple sources, including nanoparticle surface reactivity, Ag^+-mediated catalytic reactions, and mitochondrial electron transport chain disruption.

Transcriptomic studies across diverse models consistently demonstrate the activation of redox-sensitive defense pathways following AgNP exposure. The upregulation of genes such as HMOX1, NQO1, GCLC, and metallothioneins (MT1, MT2) reflects an adaptive response aimed at restoring redox balance [48–50]. In neural cells, however, sustained or excessive ROS generation can overwhelm antioxidant defenses, leading to oxidative damage to lipids, proteins, and nucleic acids.

Neurons appear particularly vulnerable due to their limited capacity for glutathione regeneration and reliance on astrocytes for redox support. In contrast, glial cells – especially astrocytes and microglia – exhibit a dual role, initially mounting antioxidant and inflammatory responses that may become neurotoxic under chronic or high-dose exposure conditions [51].

3.2 Mitochondria as primary targets of AgNP-induced toxicity

Mitochondria are both a major source and a primary target of AgNP-induced oxidative stress. Silver nanoparticle exposure disrupts mitochondrial membrane potential, impairs electron transport chain function, and reduces ATP synthesis [52]. These effects have been observed in neuronal cultures, brain endothelial cells, and *in vivo* vertebrate models, underscoring their conserved nature.

Transcriptomic analyses from developmental zebrafish models reveal significant downregulation of genes involved in oxidative phosphorylation, including components of complexes I–IV, shortly after AgNP exposure [53, 54]. Similar mitochondrial gene suppression has been reported in mammalian neural tissues, suggesting that mitochondrial vulnerability is a key determinant of neurotoxicity across species [55].

In neurons, mitochondrial dysfunction has direct functional consequences, including impaired axonal transport, synaptic failure, and altered neurotransmitter release. In glial cells, mitochondrial stress may promote pro-inflammatory phenotypes, thereby amplifying neuroinflammatory signaling and secondary neuronal injury.

3.3 Mitochondrial–nuclear crosstalk and stress signaling

Mitochondrial dysfunction induced by AgNPs triggers retrograde signaling pathways that alter nuclear gene expression. Activation of transcription factors such as NRF2, AP-1, and p53 links mitochondrial stress to broader cellular responses, including antioxidant defense, cell cycle regulation, and apoptosis [52, 56].

Importantly, transcriptomic evidence suggests that these stress-response programs are not transient. Persistent mitochondrial impairment can lead to long-term reprogramming of cellular metabolism and inflammatory signaling, particularly in post-mitotic neurons, where recovery capacity is limited [57].

3.4 Functional consequences for neural integrity

The convergence of oxidative stress and mitochondrial dysfunction has significant implications for neural integrity. Oxidative damage to synaptic proteins and cytoskeletal elements disrupts neuronal connectivity, while ATP depletion compromises ion homeostasis and membrane excitability [58].

In developing nervous systems, such as those studied in zebrafish embryos, mitochondrial disruption during critical windows of neurodevelopment may result in lasting structural and functional abnormalities [59]. In adult brains, chronic low-level mitochondrial stress may contribute to neurodegenerative processes, particularly when combined with aging-related declines in mitochondrial resilience.

3.5 Context-dependent and modifiable nature of oxidative toxicity

Despite the centrality of oxidative stress in AgNP neurotoxicity, it is increasingly evident that this pathway is highly context-dependent. Particle size, surface coating, dose, and exposure duration critically modulate ROS generation and downstream outcomes [25]. Several studies demonstrate that appropriately coated or larger AgNPs induce significantly attenuated oxidative responses, supporting the feasibility of safer-by-design approaches [44].

Moreover, antioxidant supplementation and modulation of cellular redox pathways have been shown to partially mitigate AgNP-induced toxicity in experimental systems, further underscoring the causal role of oxidative stress while highlighting potential intervention strategies [60].

4. Genetic and transcriptomic alterations induced by AgNPs in neural and glial systems

Advances in transcriptomic and genomic profiling have provided critical insights into the molecular mechanisms underlying AgNP-induced neurotoxicity. Unlike traditional cytotoxicity assays, omics-based approaches capture early and system-wide perturbations in gene regulation, enabling identification of conserved stress pathways as well as cell-type-specific vulnerabilities within the nervous system (see **Table 1**) [60, 61].

Table 1 summarizes conserved genetic responses to AgNP exposure across human endothelial cells, human immune cells, and zebrafish embryos. Shared molecular signatures include oxidative stress-related genes (HMOX1, NQO1, MT1/MT2), stress and chaperone proteins (HSP70), inflammatory cytokines (IL-8, IL-11), and DNA damage and cell cycle regulators (TP53, CDKN1A), indicating conserved toxicity pathways across species, as illustrated in **Figure 4**.

Gene cluster	Human endothelial cells	Human immune cells	Zebrafish embryos
Oxidative stress genes	HMOX1, NQ01, MT1/MT2	HMOX1, NQ01, MT1/MT2	HMOX1, NQ01, MT1/MT2
Stress and chaperone response	HSP70	HSP70	HSP70
Inflammatory cytokines	IL-8, IL-11	IL-8, IL-11	IL-8, IL-11
DNA damage and cell cycle control	TP53, CDKN1A(p21)	TP53, CDKN1A(p21)	TP53, CDKN1A(p21)

Table 1.
Conserved AgNP-responsive genes across biological models.

4.1 Transcriptomic signatures in neuronal models

Studies employing neuronal cell lines, primary neurons, and neuron-enriched cultures consistently reveal that AgNP exposure induces pronounced alterations in genes associated with oxidative stress, synaptic function, and neuronal survival [62]. Among the most frequently affected pathways are redox homeostasis, mitochondrial metabolism, calcium signaling, and synaptic vesicle cycling.

Key oxidative stress-responsive genes, such as HMOX1, TXNRD1, and SOD2, are robustly upregulated following AgNP exposure, reflecting activation of NRF2-mediated defense programs [11]. Simultaneously, transcriptomic suppression of genes involved in synaptic transmission – including those regulating neurotransmitter release, postsynaptic density organization, and axonal transport – suggests that neuronal communication may be compromised even in the absence of overt cell death [63].

Notably, neuronal cells exhibit heightened sensitivity to AgNP-induced DDR. Activation of TP53, CDKN1A (p21), and DNA repair genes indicates engagement of checkpoint signaling, which may lead to cell cycle arrest in neural progenitors or apoptosis in post-mitotic neurons under severe stress conditions [64].

4.2 Glial cell responses: Astrocytes and microglia

Glial cells play a central role in maintaining neural homeostasis and orchestrating inflammatory responses. Transcriptomic analyses demonstrate that astrocytes and microglia respond to AgNP exposure with distinct, yet interconnected, gene expression programs [65].

Astrocytes exhibit a strong induction of antioxidant and metal-binding genes, including metallothioneins (MT1, MT2), glutathione-related enzymes, and stress chaperones [66]. While these responses may initially serve a neuroprotective function, sustained activation can disrupt metabolic support to neurons and impair neurotransmitter recycling.

Microglial cells, by contrast, show pronounced activation of inflammatory signaling pathways. Upregulation of genes associated with NF-κB signaling, cytokine production (IL1B, TNF, IL6), and pattern recognition receptors suggests that AgNPs can act as potent triggers of neuroinflammatory responses [67]. Such activation may be further amplified by mitochondrial stress and ROS production, promoting a pro-inflammatory microglial phenotype that exacerbates neuronal injury.

4.3 Endothelial and neurovascular transcriptomic perturbations

Endothelial cells of the BBB represent a critical interface between systemic exposure and neural tissue. Transcriptomic profiling of brain endothelial models exposed to AgNPs reveals activation of oxidative stress, inflammatory signaling, and tight junction remodeling pathways [68].

Downregulation of genes encoding tight junction proteins and transporters, alongside upregulation of adhesion molecules and cytokines, suggests compromised BBB integrity. This molecular disruption may facilitate increased permeability to AgNPs themselves or to secondary inflammatory mediators, thereby amplifying neurotoxic outcomes [69].

4.4 *In vivo* transcriptomic evidence from vertebrate brains

In vivo transcriptomic studies provide essential validation of *in vitro* findings. Zebrafish embryo models exposed to AgNPs display conserved alterations in genes related to mitochondrial energy metabolism, protein synthesis, and stress responses within developing neural tissues [17]. Importantly, temporal analyses indicate that while some transcriptomic changes partially recover, others persist, suggesting lasting neurodevelopmental consequences.

Rodent studies further corroborate these findings, revealing AgNP-induced dysregulation of inflammatory and oxidative pathways in mammalian brains [70]. Notably, these alterations often occur at exposure levels below those causing overt neurodegeneration, underscoring the sensitivity of transcriptomic endpoints in detecting early neurotoxic risk.

4.5 Convergence and cell-type-specific vulnerability

Collectively, transcriptomic evidence supports a model in which AgNP exposure activates a conserved stress–inflammation–genotoxicity axis across neural cell types, while simultaneously revealing cell-type-specific vulnerabilities. Neurons exhibit pronounced sensitivity in pathways related to energy metabolism and synaptic integrity, whereas glial cells dominate inflammatory and antioxidant responses.

This division of labor within the neural transcriptomic response highlights the importance of systems-level analyses that account for cellular heterogeneity within the CNS. Disruption of intercellular signaling between neurons, astrocytes, microglia, and endothelial cells may ultimately determine the severity and persistence of AgNP-induced neurotoxicity [71].

5. *In vivo* and developmental evidence from vertebrate models

In vivo vertebrate models provide an essential bridge between cellular mechanistic findings and organism-level neurotoxic outcomes. By integrating transcriptomic alterations with developmental, anatomical, and behavioral endpoints, these models enable the evaluation of how AgNP-induced molecular perturbations translate into functional consequences within intact nervous systems [72].

5.1 Zebrafish embryos as a model for developmental neurotoxicity

Zebrafish (*Danio rerio*) embryos have emerged as a powerful model for investigating nanomaterial-induced neurotoxicity due to their rapid development, optical transparency, genetic conservation with mammals, and suitability for high-throughput transcriptomic analysis [59]. Exposure of zebrafish embryos to AgNPs consistently results in dose- and size-dependent alterations in gene expression profiles associated with mitochondrial function, oxidative phosphorylation, and cellular stress responses [73].

Transcriptomic studies demonstrate early suppression of genes encoding mitochondrial respiratory chain components, ribosomal proteins, and ATP synthesis machinery following AgNP exposure [74]. These molecular disruptions coincide with developmental abnormalities, including delayed neurogenesis, impaired neuronal differentiation, and altered neural patterning. Importantly, comparative analyses indicate that AgNPs elicit stronger and more persistent transcriptomic disturbances than bulk silver or ionic silver at equivalent exposure levels, supporting the existence of nanoparticle-specific effects [16].

Temporal profiling further reveals partial recovery of mitochondrial gene expression at later developmental stages, suggesting activation of compensatory mechanisms. However, persistent dysregulation of stress and inflammatory pathways raises concerns regarding subtle but lasting neurodevelopmental consequences that may not be immediately apparent in early phenotypic assessments [75].

5.2 Neurobehavioral and functional outcomes in zebrafish

Beyond molecular endpoints, AgNP exposure has been associated with measurable neurobehavioral alterations in zebrafish larvae. Studies report changes in locomotor activity, altered responses to visual or tactile stimuli, and impaired sensorimotor integration following developmental exposure [76]. These behavioral phenotypes correlate with transcriptomic signatures of synaptic dysfunction, neurotransmitter imbalance, and energy metabolism disruption, providing functional validation of molecular findings.

Such observations underscore the value of integrating transcriptomic data with behavioral assays to capture the full spectrum of AgNP-induced neurotoxicity.

5.3 Rodent models and mammalian relevance

Rodent models offer critical translational relevance for assessing AgNP neurotoxicity in mammalian systems. *In vivo* studies in mice and rats demonstrate that AgNPs can accumulate in brain tissue following systemic exposure, particularly under conditions of compromised BBB integrity or prolonged exposure [77].

Transcriptomic analyses of rodent brains reveal activation of oxidative stress, inflammatory signaling, and apoptosis-related pathways following AgNP exposure [77]. Upregulation of pro-inflammatory cytokines, microglial activation markers, and oxidative damage indicators has been observed in multiple brain regions, including the hippocampus and cortex [78]. These molecular alterations are often accompanied by histopathological evidence of neuroinflammation and synaptic disruption.

Importantly, several rodent studies indicate that chronic low-dose exposure can induce subtle but persistent transcriptomic and inflammatory changes without overt neuronal loss, highlighting the potential for long-term neurological effects that may evade detection by conventional toxicity endpoints [79].

5.4 Comparative insights across vertebrate models

Comparative analyses across zebrafish and rodent models reveal a high degree of conservation in AgNP-induced molecular responses, particularly with respect to mitochondrial dysfunction, oxidative stress, and inflammatory signaling [17]. This conservation supports the relevance of early developmental models for predicting mammalian neurotoxic risk, while emphasizing the importance of exposure timing, dose, and nanoparticle properties.

Notably, developmental exposure appears to confer heightened vulnerability, as disruptions during critical windows of neural differentiation and circuit formation may lead to irreversible alterations in neural architecture and function [80].

5.5 Implications for neurodevelopmental and adult neurotoxicity

Collectively, *in vivo* evidence indicates that AgNP-induced neurotoxicity is not restricted to acute cytotoxic effects but encompasses a continuum of molecular, developmental, and functional disturbances. In developing organisms, AgNP exposure may interfere with neurogenesis and circuit maturation, while in adults, chronic exposure may promote neuroinflammatory and neurodegenerative processes.

These findings reinforce the necessity of incorporating both developmental and adult exposure scenarios into neurotoxicity assessment frameworks for engineered nanomaterials.

6. Genotoxicity, DNA damage, and cell cycle dysregulation

Genotoxicity constitutes a critical downstream consequence of AgNP-induced oxidative and mitochondrial stress, providing a mechanistic bridge between early molecular perturbations and long-term neural dysfunction. Numerous experimental studies report increased DNA strand breaks, chromatin alterations, and activation of DDR pathways following AgNP exposure [81, 82].

At the molecular level, AgNPs and released Ag^{+} can directly interact with DNA and nuclear proteins or indirectly induce genotoxicity through ROS-mediated oxidative lesions. Transcriptomic analyses consistently reveal activation of key DDR regulators, including TP53, ATM, ATR, and CDKN1A (p21), indicating engagement of checkpoint signaling and cell cycle control mechanisms [83]. In neural progenitor cells, such activation may result in cell cycle arrest, thereby impairing neurogenesis during development.

In post-mitotic neurons, persistent DNA damage poses a distinct risk. Neurons exhibit limited DNA repair capacity and rely heavily on genomic integrity for long-term survival. Accumulation of unrepaired lesions may, therefore, promote apoptotic pathways or contribute to progressive neurodegenerative processes [84]. In glial cells, particularly microglia, DNA damage signaling may further potentiate inflammatory activation, amplifying neurotoxic cascades.

Importantly, genotoxic effects appear highly dose- and context-dependent. Several studies report minimal DNA damage under low-dose or well-coated AgNP exposure, underscoring the importance of physicochemical control and exposure scenarios in modulating genotoxic risk (see **Figure 2**) [85].

This schematic depicts the genotoxic cascade triggered by AgNP exposure (see **Figure 2**). AgNPs induce oxidative DNA lesions and strand breaks, leading to activation of the p53 tumor suppressor pathway. Subsequent induction of p21 (CDKN1A) causes cell cycle arrest, followed by apoptosis or cellular senescence. These processes impair neurogenesis and contribute to long-term neurological dysfunction.

7. Concluding distinct mechanisms of AgNPs inducing neurotoxicity and their experimental verification

7.1 Oxidative stress and ROS overproduction

This mechanism occurs when AgNP exposure leads to excessive generation of ROS (superoxide, hydrogen peroxide), which overwhelms neuronal antioxidant defenses and can cause lipid peroxidation, protein oxidation, and DNA damage. Hence, elevated ROS initiates downstream cascades such as apoptosis and autophagy [86].

Experimental verification of this mechanism depends on increased lipid peroxidation and decreased antioxidant enzyme activities (i.e., superoxide dismutase [SOD], catalase, glutathione peroxidase), which were measured in rat brain tissue after AgNP exposure in previous studies. Additionally, proteomic analysis in an

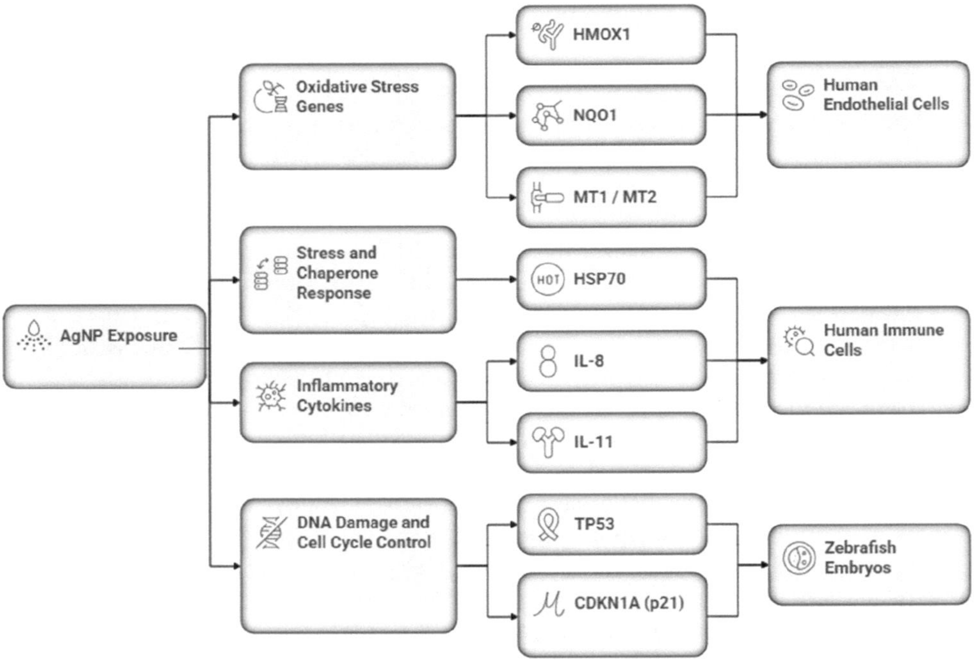

Figure 2.
DNA damage and cell cycle disruption following AgNP exposure.

in vitro human BBB model showed downregulation of antioxidative proteins and upregulation of stress pathways after AgNP treatment. Moreover, direct assays in myelin fractions demonstrated oxidative damage (increased peroxidation and SH oxidation) after AgNP exposure compared with ionic silver controls. Furthermore, experimental histologic evidence can include findings of increased lipid peroxidation (MDA), reduced antioxidant enzymes (SOD, catalase, glutathione [GSH]), elevated ROS fluorescence assays (DCFH-DA), and histological observations using stains such as hematoxylin and eosin (H&E) and Nissl staining. Common observations in rodent models also included neuronal shrinkage, that is, cell body contraction, pyknotic nuclei due to chromatin condensation, cytoplasmic eosinophilia in the form of red neurons, loss of Nissl substance, that is, chromatolysis, and reduced neuronal density in the hippocampal CA1 region [87].

7.2 Inflammation and neuroimmune activation

Experimental studies demonstrate that AgNP exposure activates microglia, astrocytes, and pro-inflammatory cytokine pathways, which produce characteristic histopathological and immunohistochemical changes in brain tissue. These structural findings are typically correlated with biochemical and molecular inflammatory markers [88].

Mechanism of this is by AgNPs that trigger neuroinflammatory responses, including the upregulation of pro-inflammatory cytokines (TNF-α, IL-1β) and the activation of glial cells (e.g., astrocytes, microglia). These inflammatory mediators further exacerbate neuronal damage and compromise neural homeostasis.

Experimental verification was performed on AgNP-treated rats, as their brains showed increased transcripts of TNF-α and IL-1β, along with elevated glial fibrillary acidic protein (GFAP) staining, indicating astrocyte activation and neuroinflammation. Additionally, *in vitro* BBB models demonstrated initial pro-inflammatory cascades in endothelial cells exposed to AgNPs (see **Table 2**) [89].

7.2.1 Microglial activation (primary neuroimmune response)

This can be experimentally evidenced by increased expression of inflammatory cytokines (TNF-α, IL-1β, IL-6), upregulation of microglial markers (Iba-1, CD68), and increased ROS and NF-κB signaling. While histological and immunohistochemical findings include Iba-1 immunostaining, an increased number of Iba-1–positive cells, a morphological shift from ramified (resting) to

Neural toxicity event	Histological evidence	Method
Microglial activation	Amoeboid microglia, increased cell density	Iba-1 immunostaining
Astrocytosis	Hypertrophic astrocytes	GFAP immunostaining
Cytokine-mediated injury	Neuronal pyknosis, degeneration	H&E/Nissl
BBB-associated inflammation	Perivascular edema/infiltration	H&E
Inflammation-induced apoptosis	TUNEL-positive neurons	TUNEL assay
Phagocytic activity	Microglial phagolysosomes	TEM

Table 2.
Experimental histologic events and evidence of AgNPs neural toxicity.

amoeboid (activated) microglia, and clustering of microglia around degenerating neurons. Finally, H&E morphology reveals microglial nodules and focal inflammatory aggregates [89].

7.2.2 Astrocyte activation (astrogliosis)

Astrocytes respond rapidly to nanoparticle-induced oxidative and inflammatory stress, which can be evidenced by the upregulation of GFAP gene and protein expression, as well as by increased cytokine production from astrocytes. Histologically, upon GFAP immunostaining, increased GFAP-positive astrocytes, hypertrophic astrocytic processes, and astrocytic network thickening will be observed. Additionally, morphological features such as enlarged astrocyte cell bodies and dense astrocytic processes surrounding damaged neurons can be seen. Thus, astrogliosis is a hallmark of chronic neuroinflammatory injury [90].

7.2.3 Perivascular inflammation and BBB

Silver nanoparticle exposure often produces inflammatory changes associated with endothelial stress that can be evidenced by increased endothelial inflammatory markers. Histological findings can include perivascular edema, perivascular inflammatory cell infiltration, and mild vascular congestion [91].

7.2.4 Neuronal damage secondary to neuroinflammation

It is known that neuroinflammatory mediators contribute to neuronal degeneration. This can be evidenced histologically by using H&E and Nissl staining to identify neuronal shrinkage, nuclear pyknosis, chromatolysis (i.e., loss of Nissl substance), as well as reduced neuronal density in the hippocampal CA1 and dentate gyrus. Notably, these changes occur in parallel with microglial and astrocytic activation, as mentioned above in Section 7.2.1 [92].

7.3 Mitochondrial dysfunction

Mechanism of toxicity here is when AgNPs disrupt mitochondrial structure and function, impair the electron transport chain, reduce ATP production, and trigger mitochondrial-mediated apoptosis via altered membrane potential and release of pro-apoptotic factors.

Experimental verification was performed in previous studies by exposing rat cortical neurons to AgNPs, which showed mitochondrial dysfunction with loss of membrane potential and morphological abnormalities. Additionally, *in vivo* studies reported mitochondrial dysfunction and induction of autophagy in rat brains exposed to low doses of AgNPs [93].

7.4 Apoptosis and cell death pathways

Mechanism of this is through ROS, inflammation, and mitochondrial failure that converge to activate intrinsic apoptotic pathways, such as BAX upregulation, caspase activation, and DNA fragmentation, which lead to neuronal and glial cell death.

Experimental verification was conducted by assessing the upregulation of apoptosis-related genes (Bax, caspase-3, caspase-9), and histological evidence of neuronal degeneration was observed in rodent models following AgNP exposure. Besides, comparative studies *in vitro* showed increased caspase activity, which is indicative of apoptosis in astrocytes treated with AgNPs [94].

7.5 Synaptic and cytoskeletal disruption

Mechanism of this toxicity concept is when AgNPs impair the structural integrity of neurons by disrupting cytoskeletal proteins such as β-tubulin and F-actin and by reducing synaptic protein levels such as synaptophysin and PSD-95, which compromises neurite outgrowth and synaptic transmission.

Experimental verification was conducted through primary rat cortical cell cultures; AgNPs induced degeneration of cytoskeletal elements and markedly decreased synaptic protein clustering [95].

7.6 BBB effects and neuronal signaling

Mechanism of toxicity through the BBB is that AgNPs may alter gene expression associated with neuronal signaling pathways (e.g., mitogen-activated protein kinase [MAPK], calcium signaling) and affect BBB functional protein expression, potentially contributing to neurotoxicity without necessarily increasing physical BBB permeability.

Experimental verification of this concept was conducted through *in vivo* IV administration of AgNPs in rats, which modulated MAPK signaling, elevated apoptotic genes, and altered calcium channel and ligand receptor gene expression in the hippocampus [96].

7.7 Genotoxicity/DNA damage

Mechanism of toxicity here is through oxidative stress and direct interactions, where AgNPs can cause DNA strand breaks, chromosomal aberrations, and influence the expression of genes involved in the cell cycle and apoptosis. Although less extensively studied in neuronal models, genotoxic effects have been reported in related cell types.

Experimental verification was performed in previous nonneural studies, which reported enrichment of pathways related to DNA replication, oxidative phosphorylation, and mitochondrial-linked apoptotic gene regulation after AgNP exposure, suggesting genotoxic stress that would likely extrapolate to neural cells [97].

8. Microbiological interactions and the gut–brain axis

Beyond direct neural and vascular effects, AgNPs exert potent antimicrobial activity that can perturb commensal microbial communities. Disruption of the gut microbiota introduces an indirect but potentially significant pathway for neurotoxicity via the gut–brain axis [98].

The AgNP-induced dysbiosis has been associated with altered microbial diversity and shifts in metabolite production, including short-chain fatty acids

(SCFAs), bile acid derivatives, and microbial-derived neurotransmitter precursors [99]. These metabolites play crucial roles in immune regulation, BBB integrity, and microglial maturation. Such perturbations may, therefore, exacerbate systemic inflammation and increase susceptibility to neuroinflammatory signaling within the CNS.

Mechanistically, microbiome disruption can elevate circulating endotoxins and pro-inflammatory cytokines, promoting oxidative stress and microglial activation in the brain [100]. This systemic inflammatory milieu may synergize with direct AgNP-induced oxidative and mitochondrial stress, lowering the threshold for neurotoxic outcomes.

Although direct causal links between AgNP-induced dysbiosis and neurological dysfunction remain underexplored, emerging evidence supports this axis as a critical modifier of neurotoxic risk, particularly under chronic exposure conditions (see **Figure 3**) [101]. For instance, Javurek et al. [102] suggested that exposure to two types of AgNPs can lead to changes in both behavior and the gut microbiome in a rat model.

Figure 3 depicts the indirect neurotoxic effects of AgNPs mediated through microbiome disruption. The antimicrobial activity of AgNPs alters gut microbial composition and diversity, leading to immune dysregulation and systemic inflammation. These secondary effects influence the gut–brain axis, increasing neuroinflammatory susceptibility and contributing to neurological impairment.

9. An integrative systems toxicology perspective

The converging evidence presented across cellular, molecular, and organismal levels supports a systems toxicology framework for understanding AgNP-induced neurotoxicity. Within this framework, physicochemical properties and protein corona formation define nanoparticle bioidentity, governing cellular uptake, biodistribution, and BBB interactions.

Subsequent intracellular events – oxidative stress, mitochondrial dysfunction, and genotoxic signaling – activate conserved transcriptional programs across neurons, glia, and endothelial cells. Intercellular crosstalk among these cell types, coupled with systemic modifiers such as microbiome-derived signals, ultimately shapes neurotoxic outcomes.

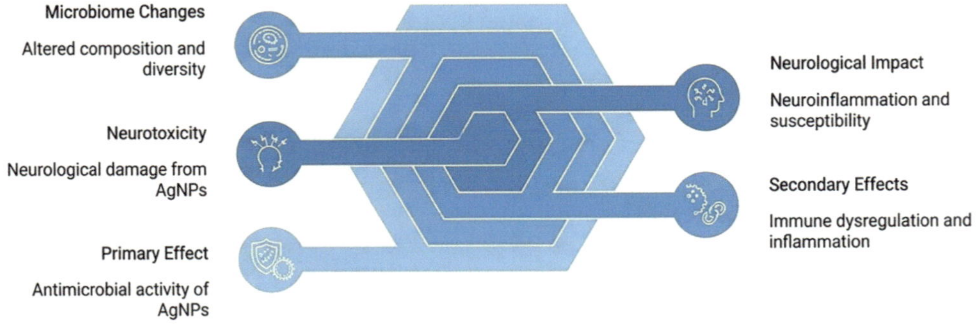

Figure 3.
AgNP-induced microbiome and gut–brain axis effects.

Integration of multi-omics datasets enables the identification of conserved pathways and vulnerability nodes, facilitating cross-species extrapolation and predictive modeling [90]. Such approaches reduce reliance on single-endpoint toxicity assays and support mechanism-based risk assessment strategies.

This integrative diagram (see **Figure 4**) illustrates conserved transcriptomic signatures induced by AgNP exposure. Oxidative stress, cellular stress responses, inflammatory signaling, and DNA damage pathways converge on a set of key regulatory genes that are consistently dysregulated across human and vertebrate models, highlighting shared molecular mechanisms underlying AgNP-induced neurotoxicity.

9.1 Limitations of current evidence

Despite the growing body of experimental and omics-based evidence on AgNP neurotoxicity, several limitations should be acknowledged. First, variability in nanoparticle physicochemical properties, exposure doses, and experimental models across studies limits direct quantitative comparisons. Second, most available transcriptomic datasets represent acute or subacute exposure scenarios, whereas chronic low-dose human exposure remains underexplored. Third, although associations between microbiome perturbation and neurotoxicity are increasingly reported, direct causal links along the gut–brain axis remain insufficiently validated. Addressing these limitations will require standardized exposure frameworks, longitudinal studies, and integrative multi-omics approaches. Additionally, another limitation of this chapter is the lack of investigation into heterogeneity in AgNP synthesis, stabilization, and surface chemistry across studies, which could be a suggestion for future review work.

10. Implications for nanomedicine and risk mitigation

The widespread use of AgNPs in biomedical applications necessitates a careful balance between therapeutic benefits and neurotoxic risks. Insights from genetic and transcriptomic studies provide actionable guidance for safer nanomaterial design.

Key mitigation strategies include the following:

- Particle size control, avoiding ultra-small AgNPs (<20 nm) where possible.
- Surface functionalization (e.g., PEGylation) to modulate protein corona composition and reduce nonspecific cellular uptake.
- Controlled dissolution kinetics to limit excessive Ag^+ release.
- Advanced *in vitro* BBB models and neurovascular coculture systems for preclinical screening [41].

Incorporating transcriptomic biomarkers into safety pipelines may enable early detection of neurotoxic signatures, improving translational predictability and regulatory decision-making.

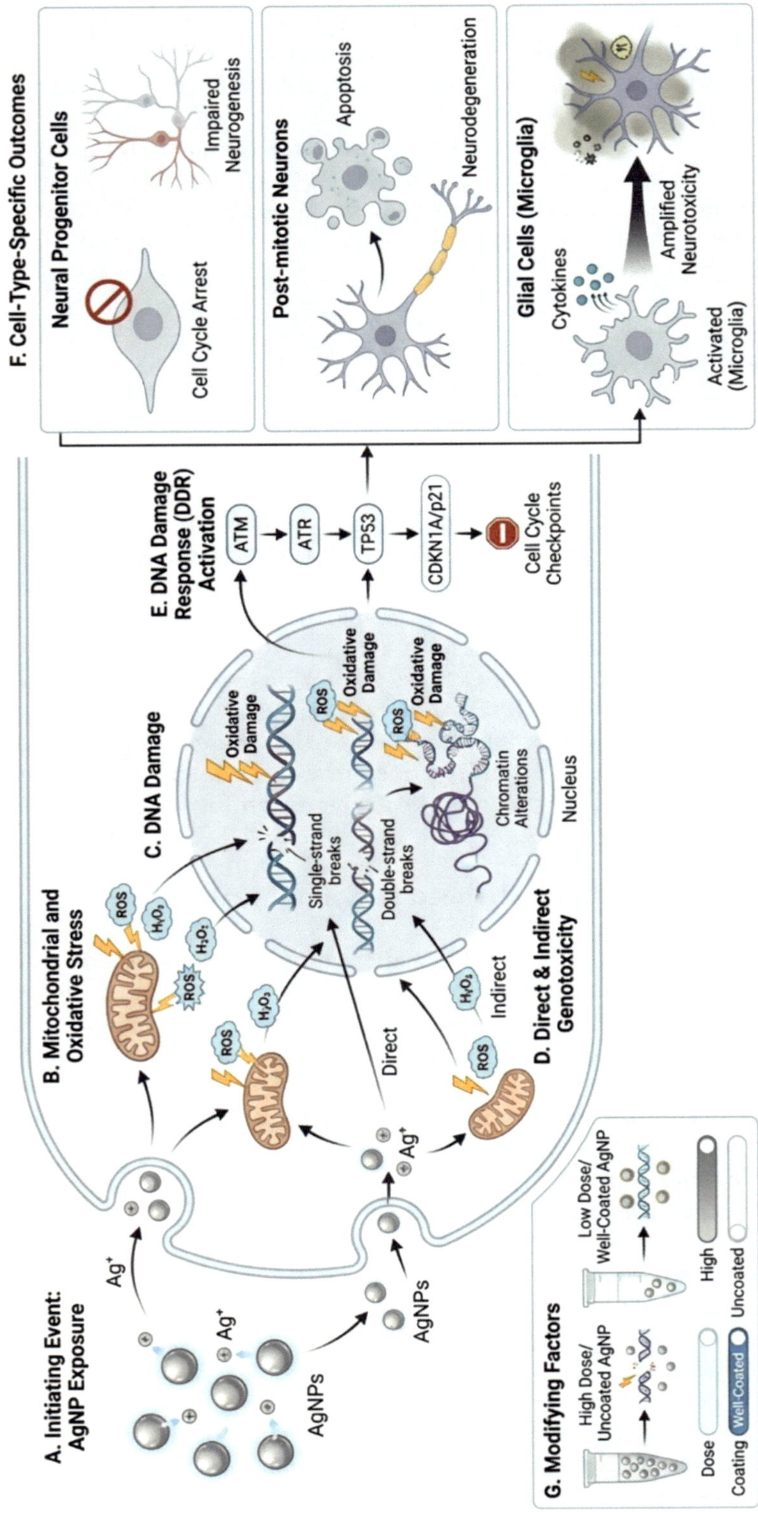

Figure 4.
Conserved genetic and transcriptomic signatures induced by AgNP exposure.

10.1 Regulatory and translational implications

The mechanistic insights presented in this chapter have important implications for regulatory toxicology and the safe development of nanomedicine. Regulatory frameworks increasingly recognize the value of omics-based biomarkers as early warning tools for toxicity assessment. Incorporating transcriptomic and mitochondrial stress signatures into safety evaluation pipelines may improve risk prediction for AgNPs. Furthermore, a "safer-by-design" strategy – considering particle size, surface coating, and dissolution behavior – can mitigate neurotoxic risks while preserving therapeutic efficacy. Such integrative approaches support evidence-based regulation and responsible clinical translation of AgNP-based technologies.

11. Conclusions and future perspectives

Silver nanoparticles induce neurotoxicity through a multifactorial and interconnected network of mechanisms involving oxidative stress, mitochondrial dysfunction, genotoxicity, neuroinflammation, and microbiome-mediated systemic effects. Genetic and transcriptomic evidence reveals conserved molecular pathways across neural cell types and vertebrate models, while also highlighting context-dependent vulnerability shaped by physicochemical properties and exposure conditions.

Future research should prioritize the following:

- Origin–pathway–outcome studies linking defined nanoparticle features (e.g., size < 20 nm) to specific molecular pathways (e.g., NLRP3 inflammasome activation) and neural outcomes.

- Chronic low-dose exposure models to complement acute toxicity paradigms.

- Advanced *in vitro* systems, including BBB-on-chip and multicellular neural models.

- Single-cell and spatial transcriptomics to resolve cell-type-specific responses within complex neural tissues.

- A mechanistic, systems-level understanding of AgNP neurotoxicity is essential for harnessing their biomedical potential while minimizing unintended neurological harm.

Acknowledgments

Authors are thankful to AL-Rayan Colleges, Al Madinah Al Munawwarah, Saudi Arabia, for their support.

The authors also acknowledge the use of "Text Guard AI"; the tool was utilized for language polishing of the chapter.

Conflict of Interest

The authors declare no conflict of interest.

Author details

Abubaker M. Hamad[1*], Mohammed Hashim Albashir[2], Abrar Khalid Aloufi[3] and Sara E. Ibrahim[4]

1 Department of General Sciences, Al-Rayan National College of Nursing, Al-Rayan National Colleges, Al Madinah Al Munawwarah, Saudi Arabia

2 Department of General Sciences, Al-Rayan National College of Health Sciences, Al-Rayan National Colleges, Al Madinah Al Munawwarah, Saudi Arabia

3 Department of Basic Medical Sciences, Al-Rayan National College of Medicine, Al-Rayan National Colleges, Al Madinah Al Munawwarah, Saudi Arabia

4 Department of Clinical Pharmacy, Al-Rayan National College of Health Sciences, Al-Rayan National Colleges, Al Madinah Al Munawwarah, Saudi Arabia

*Address all correspondence to: ab.hamad@amc.edu.sa; abkr.hamad@gmail.com

References

[1] Burduşel AC, Gherasim O, Grumezescu AM, Mogoantă L, Ficai A, Andronescu E. Biomedical applications of silver nanoparticles: An up-to-date overview. Nanomaterials (Basel). 2018;**8**(9):681. DOI: 10.3390/nano8090681

[2] González-Fernández S, Blanco-Agudín N, Rodríguez D, Fernández-Vega I, Merayo-Lloves J, Quirós LM. Silver nanoparticles: A versatile tool against infectious and non-infectious diseases. Antibiotics. 2025;**14**(3):289. DOI: 10.3390/antibiotics14030289

[3] Sati A, Ranade TN, Mali SN, Ahmad Yasin HK, Pratap A. Silver nanoparticles (AgNPs): Comprehensive insights into bio/synthesis, key influencing factors, multifaceted applications, and toxicity—a 2024 update. ACS Omega. 2025;**10**(8):7549–7582. DOI: 10.1021/acsomega.4c11045

[4] Mittal KR, Pharasi N, Sarna B, et al. Nanotechnology-based drug delivery for the treatment of CNS disorders. Translational Neuroscience. 2022;**13**(1):527–546. DOI: 10.1515/tnsci-2022-0258

[5] Fathian-Nasab MH, Manavi MA, Gelivarisarshari M, Daghighi SM, Beyer C, Baeeri M, Sanadgol N. Unveiling the hidden risks of human exposure to nanomedicine: Nanoparticle-induced blood barrier disruption and tissue toxicity. Colloids and Surfaces B: Biointerfaces. 2025;**255**:114909. DOI: 10.1016/j.colsurfb.2025.114909

[6] Bernad ES, Vasilache IA, Bernad RL, et al. Extracellular vesicles and nanoparticles in regenerative and personalised medicine: Diagnostic and therapeutic roles-A narrative review. Pharmaceutics. 2025;**17**(10):1331. DOI: 10.3390/pharmaceutics17101331

[7] Skalska J, Dąbrowska-Bouta B, Frontczak-Baniewicz M, Sulkowski G, Strużyńska L. A Low dose of nanoparticulate Silver induces Mitochondrial dysfunction and autophagy in adult rat brain. Neurotoxicity Research. 2020;**38**(3):650–664. DOI: 10.1007/s12640-020-00239-4

[8] Mello DF, Maurer LL, Ryde IT, et al. In Vivo effects of silver nanoparticles on development, behavior, and mitochondrial function are altered by genetic defects in mitochondrial dynamics. Environmental Science & Technology. 2022;**56**(2):1113–1124. DOI: 10.1021/acs.est.1c05915

[9] Park SG, Lee E, Kim HY, Yoon TH. A systems biology approach to understand temporal evolution of silver nanoparticle toxicity. NPJ Systems Biology and Applications. 2025;**11**(1):80. DOI: 10.1038/s41540-025-00561-7

[10] Huang Y, Chen R, Chen Y, Lü X. Investigation of molecular mechanisms in silver nanoparticle-induced cytotoxicity from gene to metabolite level. Scientific Reports. 2025;**15**(1):26923. DOI: 10.1038/s41598-025-12477-7

[11] Ferrer JLM, Garcia RL. Antioxidant systems, lncRNAs, and tunneling nanotubes in cell death rescue from cigarette smoke exposure. Cells. 2022;**11**(15):2277. DOI: 10.3390/cells11152277

[12] Soliman MG, Martinez-Serra A, Antonello G, Dobricic M, Wilkins T, Serchi T, Fenoglio I, Monopoli MP. Understanding the role of biomolecular coronas in human exposure to nanomaterials. Environmental Science: Nano. 2024;**11**:4421–4448. DOI: 10.1039/D4EN00488D

[13] Panico S, Capolla S, Bozzer S, Toffoli G, Dal Bo M, Macor P. Biological features of nanoparticles: Protein corona formation and interaction with the immune system. Pharmaceutics. 2022;**14**(12):2605. DOI: 10.3390/pharmaceutics14122605

[14] Teixeira MI, Lopes CM, Amaral MH, Costa PC. Navigating neurotoxicity and safety assessment of nanocarriers for brain delivery: Strategies and insights. Acta Biomaterialia. 2024;**189**:25–56. DOI: 10.1016/j.actbio.2024.09.027

[15] Desai AS, Singh A, Edis Z, et al. An In Vitro and In Vivo Study of the efficacy and toxicity of plant-extract-derived silver nanoparticles. Journal of Functional Biomaterials. 2022;**13**(2):54. DOI: 10.3390/jfb13020054

[16] Lu C, Liu Y, Liu Y, et al. Silver nanoparticles cause neural and vascular disruption by affecting key neuroactive ligand-receptor interaction and VEGF signaling pathways. International Journal of Nanomedicine. 2023;**18**: 2693–2706

[17] Batir-Marin D, Boev M, Cioanca O, et al. Exploring oxidative stress mechanisms of nanoparticles using Zebrafish (Danio rerio): Toxicological and pharmaceutical insights. Antioxidants (Basel). 2025;**14**(4):489. DOI: 10.3390/antiox14040489

[18] Opris RV, Toma V, Baciu AM, et al. Neurobehavioral and ultrastructural changes induced by phytosynthesized silver-nanoparticle toxicity in an In Vivo Rat model. Nanomaterials (Basel). 2021;**12**(1):58. DOI: 10.3390/nano12010058

[19] Alaei M, Koushki K, Taebi K, Yousefi Taba M, Keshavarz Hedayati S, Keshavarz Shahbaz S. Metal nanoparticles in neuroinflammation: Impact on microglial dynamics and CNS function. RSC Advances. 2025;**15**: 5426–5451. DOI: 10.1039/D4RA07798A

[20] Prasher P, Singh M, Mudila H. Silver nanoparticles as antimicrobial therapeutics: Current perspectives and future challenges. Biotech. 2018;**8**(10):411. DOI: 10.1007/s13205-018-1436-3

[21] Mikhailova EO. Green Silver nanoparticles: An antibacterial mechanism. Antibiotics (Basel). 2024;**14**(1):5. DOI: 10.3390/antibiotics14010005

[22] Chaudhry TS, Senapati SG, Gadam S, et al. The impact of microbiota on the Gut-Brain Axis: Examining the complex interplay and implications. Journal of Clinical Medicine. 2023;**12**(16):5231. DOI: 10.3390/jcm12165231

[23] Noga M, Milan J, Frydrych A, Jurowski K. Toxicological aspects, safety assessment, and green toxicology of Silver Nanoparticles (AgNPs)-critical review: State of the Art. International Journal of Molecular Sciences. 2023;**24**(6):5133. DOI: 10.3390/ijms24065133

[24] Ferdous Z, Nemmar A. Health impact of Silver Nanoparticles: A review of the biodistribution and toxicity following various routes of exposure.

International Journal of Molecular Sciences. 2020;**21**(7):2375. DOI: 10.3390/ijms21072375

[25] Dziendzikowska K, Wilczak J, Grodzicki W, Gromadzka-Ostrowska J, Węsierska M, Kruszewski M. Coating-Dependent neurotoxicity of Silver Nanoparticles-An In Vivo study on hippocampal oxidative stress and neurosteroids. International Journal of Molecular Sciences. 2022;**23**(3):1365. DOI: 10.3390/ijms23031365

[26] Zhang J, Wang F, Yalamarty SSK, Filipczak N, Jin Y, Li X. Nano Silver-Induced toxicity and associated mechanisms. International Journal of Nanomedicine. 2022;**17**:1851–1864. DOI: 10.2147/IJN.S355131

[27] Xu L, Wang YY, Huang J, Chen CY, Wang ZX, Xie H. Silver nanoparticles: Synthesis, medical applications and biosafety. Theranostics. 2020;**10**(20):8996–9031. DOI: 10.7150/thno.45413

[28] Rodrigues AS, Batista JGS, Rodrigues MÁV, et al. Advances in silver nanoparticles: A comprehensive review on their potential as antimicrobial agents and their mechanisms of action elucidated by proteomics. Frontiers in Microbiology. 2024;**15**:1440065

[29] Ntolia A, Chatzigiannakou T, Michailidis N, Aggeli A. A comprehensive physicochemical characterization of silver nanoparticles as a prerequisite for their successful biomedical applications. Inorganics. 2025;**13**(10):341. DOI: 10.3390/inorganics13100341

[30] Hosny S, Gaber GA, Ragab MS, et al. A comprehensive review of silver nanoparticles (AgNPs): Synthesis strategies, toxicity concerns, biomedical applications, AI-driven advancements, challenges, and future perspectives. Arabian Journal for Science and Engineering. 2025. DOI: 10.1007/s13369-025-10612-0

[31] Bélteky P, Rónavári A, Zakupszky D, et al. Are smaller nanoparticles always better? Understanding the biological effect of size-dependent silver nanoparticle aggregation under biorelevant conditions. International Journal of Nanomedicine. 2021;**16**:3021–3040

[32] Gurunathan S, Qasim M, Park C, et al. Cytotoxicity and Transcriptomic Analysis of Silver Nanoparticles in Mouse Embryonic Fibroblast Cells. International Journal of Molecular Sciences. 2018;**19**(11):3618. DOI: 10.3390/ijms19113618

[33] Mikhailova EO. Silver Nanoparticles: Mechanism of Action and Probable Bio-Application. Journal of Functional Biomaterials. 2020;**11**(4):84. DOI: 10.3390/jfb11040084

[34] Tabrizi E, Li B. Silver integrated hybrids and nanocomposites for next-generation biomedicine: Beyond antimicrobial coatings toward smart sense-response-heal platforms. Materials Today Bio. 2025;**35**:102609. DOI: 10.1016/j.mtbio.2025.102609

[35] Mosidze E, Franci G, Dell'Annunziata F, et al. Silver nanoparticle-mediated antiviral efficacy against enveloped viruses: A comprehensive review. Global Challenges. 2025;**9**(5):2400380. DOI: 10.1002/gch2.202400380

[36] Oćwieja M, Barbasz A, Wasilewska M, et al. Surface Charge-modulated toxicity of

cysteine-stabilized silver nanoparticles. Molecules. 2024;**29**(15):3629. DOI: 10.3390/molecules29153629

[37] Barbalinardo M, Chiarini F, Teti G, et al. Surface charge overrides protein corona formation in determining the cytotoxicity, cellular uptake, and biodistribution of silver nanoparticles. ACS Applied Bio Materials. 2025;**8**(6):5032–5043. DOI: 10.1021/acsabm.5c00392

[38] Mayordomo NM, Zatarain-Beraza A, Valerio F, et al. The Protein Corona Paradox: Challenges in achieving true biomimetics in nanomedicines. Biomimetics (Basel). 2025;**10**(5):276. DOI: 10.3390/biomimetics10050276

[39] Sun Y, Zhou Y, Rehman M, Wang YF, Guo S. Protein Corona of nanoparticles: Isolation and analysis. Chemical and Biological Engineering. 2024;**1**(9):757–772. DOI: 10.1021/cbe.4c00105

[40] Akhter MH, Khalilullah H, Gupta M, et al. Impact of protein corona on the biological identity of nanomedicine: Understanding the fate of nanomaterials in the biological milieu. Biomedicines. 2021;**9**(10):1496. DOI: 10.3390/biomedicines9101496

[41] Chou WC, Lin Z. Impact of protein coronas on nanoparticle interactions with tissues and targeted delivery. Current Opinion in Biotechnology. 2024;**85**:103046. DOI: 10.1016/j.copbio.2023.103046

[42] Loushambam B, Shimray MMK, Khangembam R, Krishnaswami V, Vijayaraghavalu S. Nanomedicine-Based advances in Brain Cancer Treatment—A review. Neuroglia. 2025;**6**(3):28. DOI: 10.3390/neuroglia6030028

[43] Prasanth MI, Mallya AR, Cho WC, Mundekkad D. Nano-structured strategies in combatting neurodegeneration. Frontiers in Bioengineering and Biotechnology. 2025;**13**:1638668. DOI: 10.3389/fbioe.2025.1638668

[44] Chaudhari N, Shinde S, Havelikar U, Desai S, Patel V. Recent approaches of toxicity-related attributes of NMs: A systematic review. Nano Trends. 2025;**12**:100160. DOI: 10.1016/j.nwnano.2025.100160

[45] Goshtasbi H, Hashemzadeh N, Fathi M, Movafeghi A, Barar J, Omidi Y. Mitigating oxidative stress toxicities of environmental pollutants by antioxidant nanoformulations. Nano Translational Medicine. 2025;**4**:100087. DOI: 10.1016/j.ntm.2025.100087

[46] Liu N, Liu Y, Wang Y, Feng C, Piao M, Liu M. Oxidative cell death in the central nervous system: Mechanisms and therapeutic strategies. Frontiers in Cell and Developmental Biology. 2025;**13**:1562344. DOI: 10.3389/fcell.2025.1562344

[47] Padmanaban S, Pully D, Samrot AV, et al. Rising influence of nanotechnology in addressing oxidative stress-related liver disorders. Antioxidants (Basel). 2023;**12**(7):1405. DOI: 10.3390/antiox12071405

[48] Santos I, Sousa A, Vale A, Carvalho F, Fernandes E, Freitas M. Protective effects of flavonoids against silver nanoparticles-induced toxicity. Archives of Toxicology. 2025;**99**(8):3105–3132. DOI: 10.1007/s00204-025-04068-2

[49] Jakubowska M, Costa VM, Krzeptowski W, et al. Altered NRF2 signalling in systemic redox imbalance:

Insights from non-communicable diseases. Redox Biology. 2025;**87**:103891

[50] Gambhir L, Tyagi G, Bhardwaj R, Kapoor N, Sharma G. Perturbation of Cellular Redox Status: Role of Nrf2, a Master Regulator of Cellular Redox. Biochemistry. 2022. DOI: 10.5772/intechopen.102319

[51] Zhu G, Wang X, Chen L, et al. Crosstalk between the oxidative stress and Glia Cells after stroke: From mechanism to therapies. Frontiers in Immunology. 2022;**13**:852416

[52] Dey S, Fageria L, Sharma A, et al. Silver nanoparticle-induced alteration of mitochondrial and ER homeostasis affects human breast cancer cell fate. Toxicology Reports. 2022;**9**:1977–1984

[53] Vergata C, Codogno G, De Russi G, Frigato E, Lucon-Xiccato T, Cannicci S, Bertolucci C, Fratini S. Transcriptome-wide deregulation of gene expression in zebrafish exposed to artificial light at night. Environmental Pollution. 2025;**382**:126683. DOI: 10.1016/j.envpol.2025.126683

[54] Lu C, Wu X, Meng X, Liu Y, Yang T, Zeng Y, Luo J. Silver nanoparticles exposure impairs cardiac development by suppressing the focal adhesion pathway in zebrafish. International Journal of Nanomedicine. 2024;**19**:9291–9304. DOI: 10.2147/IJN.S476168

[55] Wen H, Deng H, Li B, et al. Mitochondrial diseases: From molecular mechanisms to therapeutic advances. Signal Transduction and Targeted Therapy. 2025;**10**:9

[56] Baiskhanova D, Schäfer H. The role of Nrf2 in the regulation of mitochondrial function and ferroptosis in pancreatic cancer. Antioxidants (Basel). 2024;**13**(6):696. DOI: 10.3390/antiox13060696

[57] Zhang X, Gao Y, Zhang S, et al. Mitochondrial dysfunction in the regulation of aging and aging-related diseases. Cell Communication and Signaling. 2025;**23**(1):290. DOI: 10.1186/s12964-025-02308-7

[58] Jurcău MC, Andronie-Cioara FL, Jurcău A, et al. The Link between Oxidative Stress, Mitochondrial Dysfunction and Neuroinflammation in the Pathophysiology of Alzheimer's Disease: Therapeutic Implications and Future Perspectives. Antioxidants (Basel). 2022;**11**(11):2167. DOI: 10.3390/antiox11112167

[59] Pansera L, Mhalhel K, Cavallaro M, et al. Zebrafish as an integrative model for central nervous system research: Current advances and translational Perspectives. Life (Basel). 2025;**15**(11):1751. DOI: 10.3390/life15111751

[60] Altanam SY, Darwish N, Bakillah A. Exploring the Interplay of antioxidants, inflammation, and oxidative stress: Mechanisms, therapeutic potential, and clinical implications. Diseases. 2025;**13**(9):309. DOI: 10.3390/diseases13090309

[61] Yuan Z, Li X, Liang Z, et al. Transcriptomic analysis reveals the role of silver nanoparticles in promoting Maize germination. Plants (Basel). 2025;**14**(19):3022. DOI: 10.3390/plants14193022

[62] Chen C, Wang J, Pan D, et al. Applications of multi-omics analysis in human diseases. MedComm (2020). 2023;**4**(4):e315. DOI: 10.1002/mco2.315

[63] Kaulich E, Waselenchuk Q, Fürst N, et al. An integrated transcriptomic and proteomic map of the mouse hippocampus at synaptic resolution. Nature Communications. 2025;**16**:7942

[64] Zhang H, Xu J, Long Y, Maimaitijiang A, Su Z, Li W, Li J. Unraveling The Guardian: P53's multifaceted role in the DNA damage response and tumor treatment strategies. International Journal of Molecular Sciences. 2024;**25**(23):12928. DOI: 10.3390/ijms252312928

[65] Cibelli A, Spray DC, Mola MG. Editorial: Glial cells in homeostasis, neurodevelopment, and repair. Frontiers in Cellular Neuroscience. 2025;**19**:1575105. DOI: 10.3389/fncel.2025.1575105

[66] Yang R, Roshani D, Gao B, Li P, Shang N. Metallothionein: A Comprehensive Review of Its Classification, Structure, Biological Functions, and Applications. Antioxidants. 2024;**13**(7):825. DOI: 10.3390/antiox13070825

[67] Biswas K. Microglia mediated neuroinflammation in neurodegenerative diseases: A review on the cell signaling pathways involved in microglial activation. Journal of Neuroimmunology. 2023;**383**:578180. DOI: 10.1016/j.jneuroim.2023.578180

[68] Li X, Simo L, Zhao Q, Kim EG, Ding Y, Geng X. Endothelial cells and the Blood-Brain Barrier: Critical determinants of ineffective reperfusion in stroke. European Journal of Neuroscience. 2025;**61**(3):e16663. DOI: 10.1111/ejn.16663

[69] Gryka-Marton M, Grabowska AD, Szukiewicz D. Breaking the Barrier: The role of proinflammatory Cytokines in BBB Dysfunction. International Journal of Molecular Sciences. 2025;**26**(8):3532. DOI: 10.3390/ijms26083532

[70] Yang H, Niu S, Guo M, Xue Y. Molecular mechanisms of silver nanoparticle-induced neurotoxic injury and new perspectives for its neurotoxicity studies: A critical review. Environmental Pollution. 2024;**362**:124934. DOI: 10.1016/j.envpol.2024.124934

[71] Postel MD, Culver JO, Ricker C, Craig DW. Transcriptome analysis provides critical answers to the "variants of uncertain significance" conundrum. Human Mutation. 2022;**43**(11):1590–1608. DOI: 10.1002/humu.24394

[72] Smirnova L, Hogberg HT, Leist M, Hartung T. Revolutionizing developmental neurotoxicity testing – a journey from animal models to advanced in vitro systems. ALTEX. 2024;**41**(2):152–178. DOI: 10.14573/altex.2403281

[73] Mutalik C, Nivedita Sneka C, Krisnawati DI, Yougbaré S, Hsu-C-C, Kuo T-R. Zebrafish insights into nanomaterial toxicity: A Focused Exploration on Metallic, Metal Oxide, semiconductor, and mixed-metal nanoparticles. International Journal of Molecular Sciences. 2024;**25**(3):1926. DOI: 10.3390/ijms25031926

[74] Needs HI, Protasoni M, Henley JM, Prudent J, Collinson I, Pereira GC. Interplay between mitochondrial protein import and respiratory complexes assembly in neuronal health and degeneration. Life (Basel). 2021;**11**(5):432. DOI: 10.3390/life11050432

[75] Yang Z, Luo Y, Yang Z, et al. Mitochondrial dynamics dysfunction and neurodevelopmental disorders: From pathological mechanisms to clinical translation. Neural Regeneration Research. 2026;**21**(5):1926–1946. DOI: 10.4103/NRR.NRR-D-24-01422

[76] Maleski ALA, da Cunha, E Silva FA, Echeverry MB, Alberto-Silva C. Integrated Behavioral and Proteomic Characterization of MPP+-Induced Early Neurodegeneration and Parkinsonism in Zebrafish Larvae. International Journal of Molecular Sciences. 2025;**26**(14):6762. DOI: 10.3390/ijms26146762

[77] Safaei F, Farimaneh J, Rajabi Mohammad Abad A, et al. The effect of silver nanoparticles on learning and memory in rodents: A systematic review. Journal of Occupational Medicine and Toxicology. 2023;**18**(1):15. DOI: 10.1186/s12995-023-00381-7

[78] Maziz MNH, Chakravarthi S, Aung T, Htoo PM, Shwe WH, Gupalo S, Udayah MW, Singh H, Kabir MS, Thangarajan R, et al. Microglia-Mediated Neuroinflammation Through Phosphatidylinositol 3-Kinase signaling causes cognitive dysfunction. International Journal of Molecular Sciences. 2025;**26**(15):7212. DOI: 10.3390/ijms26157212

[79] Fang SJ, Yin ZD, Li LF, Cai Q, Zheng PF, Chen LZ. Overall effects of microplastics on brain. Frontiers in Toxicology. 2025;7:1619096. DOI: 10.3389/ftox.2025.1619096

[80] Tian C, Wu Q, Chen F, et al. Dynamic telomere length response to neurodevelopmental arsenic exposure: Insights into transcriptional regulation and neuronal morphogenesis. Environmental Health (Washington). 2025;**3**(9):1031–1042. DOI: 10.1021/envhealth.5c00026

[81] González-Vega JG, Fabian-Ortiz E, Hernández-Pérez J, Hinojosa Alvarez S, Chavez Santoscoy RA. Understanding the molecular toxicity of Metal-Based nanoparticles through nanotoxicomics. International Journal of Nanomedicine. 2025;**20**:13293–13315. DOI: 10.2147/IJN.S553276

[82] Qian J, Zhou X, Tanaka K, Takahashi A. Alteration in the chromatin landscape during the DNA damage response: Continuous rotation of the gear driving cellular senescence and aging. DNA Repair (Amsterdam). 2023;**131**:103572. DOI: 10.1016/j.dnarep.2023.103572

[83] Mohammed KR, Hussein RA, Awad HA, et al. Silver nanoparticles as a triple-action agent: Therapeutic potential in skin cancer, skin infections, and antioxidant activity. Applied Nanoscience. 2025;**15**:42

[84] Shadfar S, Brocardo M, Atkin JD. The complex mechanisms by which neurons die following DNA Damage in Neurodegenerative diseases. International Journal of Molecular Sciences. 2022;**23**(5):2484. DOI: 10.3390/ijms23052484

[85] Rogers CR, Dasari S, Patlolla AK, Tchounwou PB. Physico-chemical characterization and assessment of cytotoxic and genotoxic effects of Poly-Ethylene-Glycol coated and uncoated gold nanoparticles on Human Kidney (HK-2) cells. Austin Journal of Environmental Toxicology. 2021;7(2):1042

[86] Rohde MM, Snyder CM, Sloop J, et al. The mechanism of cell death induced by silver nanoparticles is

distinct from silver cations. Particle and Fibre Toxicology. 2021;**18**(1):37. DOI: 10.1186/s12989-021-00430-1

[87] Paciorek P, Żuberek M, Grzelak A. Products of Lipid Peroxidation as a Factor in the Toxic Effect of Silver Nanoparticles. Materials. 2020;**13**(11):2460. DOI: 10.3390/ma13112460

[88] Skóra B, Szychowski KA. Oxidative stress-driven disease-associated microglia (DAM)-like polarization in human microglia (HMC3) cells exposed to small-size silver nanoparticles in a transwell co-culture system with neurons (cholinergic differentiated SH-SY5Y) cells in vitro. Archives of Toxicology. 2026;**100**(1):207–229. DOI: 10.1007/s00204-025-04183-0

[89] García-Domínguez M. Glial Cell Dynamics in Neuroinflammation: Mechanisms, Interactions, and Therapeutic Implications. Biomedicines. 2026;**14**(1):115. DOI: 10.3390/biomedicines14010115

[90] Giovannoni F, Quintana FJ. The Role of Astrocytes in CNS Inflammation. Trends in Immunology. 2020;**41**(9):805–819. DOI: 10.1016/j.it.2020.07.007

[91] Takata F, Nakagawa S, Matsumoto J, Dohgu S. Blood-Brain Barrier Dysfunction Amplifies the Development of Neuroinflammation: Understanding of Cellular Events in Brain Microvascular Endothelial Cells for Prevention and Treatment of BBB Dysfunction. Frontiers in Cellular Neuroscience. 2021;**15**:661838. DOI: 10.3389/fncel.2021.661838

[92] Adamu A, Li S, Gao F, Xue G. The role of neuroinflammation in neurodegenerative diseases: Current understanding and future therapeutic targets. Frontiers in Aging Neuroscience. 2024;**16**:1347987. DOI: 10.3389/fnagi.2024.1347987

[93] Khan T, Waseem R, Zehra Z, et al. Mitochondrial Dysfunction: Pathophysiology and mitochondria-targeted drug delivery approaches. Pharmaceutics. 2022;**14**(12):2657. DOI: 10.3390/pharmaceutics14122657

[94] Mustafa M, Ahmad R, Tantry IQ, et al. Apoptosis: A Comprehensive overview of signaling pathways, morphological changes, and physiological significance and therapeutic implications. Cells. 2024;**13**(22):1838. DOI: 10.3390/cells13221838

[95] Skalska J, Frontczak-Baniewicz M, Strużyńska L. Synaptic degeneration in rat brain after prolonged oral exposure to silver nanoparticles. Neurotoxicology. 2015;**46**:145–154. DOI: 10.1016/j.neuro.2014.11.002

[96] Dan M, Wen H, Shao A, Xu L. Silver nanoparticle exposure induces neurotoxicity in the Rat Hippocampus Without Increasing the Blood-Brain Barrier permeability. Journal of Biomedical Nanotechnology. 2018;**14**(7):1330–1338. DOI: 10.1166/jbn.2018.2563

[97] Encinas-Gimenez M, Martin-Duque P, Martín-Pardillos A. Cellular alterations due to direct and indirect interaction of nanomaterials with nucleic acids. International Journal of Molecular Sciences. 2024;**25**(4):1983. DOI: 10.3390/ijms25041983

[98] Lazar V, Holban AM, Curutiu C, Ditu LM. Modulation of gut microbiota by essential oils and inorganic Nanoparticles: Impact in nutrition and

health. Frontiers in Nutrition. 2022;**9**:920413. DOI: 10.3389/fnut.2022.920413

[99] Wu C, Li X, Wang H, Liu Z. Altered Gut Microbial Diversity and Depletion of SCFA-Producing Taxa Associated with ASD-like Phenotypes in a Prenatal VPA Rat Model. International Journal of Molecular Sciences. 2025;**26**(18):8931. DOI: 10.3390/ijms26188931

[100] Bostick JW, Schonhoff AM, Mazmanian SK. Gut microbiome-mediated regulation of neuroinflammation. Current Opinion in Immunology. 2022;**76**:102177. DOI: 10.1016/j.coi.2022.102177

[101] Park JJ, Faustman EM. Silver nanoparticle (AgNP), neurotoxicity, and putative adverse outcome pathway (AOP): A review. Neurotoxicology. 2025;**108**:11–27. DOI: 10.1016/j.neuro.2025.02.001

[102] Ramzan U, Majeed W, Hussain AA, Qurashi F, Qamar SUR, Naeem M, Uddin J, Khan A, Al-Harrasi A, Razak SIA, et al. New insights for exploring the risks of bioaccumulation, molecular mechanisms, and cellular toxicities of AgNPs in aquatic ecosystem. Water. 2022;**14**(14):2192. DOI: 10.3390/w14142192

Section 4

Agricultural, Environmental, and Aquaculture Applications

Chapter 8

New Tools to Improve Plant Adaptation to Acute Drought Stress: Bio-Ag Nanoparticles Coated with Bacterial Elicitors. Case Study with *Pseudomonas* sp. N5.12

Svitlana Plokhovska, Ana García-Villaraco, José Antonio Lucas, Francisco Javier Gutiérrez-Mañero and Beatriz Ramos-Solano

Abstract

Forecasted increases in global population demand an increase in agronomic production, currently compromised by climate change. In this framework, the combination of plant growth-promoting rhizobacteria (PGPR)-based biofertilizers and nanotechnology has resulted in a new tool, Bio-AgNP, an Ag^0 core covered with an organic shell of bacterial metabolites endowed with biological activity; this Bio-AgNP is extremely powerful because of the active metabolites and the nanosize. This chapter summarizes the results of a two-year postdoctoral study focused on the development of an eco-friendly method for the biosynthesis of silver nanoparticles (N5.12-AgNPs) using plant growth-promoting bacteria (*Pseudomonas* sp. N5.12). The study examined the influence of physicochemical parameters, enabling the optimization of synthesis conditions and control over nanoparticle properties. Spectroscopic and microscopic analyses (UV–Vis spectroscopy, transmission electron microscopy (TEM), X-ray diffraction (XRD), Fourier transform infrared spectroscopy (FTIR), and zeta potential) confirmed the formation of stable, spherical AgNPs with strain-dependent organic coatings. The biosynthesized nanoparticles exhibited strong antimicrobial activity against human and plant pathogens, while showing no toxicity toward plants or soil microbial communities. Furthermore, they promoted plant growth, enhanced nutrient assimilation, and enhanced resilience to acute drought stress through the induction of chloroplast gene transcription and the activation of chloroplast-associated response mechanisms, among others; the response was more intense and different from the non-formulated elicitors. These findings highlight the potential of safe biogenic nanomaterials to support plant health and improve agricultural productivity under water-limited conditions.

Keywords: biological synthesis, silver nanoparticles, bacterial metabolites, antimicrobial activity, drought stress

1. Introduction

The global population is projected to approach six billion by 2050. At the same time, crop productivity is declining due to biotic and abiotic stresses and limited water resources, which slows progress in the agricultural sector. A promising approach to addressing these challenges is the application of nanotechnology [1, 2]. Nanoparticles (NPs) are extremely small particles, typically ranging from 1 nm to 100 nm in size. Their nanoscale dimensions, combined with a high surface-to-volume ratio, confer unique physical, chemical, and mechanical properties, making them particularly useful in diverse technological and agricultural applications [1]. Nanotechnology can improve agriculture and food production. Nanoparticles have a much larger surface area than regular materials, which enhances their chemical and physical interactions [3]. They are also stronger and more stable, making them more efficient for delivering nutrients, protecting crops, and improving soil and water use [4].

Nanoparticles can be synthesized through various approaches, generally categorized as top-down or bottom-up methods [1]. Among biological techniques, green synthesis using biomaterials is particularly notable, offering a rapid, cost-effective, and scalable strategy that utilizes the bioactive compounds in biomaterials to naturally reduce and stabilize NPs [5]. The use of bacterial metabolites in the synthesis of nanoparticles has emerged as a promising and sustainable strategy because these biomolecules serve as natural reducing and stabilizing agents, enabling nanoparticle formation under mild and environmentally friendly conditions. In particular, metabolites produced by plant growth-promoting bacteria (PGPB) are of growing interest, as they combine the advantages of green synthesis with biological functions that support crop productivity and stress tolerance [6]. Plant growth-promoting bacteria are known to enhance plant growth through diverse mechanisms, including phytohormone production, nutrient solubilization, and modulation of stress responses, thereby improving plant nutrient uptake and resilience in agroecosystems [7].

Metabolites from beneficial bacteria have been effectively used for the eco-friendly biosynthesis of silver nanoparticles (AgNPs), where bacterial secretions act as natural reducing and stabilizing agents. For example, AgNPs synthesized using *Pseudomonas poae* CO showed strong antifungal activity against *Fusarium graminearum*, indicating their potential for crop disease management (Ibrahim et al., 2020 [8]). AgNPs produced with metabolites from *Pseudomonas canadensis* demonstrated antibacterial effects against *Pseudomonas tolaasii*, a causative agent of mushroom brown blotch disease [9]. Additionally, the extracellular synthesis of AgNPs using *Bacillus siamensis* C1 enhanced both antimicrobial activity and plant growth promotion in rice seedlings [10]. Other PGPB, such as mixed cultures of *Lactobacillus* sp. and *Bacillus* sp., have also been reported to efficiently produce AgNPs with notable biological activity, underscoring the versatility of bacterial systems in green nanoparticle production [11].

Building on these findings, AgNPs synthesized through biological material are not only valuable for disease management but also have shown promising potential in mitigating abiotic stresses. Foliar application of AgNPs significantly alleviated drought-induced damage in *Pelargonium hortorum*, improving growth parameters and activating antioxidant defenses under water deficit conditions [12]. Seed priming with AgNPs enhanced the tolerance of *Chinese cabbage* to multiple abiotic stresses, including drought, salinity, and low temperature, by improving germination, seedling vigor, and biomass relative to hydroprimed controls [13]. In wheat, AgNP treatment increased drought resilience by promoting antioxidant activity, reducing oxidative damage, and improving photosynthetic performance through stress memory mechanisms [14]. Similarly, AgNPs have been reported to improve salinity tolerance in wheat by modulating ion homeostasis and antioxidant systems, leading to better stomatal function and stress adaptation compared with untreated plants [15]. Together, these findings indicate that AgNPs can enhance plant performance under diverse abiotic stresses, supporting their investigation in the present study using PGPB-mediated synthesis.

Therefore, metabolites from PGPB strains provide an eco-friendly route for synthesizing AgNPs with controlled size, morphology, and antimicrobial activity. Based on these data, we hypothesize that the selected *Pseudomonas* sp. N5.12 strain can efficiently produce N5.12-AgNPs with reproducible physicochemical properties, capable of protecting plants from pathogens and enhancing drought tolerance beyond the effects of the strain itself [6, 16]. To test this, optimization of synthesis conditions, characterization of the N5.12-AgNPs obtained by spectroscopic and microscopic techniques, and evaluation of their antimicrobial activity were conducted; once the most active NPs were available, toxic and physiological concentrations on plants were determined, and the physiological concentration was tested to determine effects on plant growth and stress resilience, undertaking a deep study at the metabolic and transcriptomic level. Finally, a preliminary evaluation of their impact on soil microbial communities was undertaken when delivered through the roots or leaves. These findings will help to elucidate the potential of PGPB-mediated AgNPs as safe biogenic nanomaterials for improving crop productivity under water-limited conditions.

2. *Pseudomonas*-mediated biosynthesis of AgNPs

2.1 Optimization of synthesis conditions

Optimization of synthesis parameters is critical for obtaining reproducible, stable, and functionally consistent biosynthesized NPs. Even minor deviations in reaction conditions can significantly influence NP formation, leading to variations in size, morphology, and physicochemical characteristics. The extracellular synthesis strategy was selected because it allows direct interaction between secreted metabolites and Ag^+ ions, facilitating efficient reduction and simplifying NP recovery without the need for cell disruption [17].

The beneficial strain to be used, *Pseudomonas* N5.12, has an excellent background in stimulating plant protection against pathogens and aiding plant adaptation to drought conditions [6]. With the hypothesis that bacterial metabolites would be able to reduce Ag^+ ions, the synthesis of N5.12-AgNPs was optimized by considering

Parameter	Tested range	Observed effect	Optimal
$AgNO_3$ concentration	1–5 mM	Higher concentrations increased yield but led to aggregation	1 mM
Ratio (ML:$AgNO_3$)	5:1–1:5	Variation in color intensity and UV–Vis absorption peak position	2:4
pH	5–9	Faster reduction under alkaline conditions	pH 9
Temperature	28°C, 37°C	Increased synthesis rate at higher temperatures	37°C
Incubation time	24 –72 hours	Complete reduction after one day	24 hours
Light exposure	Dark/light	No significant difference observed	Light

Table 1.
Optimization of extracellular biosynthesis conditions of N5.12-AgNPs.

several key parameters influencing nanoparticle formation, including $AgNO_3$ concentration, the ratio of bacterial supernatant (metabolites liquid [ML]) to metal precursor ($AgNO_3$), pH, temperature, and incubation time [18]. A preliminary study was carried out to determine the best incubation period to obtain bacterial metabolites (24 , 48 , and 72 hours), as metabolites produced during bacterial fermentation change depending on whether the bacteria are in the exponential or stationary phase, with 24 hours being the best incubation period.

The evaluated parameters and their corresponding effects on N5.12-AgNP synthesis are summarized in **Table 1**. Increasing the concentration of Ag^+ generally accelerates the reduction process and enhances NP yield. Several studies also indicate that higher $AgNO_3$ concentrations are associated with larger AgNPs, suggesting that precursor availability influences particle growth [19, 20]. However, when the concentration becomes too high, the stabilizing capacity of biological molecules may be exceeded, causing uncontrolled nucleation and particle aggregation, which negatively affects size uniformity and stability [21]. In our experiments, higher $AgNO_3$ concentrations (5 mM) led to noticeable particle clustering, suggesting that metabolite availability became a limiting factor for effective stabilization. Therefore, 1 mM was selected as the optimal stock concentration, providing a balance between efficient reduction and satisfactory colloidal stability [18].

Similarly, the supernatant (ML) to precursor ($AgNO_3$) ratio influenced the balance between reducing and capping biomolecules. A visible color shift from pale yellow to dark brown, along with a characteristic UV–Vis peak at 420–430 nm, confirmed Ag^+ reduction [22]. The optimal ratio (2:4, v/v) provided sufficient metabolites for efficient reduction while preventing excessive organic coating that could hinder crystallization [18]. Our data correlate with previous reports demonstrating that the bacterial extracts-to-precursor ratio is a key factor in balancing reduction efficiency and NP stabilization [23].

pH strongly affects the redox potential of biomolecules and the ionization of functional groups involved in reduction. Alkaline conditions have been shown to accelerate Ag^+ reduction in several bacterial-mediated syntheses, sometimes yielding smaller or more uniformly distributed particles [9, 24]. In contrast, acidic conditions slow the reaction and produce larger, less stable aggregates. Based on these findings, pH 9 was selected as the optimal condition in our study, as it provided fast Ag^+ reduction and produced stable, uniformly dispersed N5.12-AgNPs [18].

Temperature also played an important role in NP formation. Previous studies on microbial-mediated AgNP synthesis have reported that moderate increases in temperature enhance electron transfer, accelerate nucleation, and result in faster NP formation due to elevated metabolic or enzymatic activity [25]. In our study, at 37°C, the reduction process proceeded more rapidly and produced a stronger surface plasmon resonance signal compared to 28°C [18]. However, as noted in the literature, temperature should remain below levels where biomolecules may become denatured, which could negatively affect NP stability and synthesis efficiency [24].

Incubation time proved essential for completing the reduction process. The reaction reached equilibrium after approximately 24 hours, with no further increase in intensity [18]. Longer incubation occasionally caused minor aggregation, consistent with reports that extended reaction time can destabilize biosynthesized AgNPs [26, 27]. Thus, 24 hours was identified as the optimal incubation time, supporting complete reduction and stable N5.12-AgNP formation, particularly under light conditions.

In summary, the optimal conditions for the extracellular biosynthesis of N5.12-AgNPs using *Pseudomonas* sp. N5.12 were determined as 1 mM $AgNO_3$, a 2:4 (v/v) ML to $AgNO_3$ ratio, pH 9, incubation at 37°C, and a reaction time of 24 hours under light conditions. These optimized parameters enabled efficient and controlled AgNP synthesis, demonstrating the suitability of this strain for sustainable NP production [18].

2.2 Characterization of biosynthesized AgNPs

The N5.12-AgNPs were characterized to determine their key physicochemical properties. The characterization included analysis of particle morphology and size distribution, crystal structure, surface functional groups, and colloidal stability. When the synthesis was performed under neutral and alkaline conditions (pH 7 and pH 9), the UV–Vis spectra exhibited a distinct surface plasmon resonance band within the 400–460 nm range, confirming the successful formation of AgNPs [22, 28] . To further evaluate how pH and precursor-to-metabolite ratios influence NP architecture, TEM analysis was carried out across all tested $AgNO_3$ proportions (ML:$AgNO_3$ (v:v); S1 – 5:1, S2 – 4:2, S3 – 3:3, S4 – 2:4, and S5 – 1:5). The synthesized N5.12-AgNPs exhibited predominantly spherical morphology, with mean particle diameters ranging from 14.2 ± 0.34 nm to 16.68 ± 0.38 nm at pH 7 and from 13.75 ± 0.47 nm to 20.71 ± 0.43 nm at pH 9 [18]. Similar control over NP size and morphology has been observed in other biologically proven synthesis systems, including AgNPs produced by *Pseudomonas* spp., where mainly spherical and well-dispersed particles were reported [9, 29, 30]. The highest precursor ratio (S5) was unstable and quickly precipitated, so it was not included in further experiments. Although both pH values produced N5.12-AgNPs of comparable size, pH 9 showed clearer optical spectra; therefore, alkaline synthesis was selected. Samples S1 and S4 (see **Figure 1a, b**), representing the lowest and highest $AgNO_3$ ratios at pH 9, were selected for further physicochemical characterization.

Fourier transform infrared spectroscopy confirmed the presence of multiple functional groups associated with biomolecule-mediated synthesis. Overlapping NP profiles with metabolites from the N5.12 strain (ML) suggest a unique coating on the AgNPs (see **Figure 1c**). The similarity of the spectral characteristics and the shifts of the bands corresponding to the functional groups indicate the presence of

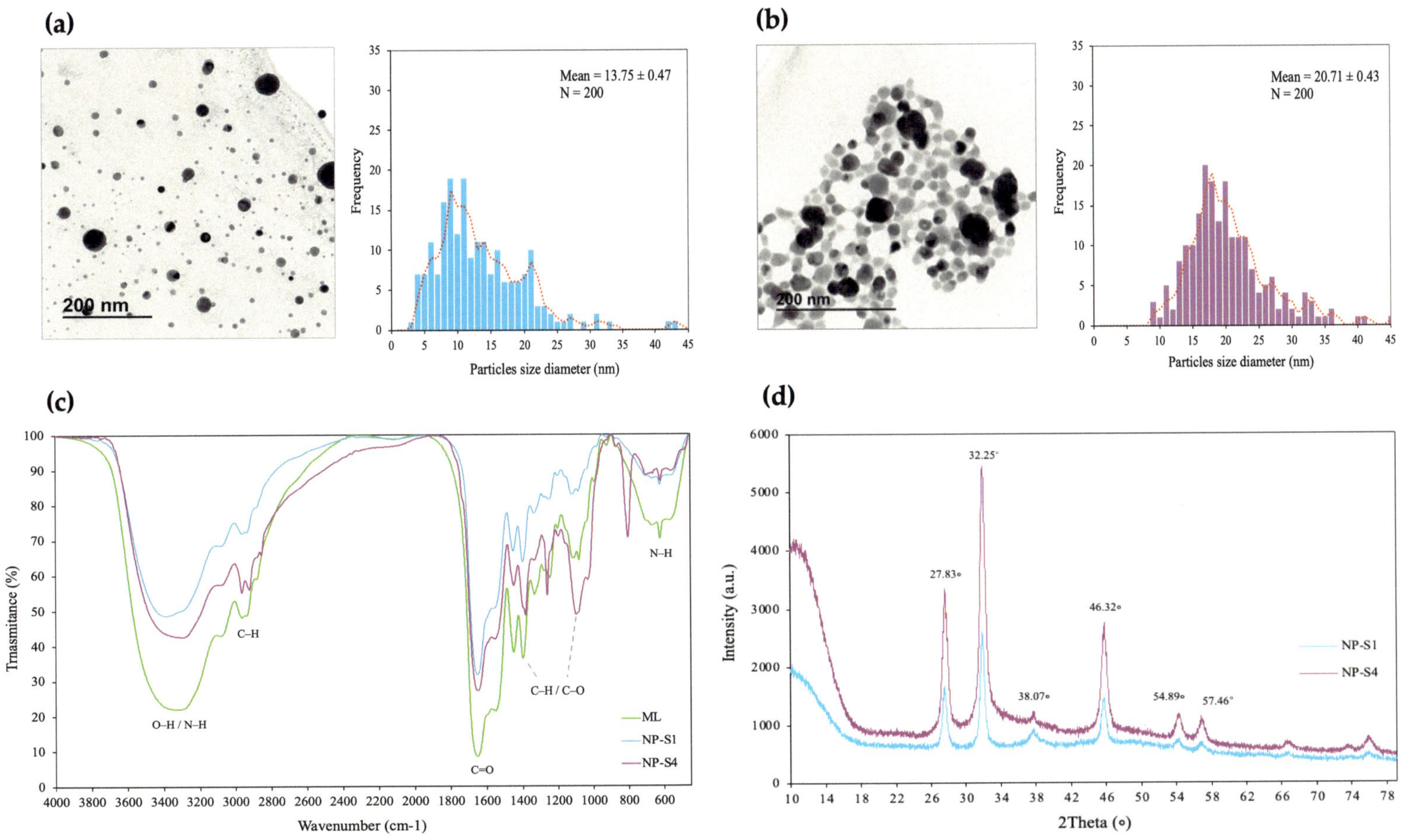

Figure 1.
TEM morphology, surface chemistry, and crystallinity of biosynthesized N5.12-AGNPs: (a–b) TEM micrographs and corresponding particle size distribution histograms of AgNPs synthesized at pH 9 for samples S1 (a) and S4 (b). The particle size distribution was statistically evaluated (n = 200) and expressed as mean ± SD. Scale bar: 200 nm. (c) FTIR spectra of ML and AgNP samples S1 and S4 showing characteristic functional groups associated with reducing and capping biomolecules. (d) XRD diffraction patterns of AgNP samples S1 and S4 confirming crystalline structure and the presence of Ag^0 *with additional AgCl phases. Source: Modified from the study by Plokhovska et al. [18], https://doi.org/10.5281/zenodo.14051744).*

biomolecules associated with the NP surface. Both S1 and S4 exhibited broad O–H stretching bands (3298–3391 cm^{-1}), strong amide-related C=O peaks (1646–1649 cm^{-1}), and characteristic C–H and C–O signals (1401–1451 and 1098–1386 cm^{-1}). A shared peak at 620 cm^{-1} indicated N–H bending, supporting the role of proteins and polysaccharides in N5.12-AgNP stabilization [18]. Similar FTIR patterns are commonly reported for biogenic AgNPs, where O–H/N–H, amide, and C–O/C–N vibrations indicate the presence of organic molecules acting as both reducing and capping agents [29, 31, 32]. These functional groups suggest the formation of a stabilizing crown, contributing to NP stability and dispersibility.

X-ray diffraction (XRD) analysis confirmed the crystalline structure of our N5.12-AgNPs, showing peaks corresponding to face-centered cubic silver (see **Figure 1d**), as well as additional signals attributed to AgCl, likely due to chloride ions present in the culture medium [18]. Comparable XRD results have also been reported in previous studies involving the microbial synthesis of AgNPs [29, 33]. AgCl phases have also been reported in biosynthesized AgNPs [32], and their presence may affect ion-release dynamics, which is important for antimicrobial activity.

Zeta potential measurements showed a negative surface charge of -26.87 ± 0.84 mV, indicating sufficient electrostatic repulsion and stable colloidal behavior. The observed charge suggests that biomolecular capping contributes to N5.12-AgNP stabilization by preventing aggregation. A zeta potential value below -25 mV is generally considered indicative of good electrostatic stabilization [34, 35], suggesting that the NPs are unlikely to aggregate under aqueous conditions.

Based on the combined characterization results, it can be concluded that the biosynthesized N5.12-AgNPs are nanosized, predominantly spherical, crystalline, and colloidally stable, with a biological organic coating [18]. These findings show that the synthesis method is reliable and produces stable NPs that may be useful for antimicrobial and other biological applications.

3. Antimicrobial activity

AgNPs are among the most extensively studied antimicrobial agents, demonstrating activity against a wide range of bacterial and fungal pathogens important to both human and plant health [36]. Two N5.12-AgNP samples were evaluated for their antimicrobial activity against human pathogenic bacterial strains, including *Staphylococcus aureus*, *S. epidermidis*, *Enterococcus* sp., *Salmonella* sp., *P. aeruginosa*, and *E. coli* [18]. The antibacterial effect was determined by measuring the inhibition zones at different NP concentrations. The inhibitory effect depended on the concentration and varied across bacterial species. S4 consistently showed slightly stronger activity than S1 (see **Figure 2a**). At the highest tested concentration, inhibition ranged from 13.3–21.6% for S1 and 11.6–31.7% for S4, depending on the strain. Comparison between Gram-positive and Gram-negative groups indicated stronger susceptibility among Gram-negative strains, with *P. aeruginosa* demonstrating the highest sensitivity (14.9 ± 0.3 mm). This conclusion is based on experimentally observed antibacterial effects (inhibition zone diameters) and not on direct mechanistic confirmation. These results agree with previous findings showing that biosynthesized AgNPs can inhibit a wide range of microorganisms, although sensitivity differs between species [37]. Gram-negative bacteria, particularly *P. aeruginosa*, often appear to be more sensitive to N5.12-

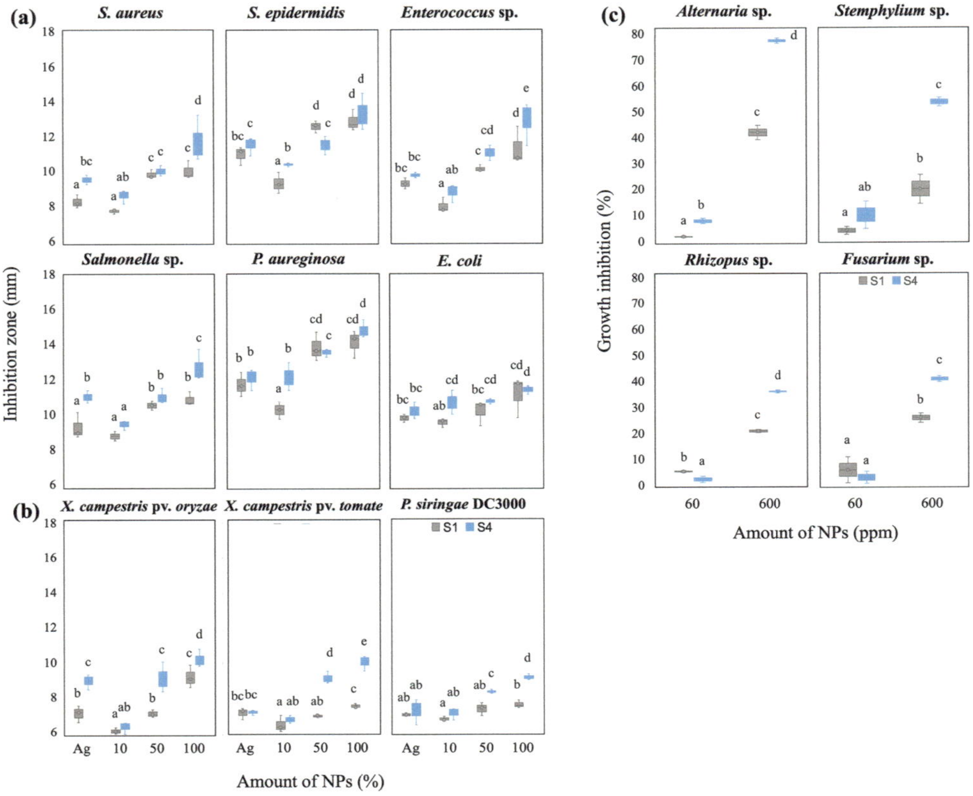

Figure 2.
Antimicrobial activity of biosynthesized N5.12-AgNPs (S1 and S4) against different pathogens: (a) human bacterial pathogens; (b) plant pathogenic bacteria; and (c) phytopathogenic fungi. Values represent mean ± SD from triplicate experiments. Different letters indicate statistically significant differences according to ANOVA (LSD, $p < 0.05$). Source: Modified from the study by Plokhovska [18], https://doi.org/10.5281/zenodo.14051744).

AgNPs, likely due to a thinner peptidoglycan layer and the presence of negatively charged lipopolysaccharides, which promote stronger interaction and uptake of NPs. The concentration-dependent pattern observed in this study is also consistent with the known dose-response relationship of metal NPs, where higher levels of AgNPs typically enhance membrane disruption, generation of reactive oxygen species (ROS), and release of Ag^+ ions [38, 39].

In addition to human pathogens, both samples N5.12-AgNP (S1 and S4) were also tested against phytopathogenic bacteria (*Xanthomonas campestris* pv. *oryzae*, *X. campestris* pv. *tomato*, and *P. syringae* DC3000), where a similar dose-dependent trend was observed, and S4 again demonstrated superior performance (see **Figure 2b**). At the highest tested concentration, sample S4 inhibited bacterial growth by 13.3% for *X. campestris* pv. *oryzae*, 39.2% for *X. campestris* pv. *tomato*, and 25.5% for *P. syringae* DC3000 relative to the silver control [18]. These results are consistent with previous studies showing that AgNPs are effective in controlling plant bacterial diseases caused by *X. oryzae*, *X. phaseoli*, *Ralstonia solanacearum*, and *Clavibacter michiganensis* [40]. Our results demonstrate species-specific differences in sensitivity, consistent with previous reports for *P. carotovorum*, *P. syringae*, and *X. campestris* [41], indicating that susceptibility to AgNPs varies among these species.

The antifungal properties of the N5.12-AgNPs were also examined against four phytopathogenic fungi (*Alternaria* sp., *Stemphylium* sp., *Rhizopus* sp., and *Fusarium* sp.). Growth inhibition varied depending on the NP type, fungal species, and concentration. Minimal effects were observed at 60 ppm (<10% inhibition), whereas strong suppression occurred at 600 ppm, particularly for sample S4 (see **Figure 2c**). Thus, *Alternaria* sp. was the most sensitive (75.7%), followed by *Stemphylium* sp. (52.6%), *Fusarium* sp. (40.5%), and *Rhizopus* sp. (35.9%) [18]. These results are consistent with previous studies that have shown that different fungi differ in their sensitivity to AgNPs. For example, *A. solani* was strongly inhibited by AgNPs, while *F. oxysporum* showed a much lower level of growth inhibition [42]. Almost complete inhibition of *A. solani* was observed at a concentration of 100 ppm AgNPs [43], while *F. solani* and *Rhizopus stolonifer* were completely inhibited at a concentration of 1000 ppm AgNPs [44]. The particularly high sensitivity of *Alternaria* sp. in this and other studies may reflect differences in cell wall architecture or metabolic activity compared to more resistant species like *Rhizopus* sp. In general, the results show that N5.12-AgNPs have strong potential as antimicrobial agents against a range of human and plant pathogens.

4. Impact on tomato plants and soil microbiota

Once the Bio-AgNPs were available, we explored safety conditions for plants and soils. In order to determine the toxic and physiological dose in plants, cytoskeleton analysis on *Arabidopsis* mutants was carried out, selecting a physiological concentration that was tested on tomato. On the other hand, as the Bio-AgNPs had antimicrobial and antifungal activity in vitro, it raised some concerns about the potential to alter soil microbial communities when delivered to plants. So, it was of paramount importance to confirm this before further studies were carried out toward a safe agronomic product. Therefore, we conducted an experiment with this purpose, delivering NPs to the roots and leaves to evaluate the impact on soil communities.

4.1 Phytotoxicity evaluation in plants

The cytoskeleton is a sensitive marker of biological activity because it quickly changes its structure in response to even small internal or external stresses [45]. Silver nanoparticles can affect the structure of microtubules and actin filaments depending on their size, shape, concentration, and exposure duration [46]. It has been shown that AgNPs reduce the number and growth rate of microtubules and may induce their aggregation [45]. Low concentrations of stabilized N5.12-AgNPs do not cause significant changes, whereas higher concentrations lead to shortening and destabilization of cytoskeleton structures, disrupting cell division and growth (see **Figure 3a**). Based on the cytoskeleton response, a physiologically active concentration of N5.12-AgNPs was determined for *Arabidopsis* [47]. The concentrations of 30 and 60 ppm were not toxic to the plant. Tomato plants exposed to N5.12-AgNPs at a concentration of 30 ppm showed generally positive physiological responses (see **Figure 3b**). Photosynthetic efficiency parameters did

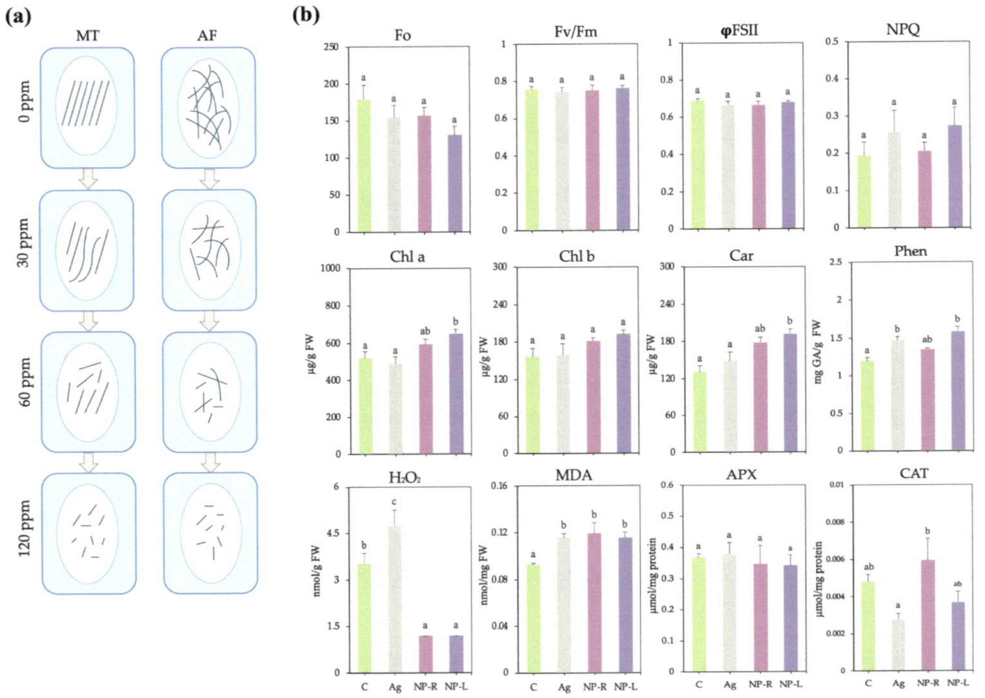

Figure 3.
Plant cellular and physiological responses to N5.12-AgNP treatment: (a) schematic visualization of cytoskeletal organization (MT and AF) in arabidopsis thaliana *cells exposed to different concentrations of NP; (b) effects of AgNPs on tomato plants, including changes in stress markers and oxidative balance. C – control; Ag – treated with 1 mM $AgNO_3$; NP–R – AgNP applied via soil drench; NP–L – AgNP applied via foliar spray. Data are presented as mean ± SD of three replicates. Different letters indicate significant differences according to ANOVA (LSD, $p < 0.05$). Source: Data are based on results from the study by Plokhovska et al. [47], https://doi.org/10.5281/zenodo.14626361.*

not differ significantly between treatments, although N5.12-AgNPs exposure led to a noticeable, non-significant increase in NPQ. Both delivery methods enhanced photosynthetic pigment content; chlorophyll a increased by 14.2% (root) and 25.2% (leaves), while carotenoids rose by 36% and 46.6%, respectively. Chlorophyll b remained statistically unchanged, and silver nitrate did not affect pigment accumulation [47]. Our results are consistent with previous studies reporting that AgNPs increase carotenoids and non-photochemical quenching (NPQ) in cotton, indicating photoprotective responses [48]. Long-term treatment of tomato plants with AgNPs (combined with mineral nutrition) altered chloroplast ultrastructure and improved Photosystem II (PSII) activity, suggesting enhanced light-use efficiency [49]. In rice seedlings, AgNPs increased chlorophyll a and carotenoid levels, which was interpreted as a mild stress response involving antioxidative mechanisms [50].

N5.12-AgNP treatment also improved redox balance in tomato, reducing H_2O_2 levels by approximately 66% compared to the control, whereas silver nitrate caused an increase of about 35% (see **Figure 3b**). Despite this reduction in oxidative stress, ascorbate peroxidase (APX) and catalase (CAT) enzyme activities remained similar to the control [47]. Silver nanoparticles have been shown in other plant systems to reduce H_2O_2 levels while increasing antioxidant enzyme activities. For example, AgNP treatment decreased H_2O_2 and enhanced APX and CAT activities in *Brassica juncea* [51]. On the other hand, low-dose AgNPs reduced ROS and H_2O_2 in *Oryza sativa* without strongly affecting superoxide dismutase (SOD), while increasing APX and glutathione reductase (GR), suggesting a reliance on the ascorbate–glutathione cycle [50].

In addition, the total phenolic content increased significantly, especially in plants treated on leaves with N5.12-AgNPs (32.5%), reaching levels similar to those seen with silver nitrate treatment [47]. These results are in agreement with other studies showing that AgNPs can boost phenolic content. For instance, in *Clinopodium nepeta*, AgNPs raised rosmarinic acid and total phenolics by 2.5- to 4.5-fold [52]. Similarly, in hairy root cultures of *Cucumis anguria*, AgNPs elicited markedly higher total phenolic, flavonoid, and specific hydroxycinnamic and hydroxybenzoic acids than Ag^+ ions [53].

On the other hand, these N5.12-AgNPs are effective even during the postharvest period, also increasing phenolic compounds of different nature. For example, when delivered to *Rosmarinus* stems after harvest, a strong metabolic remodeling consisting of increased rosmarinic acid and the triterpenes ursolic and betulinic acid was detected, associated with enhanced anti-inflammatory effects [54].

In summary, low-dose N5.12-AgNPs promote physiological and biochemical processes in tomato plants, enhancing photosynthetic pigments, photoprotective mechanisms, and phenolic compound accumulation. These results indicate that N5.12-AgNPs can act as mild biostimulants, supporting plant growth and metabolism without causing adverse effects. Furthermore, they can be used as tools to enhance bioactive contents with an impact on human health.

4.2 Effects on soil microbial activity

Evaluating rhizosphere bacterial metabolic activity helps to understand how N5.12-AgNPs influence microbial communities. The average well color development (AWCD) obtained from Biolog Eco plates indicated similar metabolic activity of the rhizosphere microbiota across all treatments, including those exposed to NPs applied to roots and leaves [47]. Carbon source utilization patterns were largely comparable, with minor differences observed only in the $AgNO_3$ treatment for a few substrates. Major compound groups were used at similar levels in all treatments, and significant variation in metabolic profiles occurred only between $AgNO_3$ and the control (see **Figure 4a**). The Shannon index confirmed no significant changes in metabolic diversity among N5.12-AgNP treatments.

These findings are consistent with recent studies reporting limited or dose-dependent effects of AgNPs on soil microbial functioning, where moderate exposure did not significantly alter microbial activity or diversity [55]. However, some evidence suggests that higher concentrations may disrupt metabolic pathways and carbon cycling processes [56]. In summary, the N5.12-AgNPs had minimal impact on bacterial function, likely due to the stability of the NP conferred by the organic matter crown preventing Ag oxidation, which would affect microbial communities, as shown by $AgNO_3$, which negatively affected the metabolic profile. Overall, at the tested concentrations, these N5.12-AgNPs appear to have limited ecological impact.

The structural diversity of rhizosphere microbial communities was assessed across treatments [47]. Bacterial alpha diversity was stable under N5.12-AgNP

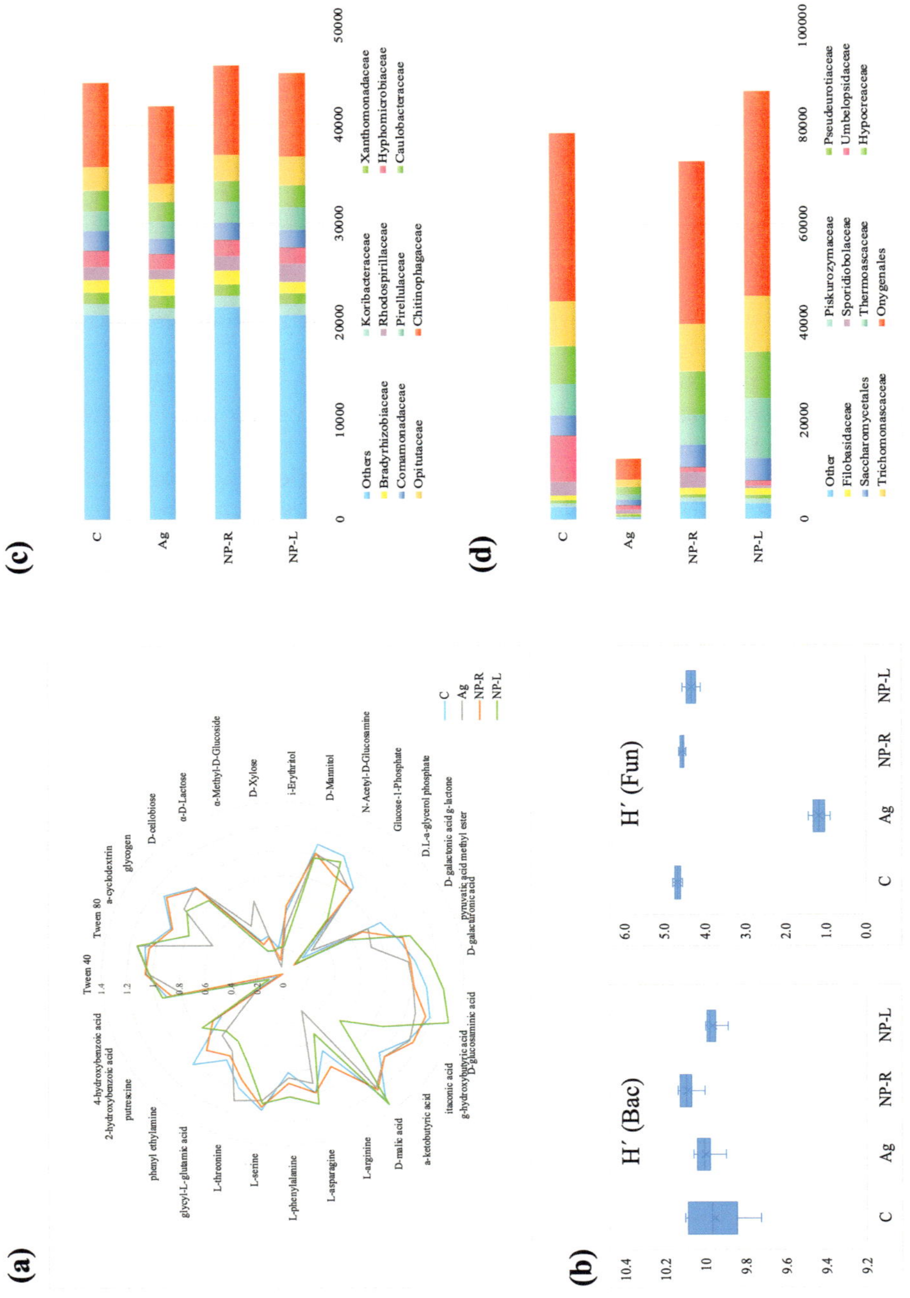

Figure 4.
Effects of N5.12-AgNP treatments on microbial community structure and functional profiles: (a) metabolic profile of the bacterial communities visualized as a radar plot, showing the mean absorbance values for the substrates in each category; (b) Shannon diversity index (H′) of bacterial and fungal communities; (c) relative abundance (%) of the top 10 bacterial families; (d) relative abundance (%) of the top 10 fungal families. Source: Modified from the study by Plokhovska et al. [47], https://doi.org/10.5281/zenodo.14626361.

treatments, while fungal diversity declined under $AgNO_3$, whereas N5.12-AgNP and controls showed similar levels (see **Figure 4b**); beta diversity showed no significant differences. The top 10 bacterial families made up 53% of the community, with *Chitinophagaceae* being the most abundant (19%), and their abundance did not differ between treatments (see **Figure 4c**). Fungal communities were dominated by the top 10 families (96%), with *Onygenales* (Fam. incertae sedis) at 43%. Ag treatment significantly reduced the abundance of all fungal families by 76–88%, whereas NP treatments (leaves or roots) had no significant negative effect; in contrast, they slightly increased abundance (see **Figure 4d**). These results are consistent with previous reports of minor effects of AgNPs on plant growth and rhizosphere bacterial structure [57], while fungi are more sensitive, showing reduced numbers and changes in their community structure in maize [58].

5. Impact of *Pseudomonas*-AgNPs on plant growth and drought tolerance

Once the safety of the N5.12-NP was determined, we explored the ability of the bacterial metabolites formulated in NP to reproduce the plant protection effects of the non-formulated metabolites (metabolic liquid-ML) against an acute drought stress event. The expected net outcome of the experimental set up was to obtain a formulation with a more potent biostimulant effect than bacterial strains or non-formulated metabolites, which would have more chances to be efficient in agronomic fields, avoiding risks associated with biostimulants based on living microorganisms, such as root colonization and alteration of native microbial communities [59].

Drought during the cropping season can be constitutive, that is, no water availability, or overcome, that is, insufficient rainfall or, in irrigated systems, punctual failure of watering systems. Irrespective of its nature, drought causes significant suppression of plant growth and yield [60]. In our experiment, this effect was reflected in reduced shoot length and lower biomass accumulation, while a significant decline in transpiration rate confirmed an early physiological response to water deficit. The stress condition also led to a significant increase in hydrogen peroxide levels and altered the activity of key antioxidant enzymes, indicating the activation of plant defense mechanisms under drought.

N5.12-AgNPs synthesized from *Pseudomonas* N5.12 at 60 ppm concentration (NP60) were tested for their potential to enhance plant resilience to drought stress. Preliminary results indicate that the ML and NP60 improve plant performance under a six-day drought period (unpublished data). It was found that both treatments improved the growth parameters of drought-stressed plants, including plant height as well as fresh and dry biomass (see **Figure 5**). Our results are consistent with the literature, which shows that NPs can positively influence plant growth by modulating key physiological and biochemical processes [61], and their protective effects under drought conditions have also been reported [62, 63].

Non-formulated bacterial metabolites (ML) induced a minor increase in photosynthetic pigment content (see **Figure 5**), whereas NP60 enhanced plant physiological activity by increasing transpiration and activating stress markers (H_2O_2 and malondialdehyde [MDA]). The accompanying reduction in water use efficiency (WUE) may represent an adaptive response that helps maintain gas exchange and metabolic activity under drought stress conditions. In addition, NP60

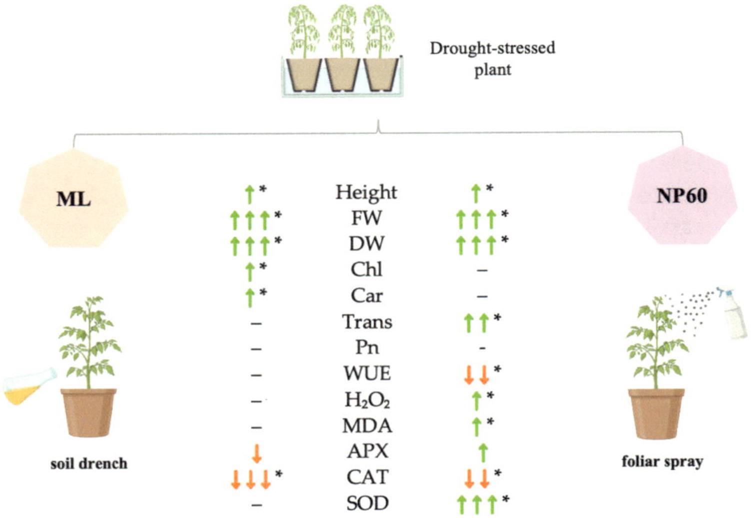

Figure 5.
Effect of ML and NP60 on plant resilience to drought stress. Arrows indicate generalized trends relative to drought-stressed control plants: ↑/↓ – weak increase or decrease (10–25% change); ↑↑/↓↓ – moderate change (25–40%); ↑↑↑/↓↓↓ – pronounced change (>40%); – no substantial change (< 10 %). Asterisks indicate significant differences according to the ANOVA test ($p < 0.05$).

treatment significantly enhanced SOD activity, while CAT activity was reduced (see **Figure 5**). These results are consistent with previous studies, demonstrating that AgNPs enhance wheat growth, antioxidant activity, and drought tolerance while mitigating oxidative stress [64]. Low doses of green-synthesized AgNPs increased SOD and glutathione peroxidase (GPOx) activities in drought-tolerant *Pimpinella anisum* L., while CAT remained unchanged, showing safe effects on antioxidant defenses [65].

As the bacterial strain had demonstrated its ability to enhance plant nutrient absorption in tomato and blueberry (patent ES3032895-A1), nutrient contents were also analyzed in this experiment. Preliminary analysis of nutrient contents in tomato under drought stress revealed that ML and NP60 treatments improved the accumulation of key macro- and micronutrients, with NP60 showing the most pronounced positive effect. The strongest responses were observed for nitrogen and phosphorus, while other essential elements, including potassium, sulfur, and calcium, exhibited consistent but less intensive increases. Magnesium and calcium also showed positive shifts, although their dynamics were comparatively moderate. This improvement in nutrient status suggests enhanced plant resilience under moisture limitation. These results align with previous studies showing that AgNPs increased phosphorus and potassium in geranium leaves, while higher concentrations and drought stress reduced nitrogen and potassium levels [12].

In addition, preliminary RNA-seq analysis in tomato under drought stress indicates differential gene expression under ML and NP60 treatments, with a group of genes uniquely responsive to these treatments but not expressed in controls. Functional categorization suggests involvement in stress response, metabolism, photosynthesis, and protein synthesis, highlighting potential molecular mechanisms underlying the observed physiological effects (unpublished data). Notably, about half of NP60-responsive genes are chloroplast- or mitochondrion-encoded, revealing a novel finding that nanoparticles can reach organelle genomes and highlighting the unique effects of NP60 (https://doi.org/10.5281/zenodo.15089625).

Proteomic analyses have revealed multiple drought-responsive proteins, such as Rubisco, chlorophyll a-/b-binding proteins, sucrose synthase, and uridine diphosphate (UDP)-glucose-6-dehydrogenase, all linked to carbon metabolism regulation. These proteins contribute to drought adaptation in *Arabidopsis*, rice, and maize [66, 67].

6. Conclusions

The eco-friendly method for AgNP synthesis using metabolites of *Pseudomonas* sp. N5.12 was optimized to allow reliable and controllable production of stable, spherical nanoparticles with natural strain-specific surface coatings. The N5.12-AgNPs are highly stable in colloids and compatible with biological systems and effectively inhibit a wide range of human and plant pathogens *in vitro*. The N5.12-AgNPs are safe for plants and soil microorganisms, do not disturb the soil microbiome, and, at the same time, act as bio-based stimulants that support plant growth, improve nutrient uptake, and increase tolerance to drought. This effect is linked to a strong chloroplast-related stress response, which differs in intensity and mechanism from responses induced by non-formulated elicitors and helps maintain photosynthesis and plant water balance under stress. Overall, the results demonstrate that N5.12-AgNPs can be used as a safe and practical tool for sustainable agriculture to protect crops, stimulate development, and strengthen plant stress resistance in environments with limited water.

Acknowledgments

The authors gratefully acknowledge the support that made this research possible. This research was funded by the European Union through HORIZON EUROPE Marie Skłodowska-Curie Actions (MSCA4Ukraine project, Grant No. 101101923).

The author acknowledges the use of ChatGPT tool for language polishing of the chapter.

Conflict of Interest

The authors declare no conflict of interest.

Author details

Svitlana Plokhovska[1,2]*, Ana García-Villaraco[1], José Antonio Lucas[1], Francisco Javier Gutiérrez-Mañero[1] and Beatriz Ramos-Solano[1]

1 Facultad de Farmacia, Universidad San Pablo-CEU, CEU-Universities, Madrid, Spain

2 Institute of Food Biotechnology and Genomics, NAS of Ukraine, Kyiv, Ukraine

*Address all correspondence to: svetaplohovska@gmail.com; bramsol@ceu.es

References

[1] Baig N, Kammakakam I, Falath W. Nanomaterials: A review of synthesis methods, properties, recent progress, and challenges. Materials Advances. 2021;**2**(6):1821–1871. DOI: 10.1039/D0MA00807A

[2] Duhan JS, Kumar R, Kumar N, Kaur P, Nehra K, Duhan S. Nanotechnology: The new perspective in precision agriculture. Biotechnology Reports. 2017;**15**:11–23. DOI: 10.1016/j.btre.2017.03.002

[3] Vijayakumar MD, Surendhar GJ, Natrayan L, Patil PP, Ram PMB, Paramasivam P. Evolution and recent scenario of nanotechnology in agriculture and food industries. Journal of Nanomaterials. 2022;**2022**. DOI: 10.1155/2022/1280411

[4] Kale SS, Chauhan R, Nigam B, Gosavi S, Chaudhary IJ. Effectiveness of nanoparticles in improving soil fertility and eco-friendly crop resistance: A comprehensive review. Biocatalysis and Agricultural Biotechnology. 2024;**56**:103066. DOI: 10.1016/j.bcab.2024.103066

[5] Fahim M, Shahzaib A, Nishat N, Jahan A, Bhat TA, Inam A. Green synthesis of silver nanoparticles: A comprehensive review of methods, influencing factors, and applications. JCIS Open. 2024;**16**:100125. DOI: 10.1016/j.jciso.2024.100125

[6] Lucas JA, Garcia-Villaraco A, Montero-Palmero MB, Montalban B, Ramos Solano B, Gutierrez-Mañero FJ. Physiological and genetic modifications induced by Plant-Growth-Promoting Rhizobacteria (PGPR) in tomato plants under moderate water stress. Biology. 2023;**12**(7):901. DOI: 10.3390/biology12070901

[7] Bianco C. Plant-Growth-Promoting Bacteria. Plants. 2024;**13**(10):1323. DOI: 10.3390/plants13101323

[8] Ibrahim E, Zhang M, Zhang Y, Hossain A, Qiu W, Chen Y, Wang Y, Wu W, Sun G, Li B. Green-synthesization of silver nanoparticles using endophytic bacteria isolated from garlic and its antifungal activity against wheat Fusarium head blight pathogen Fusarium graminearum. Nanomaterials. 2020; **10**(2):219. DOI: 10.3390/nano10020219

[9] Ghasemi S, Harighi B, Ashengroph M. Biosynthesis of silver nanoparticles using *Pseudomonas canadensis*, and its antivirulence effects against *Pseudomonas tolaasii*, mushroom brown blotch agent. Scientific Reports. 2023;**13**:3668. DOI: 10.1038/s41598-023-30863-x

[10] Ibrahim E, Fouad H, Zhang M, Zhang Y, Qiu W, Yan C, et al. Biosynthesis of silver nanoparticles using endophytic bacteria and their role in inhibition of rice pathogenic bacteria and plant growth promotion. RSC Advances. 2019;**9**(50):29293–29299. DOI: 10.1039/C9RA04246F

[11] Al-Asbahi MGSS, Al-Ofiry BA, Saad FAA, Alnehia A, Al-Gunaid MQA. Silver nanoparticles biosynthesis using mixture of *Lactobacillus* sp. and *Bacillus* sp. growth and their antibacterial activity. Scientific Reports. 2024;**14**(1):10224. DOI: 10.1038/s41598-024-59936-1

[12] Babarabie M, Zolfaghary P, Sardoei AS, Fotoohiyan Z,

Ghorbanzadeh A, Danyaei A, et al. The effect of foliar application of silver nanoparticles synthesized by *Moringa oleifera* on improving the yield and quality of *Pelargonium hortorum* under drought stress. Scientific Reports. 2025;**15**:23239

[13] Zhuang D, Zhou X, Zhao D, Wang J, Wang Y, Chen M, et al. AgNPs seeds nanopriming enhanced the tolerance of *Chinese cabbage* to diverse abiotic stresses. ACS Agricultural Science & Technology. 2025;**5**(8): 1722–1734. DOI: 10.1021/acsagscitech.5c00376

[14] Ding S, Zheng L, Tao T, Li Q, Cai J, Zhou Q, et al. Silver nanoparticles priming for drought tolerance in wheat: Insights from antioxidant system activation and stress memory. Chemical and Biological Technologies in Agriculture. 2025;**12**:57

[15] Wahid I, Kumari S, Ahmad R, Hussain SJ, Alamri S, Siddiqui MH, et al. Silver nanoparticle regulates salt tolerance in wheat through changes in ABA concentration, ion homeostasis, and defense systems. Biomolecules. 2020;**10**(11):1506. DOI: 10.3390/biom10111506

[16] Martin-Rivilla H, Gutierrez-Mañero FJ, Gradillas A, P. Navarro MO, Andrade G, Lucas JA. Identifying the compounds of the metabolic elicitors of *Pseudomonas fluorescens* N 21.4 responsible for their ability to induce plant resistance. Plants. 2020;**9**(8):1020. DOI: 10.3390/plants9081020

[17] Shah AH, Rather MA. Intracellular and extracellular microbial enzymes and their role in nanoparticle synthesis. In: Ansari MA, Rehman S, editors. Microbial Nanotechnology: Green Synthesis and Applications. Singapore: Springer; 2021. p. 41–59. DOI: 10.1007/978-981-16-1923-6_3

[18] Plokhovska S, García-Villaraco A, Lucas JA, Gutierrez-Mañero FJ, Ramos-Solano B. Silver nanoparticles coated with metabolites of *Pseudomonas* sp. N5.12 inhibit bacterial pathogens and fungal phytopathogens. Scientific Reports. 2025;**15**(1):1522. DOI: 10.1038/s41598-024-84503-z

[19] Bamsaoud SF, Basuliman MM, Bin-Hameed EA, Balakhm SM, Alkalali AS. The effect of volume and concentration of $AgNO_3$ aqueous solutions on silver nanoparticles synthesized using *Ziziphus Spina–Christi* leaf extract and their antibacterial activity. Journal of Physics: Conference Series. 2021;**1900**(1):012005. DOI: 10.1088/1742-6596/1900/1/012005

[20] Htwe YZN, Chow WS, Suda Y, Mariatti M. Effect of silver nitrate concentration on the production of silver nanoparticles by green method. Materials Today: Proceedings. 2019;**17**(3):568–573. DOI: 10.1016/j.matpr.2019.06.336

[21] Elnaggar M, Emam H, Fathalla M, Abdel-Aziz M, Zahran M. Chemical synthesis of silver nanoparticles in its solid state: Highly efficient antimicrobial cotton fabrics for wound healing properties. Egyptian Journal of Chemistry. 2021;**64**(5):2697–2709. DOI: 10.21608/ejchem.2021.57667.3236

[22] Zhang XF, Liu ZG, Liu ZG, Shen W, Gurunathan S, Zhang XF, Liu ZG, et al. Silver nanoparticles: Synthesis, characterization, properties, applications, and therapeutic approaches. International Journal of Molecular Sciences. 2016;**17**(9):1534. DOI: 10.3390/ijms17091534

[23] Solís-Sandí I, Cordero-Fuentes S, Pereira-Reyes R, Vega-Baudrit JR, Batista-Menezes D, Montes de Oca-vásquez G. Optimization of the biosynthesis of silver nanoparticles using bacterial extracts and their antimicrobial potential. Biotechnology Reports (Amsterdam, Netherlands). 2023;**40**:e00816. DOI: 10.1016/j.btre.2023.e00816

[24] Peiris MK, Gunasekara CP, Jayaweera PM, Arachchi ND, Fernando N. Biosynthesized silver nanoparticles: Are they effective antimicrobials? Memórias Do Instituto Oswaldo Cruz. 2017;**112**(8):537–543. DOI: 10.1590/0074-02760170023

[25] Bamal D, Singh A, Chaudhary G, Kumar M, Singh M, Rani N, et al. Silver nanoparticles biosynthesis, characterization, antimicrobial activities, applications, cytotoxicity and safety issues: An updated review. Nanomaterials (Basel, Switzerland). 2021;**11**(8):2086. DOI: 10.3390/nano11082086

[26] Bélteky P, Rónavári A, Igaz N, Szerencsés B, Tóth IY, Pfeiffer I, et al. Silver nanoparticles: Aggregation behavior in biorelevant conditions and its impact on biological activity. International Journal of Nanomedicine. 2019;**14**:667–687

[27] Kumar Sahu D, Sarkar P, Singha D, Sahu K. Protein-activated transformation of silver nanoparticles into blue and red-emitting nanoclusters. RSC Advances. 2019;**9**(67):39405–39409. DOI: 10.1039/c9ra06774d

[28] Iwuji C, Saha H, Ghann W, Dotson D, Bhuiya M, Parvez M, et al. Synthesis and characterization of silver nanoparticles and their promising antimicrobial effects. Chemical Physics Impact. 2024;**9**:100758

[29] Hossain A, Hong X, Ibrahim E, Li B, Sun G, Meng Y, et al. Green synthesis of silver nanoparticles with culture supernatant of a bacterium *Pseudomonas rhodesiae* and their antibacterial activity against soft rot pathogen *Dickeya dadantii*. Molecules. 2019;**24**(12):2303. DOI: 10.3390/molecules24122303

[30] Huseen DA, Ahmed ME, Hamid MK. Biosynthesis of silver nanoparticles by *Pseudomonas aeruginosa* and evaluation of their antibacterial activity and cytotoxicity assay properties. Iraqi Journal of Medical Sciences. 2024;**22**(1):154–163. Available from: https://iraqijms.net/index.php/jms/article/view/983

[31] Kalaimurugan D, Vivekanandhan P, Sivasankar P, Durairaj K, Senthilkumar P, Shivakumar MS, et al. Larvicidal activity of silver nanoparticles synthesized by *Pseudomonas fluorescens* YPS3 isolated from the Eastern Ghats of India. Journal of Cluster Science. 2019;**30**(1):225–233. DOI: 10.1007/s10876-018-1478-z

[32] Pernas-Pleite C, Conejo-Martínez AM, Marín I, Abad JP. Green extracellular synthesis of silver nanoparticles by *Pseudomonas alloputida*, their growth and biofilm-formation inhibitory activities and synergic behavior with three classical antibiotics. Molecules. 2022;**27**(21):7589. DOI: 10.3390/molecules27217589

[33] Ahsan T. Biofabrication of silver nanoparticles from *Pseudomonas fluorescens* to control tobacco mosaic virus. Egyptian Journal of Biological

Pest Control. 2020;**30**(1):66. DOI: 10.1186/s41938-020-00268-3

[34] Bhattacharjee S. DLS and zeta potential – What they are and what they are not? Journal of Controlled Release. 2016;**235**:337–351. DOI: 10.1016/j.jconrel.2016.06.017

[35] Zając M, Kotyńska J, Zambrowski G, Breczko J, Deptuła P, Cieśluk M, et al. Exposure to polystyrene nanoparticles leads to changes in the zeta potential of bacterial cells. Scientific Reports. 2023;**13**(1):9552. DOI: 10.1038/s41598-023-36603-5

[36] Kutawa AB, Ahmad K, Ali A, Hussein MZ, Abdul Wahab MA, Adamu A, et al. Trends in nanotechnology and its potentialities to control plant pathogenic fungi: A review. Biology. 2021;**10**(9):881. DOI: 10.3390/biology10090881

[37] Saravanan M, Arokiyaraj S, Lakshmi T, Pugazhendhi A. Synthesis of silver nanoparticles from *Phenerochaete chrysosporium* (MTCC-787) and their antibacterial activity against human pathogenic bacteria. Microbial Pathogenesis. 2018;**117**:68–72. DOI: 10.1016/j.micpath.2018.02.008

[38] Menichetti A, Mavridi-Printezi A, Mordini D, Montalti M. Effect of size, shape and surface functionalization on the antibacterial activity of silver nanoparticles. Journal of Functional Biomaterials. 2023;**14**(5):244. DOI: 10.3390/jfb14050244

[39] Zheng K, Setyawati MI, Leong DT, Xie J. Antimicrobial silver nanomaterials. Coordination Chemistry Reviews. 2018;**357**:1–17. DOI: 10.1016/j.ccr.2017.11.019

[40] Tariq M, Mohammad KN, Ahmed B, Siddiqui MA, Lee J, Tariq M, et al. Biological synthesis of silver nanoparticles and prospects in plant disease management. Molecules. 2022;**27**(15):4754. DOI: 10.3390/molecules27154754

[41] Trzcińska-Wencel J, Wypij M, Rai M, Golińska P. Biogenic nanosilver bearing antimicrobial and antibiofilm activities and its potential for application in agriculture and industry. Frontiers in Microbiology. 2023;**14**:1125685. DOI: 10.3389/fmicb.2023.1125685

[42] Vera-Reyes I, Altamirano-Hernández J, Reyes-de la Cruz H, Granados-Echegoyen CA, Loera-Alvarado G, López-López A, et al. Inhibition of phytopathogenic and beneficial fungi applying silver nanoparticles *in vitro*. Molecules (Basel, Switzerland). 2022;**27**(23):8147. DOI: 10.3390/molecules27238147

[43] Khatoon J, Mehmood A, Khalid AUR, Khan MAR, Ahmad KS, Amjad MS, et al. Green-fabricated silver nanoparticles from *Quercus incana* leaf extract to control the early blight of tomatoes caused by *Alternaria solani*. BMC Plant Biology. 2024;**24**(1):302. DOI: 10.1186/s12870-024-05008-5

[44] Moreno-Vargas JM, Echeverry-Cardona LM, Moreno-Montoya LE, Restrepo-Parra E, Moreno-Vargas JM, Echeverry-Cardona LM, et al. Evaluation of antifungal activity of Ag nanoparticles synthetized by green chemistry against *Fusarium solani* and *Rhizopus stolonifera*. Nanomaterials. 2023;**13**(3):548. DOI: 10.3390/nano13030548

[45] Angelini J, Klassen R, Široká J, Novák O, Záruba K, Siegel J, et al. Silver

nanoparticles alter microtubule arrangement, dynamics and stress phytohormone levels. Plants. 2022;**11**(3):313. DOI: 10.3390/plants11030313

[46] Sharma P, Chauhan NS. Effect of nanoparticles on plant cell morphology, physiology, and metabolism. In: Chauhan NS, Gill SS, editors. The Impact of Nanoparticles on Agriculture and Soil (Nanomaterial-plant Interactions) Amsterdam, The Netherlands. p. 95–113. Elsevier (Academic Press); 2023. DOI: 10.1016/B978-0-323-91703-2.00004-X

[47] Plokhovska S, García-Villaraco A, Lucas JA, Gutiérrez-Mañero FJ, Ramos-Solano B. *Pseudomonas* sp. N5.12 metabolites formulated in AgNPs enhance plant fitness and metabolism without altering soil microbial communities. Plants. 2025;**14**(11):1655. DOI: 10.3390/plants14111655

[48] Yang Y, Du P, Lai W, Yin L, Ding Y, Li Z, et al. Changes in primary metabolites and volatile organic compounds in cotton seedling leaves exposed to silver ions and silver nanoparticles revealed by metabolomic analysis. PeerJ. 2022;**10**:e13336

[49] Kulkov L, Arkhipov R, Abramova A, Vereshchagin M, Voronkov A, Khalilova L, et al. Long-term effects of silver nanoparticles and mineral nutrition components on the photosynthetic processes, chloroplast ultrastructure and productivity of *Solanum lycopersicum* plants. Journal of Photochemistry and Photobiology B: Biology. 2024;**260**:113038

[50] Gupta SD, Agarwal A, Pradhan S. Phytostimulatory effect of silver nanoparticles (AgNPs) on rice seedling growth: An insight from antioxidative enzyme activities and gene expression patterns. Ecotoxicology and Environmental Safety. 2018;**161**:624–633. DOI: 10.1016/j.ecoenv.2018.06.023

[51] Sharma P, Bhatt D, Zaidi MGH, Saradhi PP, Khanna PK, Arora S. Silver nanoparticle-mediated enhancement in growth and antioxidant status of *Brassica juncea*. Applied Biochemistry and Biotechnology. 2012;**167**(8):2225–2233. DOI: 10.1007/s12010-012-9759-8

[52] Bektaş E. Influence of silver nanoparticles on the in vitro growth, phenolic profile, antioxidant potential, enzyme inhibition, and essential oil composition of *Clinopodium nepeta* subsp. Scientific Reports. 2025;**15**(1):38744. DOI: 10.1038/s41598-025-22586-y

[53] Chung IM, Rajakumar G, Thiruvengadam M. Effect of silver nanoparticles on phenolic compounds production and biological activities in hairy root cultures of *Cucumis anguria*. Acta Biologica Hungarica. 2018;**69**(1):97–109. DOI: 10.1556/018.68.2018.1.8

[54] Gutierrez-Albanchez E, Fuente-González E, Plokhovska S, Gutierrez-Mañero FJ, Ramos-Solano B, Gutierrez-Albanchez E, et al. Enhanced anti-inflammatory effects of rosemary (*Salvia rosmarinus*) extracts modified with *Pseudomonas shirazensis* nanoparticles. Antioxidants. 2025;**14**(8):931. DOI: 10.3390/antiox14080931

[55] Chen H, Li S, Fan C, Cao J, Chen H, Li S, et al. Silver nanoparticles alter the diazotrophic community structure and co-occurrence patterns in maize rhizosphere. Agronomy. 2025;**15**(7):1601. DOI: 10.3390/agronomy15071601

[56] Hong X, Song Y, Cao D, Xu S, Gao F, Fan H, et al. Silver nanoparticles altered soil microbial activity and carbon dynamics: Unraveling microbial sensitivities and ecosystem responses [preprint]. SSRN. 2023

[57] Zarco-González KE, Valle-García JD, Navarro-Noya YE, Fernández-Luqueño F, Dendooven L, Zarco-González KE, et al. Silver and hematite nanoparticles had a limited effect on the bacterial community structure in soil cultivated with *Phaseolus vulgaris* L. agronomy. 2023;**13**(9):2341. DOI: 10.3390/agronomy13092341

[58] Zhao H, Liu Z, Han Y, Cao J, Zhao H, Liu Z, et al. Impact of silver nanoparticles on arbuscular mycorrhizal fungi and glomalin-related soil proteins in the rhizosphere of maize seedlings. Diversity. 2024;**16**(5):273. DOI: 10.3390/d16050273

[59] Herrmann MN, Griffin LG, John R, Mosquera-Rodríguez SF, Nkebiwe PM, Chen X, et al. Limitations of soil-applied non-microbial and microbial biostimulants in enhancing soil *P. turnover* and recycled *P. fertilizer* utilization-a study with and without plants. Frontiers in Plant Science. 2024;**15**:1465537

[60] Yang X, Lu M, Wang Y, Wang Y, Liu Z, Chen S. Response mechanism of plants to drought stress. Horticulturae. 2021;**7**(3):50. DOI: 10.3390/horticulturae7030050

[61] Khan S, Zahoor M, Khan RS, Ikram M, Islam NU. The impact of silver nanoparticles on the growth of plants: The agriculture applications. Heliyon. 2023;**9**(6):e16928. DOI: 10.1016/j.heliyon.2023.e16928

[62] Rasheed A, Li H, Tahir MM, Mahmood A, Nawaz M, Shah AN, et al The role of nanoparticles in plant biochemical, physiological, and molecular responses under drought stress: A review. Frontiers in Plant Science. 2022;**13**:976179

[63] Şener S, Sayğı H. The role of silver nanoparticles in response of *in vitro* boysenberry plants to drought stress. Horticulturae. 2023;**9**:1177. DOI: 10.3390/horticulturae9111177

[64] Sahoo SS, Sahu DP, Behera RK, Sahoo SS, Sahu DP, Behera RK. Drought stress mitigation in wheat seedlings via green-synthesized silver nanoparticle priming. Seeds. 2025;**4**(4):62. DOI: 10.3390/seeds4040062

[65] Gaber E, Ulusoy E. Impact of green synthesis of silver nanoparticles on antioxidant activity in drought-sensitive and drought-tolerant *Pimpinella anisum* L. International Journal of Life Sciences and Biotechnology. 2025;**8**(1):24–37. DOI: 10.38001/ijlsb.1581794

[66] Ahmad P, Abdel Latef AAH, Rasool S, Akram NA, Ashraf M, Gucel S. Role of proteomics in crop stress tolerance. Frontiers in Plant Science. 2016;**7**:1336. DOI: 10.3389/fpls.2016.01336

[67] Liu JX, Bennett J. Reversible and irreversible drought-induced changes in the anther proteome of rice (*Oryza sativa* L.) genotypes IR64 and moroberekan. Molecular Plant. 2011;**4**(1):59–69. DOI: 10.1093/mp/ssq039

Chapter 9

Advances in the Application of Green Silver Nanoparticles in Farmed Shrimp: Efficacy, Safety and Future Perspectives

Maribel Maldonado-Muñiz, Sonia A. Soto-Rodriguez, Martha Guadalupe Nieto Lopez and Mireya Tapia Salazar

Abstract

The use of silver nanoparticles (AgNPs) in aquaculture has gained growing attention due to their broad-spectrum antimicrobial properties and potential to improve animal health and production efficiency. In shrimp farming, AgNPs have been explored as alternatives to conventional antibiotics to prevent bacterial diseases such as *Vibrio* infections, which represent a major constraint in global production. This chapter reviews recent advances in the synthesis, characterization, and application of AgNPs in cultured shrimp, with particular emphasis on green synthesis using eco-friendly reducing agents. Experimental evidence demonstrates that optimized doses of AgNPs can enhance immune response, growth performance, and survival rates under pathogenic challenge. However, their use raises concerns regarding nanoparticle accumulation, oxidative stress, and possible ecological risks. The chapter discusses key findings on the toxicological thresholds, histopathological alterations, and microbiome shifts induced by AgNPs exposure. Finally, emerging trends in feed formulation, biosafety and environmental assessment, and possible regulatory frameworks are addressed to outline a sustainable path for nanoparticle integration in shrimp aquaculture. This synthesis highlights both the promising potential and the critical need for responsible innovation in the application of silver nanotechnology to ensure environmental safety and food security.

Keywords: silver nanoparticles, shrimp aquaculture, green synthesis, antimicrobial activity, nanotechnology applications

1. Introduction

Shrimp aquaculture has undergone rapid intensification over the last three decades, becoming one of the most economically significant sectors of global aquaculture. However, this growth has been accompanied by persistent and recurrent health challenges that continue to compromise productivity, biosecurity, and sustainability [1]. Among these challenges, infectious diseases remain the primary limiting factor for stable shrimp production worldwide. Bacterial diseases caused by *Vibrio* species have long been associated with mass mortalities in shrimp farming systems. More recently, the emergence of acute hepatopancreatic necrosis disease (AHPND), also known as early mortality syndrome (EMS), has caused severe production losses, particularly in *Penaeus vannamei* and *Penaeus monodon* cultures [2, 3]. In parallel, necrotizing hepatopancreatitis bacterium (NHP-B) continues to represent a chronic threat in intensive systems [4, 5]. These diseases not only reduce survival and growth rates but also destabilize production cycles, generating substantial economic uncertainty for producers.

Historically, antibiotics were widely used as prophylactic and therapeutic tools to control bacterial outbreaks in shrimp aquaculture. However, the indiscriminate use of these compounds has led to well-documented consequences, including the emergence of antimicrobial-resistant bacterial strains, accumulation of drug residues in edible tissues, and negative impacts on surrounding aquatic environments. As a result, many countries have implemented increasingly strict regulations limiting or banning the use of antibiotics in aquaculture, particularly for export-oriented production. These regulatory pressures, combined with growing consumer demand for residue-free seafood, have intensified the search for alternative disease management strategies [6, 7]. Within this context, silver nanoparticles (AgNPs) have emerged as a promising technological response rather than a transient trend. Their broad-spectrum antimicrobial activity, effectiveness at low concentrations, and reduced propensity for inducing classical resistance mechanisms distinguish them from conventional antibiotics [8, 9]. Early studies demonstrated that green-synthesized AgNPs could inhibit pathogenic *Vibrio* species and enhance shrimp survival under experimental challenge conditions, providing proof of concept for their application in aquaculture systems [10]. Importantly, the adoption of green synthesis routes, using biological extracts as reducing and stabilizing agents, addresses additional concerns related to environmental safety and biocompatibility. Consequently, the interest in green-synthesized AgNPs within shrimp farming should be understood as a response to concrete sanitary, regulatory, and technological needs. Their development reflects an effort to reconcile effective disease control with sustainability, regulatory compliance, and long-term ecosystem health, setting the foundation for the exploration of nano-enabled strategies in modern shrimp aquaculture.

This chapter presents studies reported over time to discuss the evolution of green AgNPs applications in shrimp aquaculture. In parallel, it adopts a cross-sectional analytical approach, examining plant, algal, and bacterial-mediated synthesis strategies by integrating key physicochemical characteristics with their *in vitro* applications, exposure pathways, and the scope of biological outcomes observed *in vivo*. Emphasis is placed on how these findings relate to critical toxicological and environmental biosafety thresholds. This combined chronological and mechanistic perspective ensures balanced analytical depth across sections while preserving coherence in the translational evaluation of silver-based nanomaterials as therapeutic alternatives for bacterial disease management in aquaculture systems and within emerging regulatory frameworks.

2. State of the art: Green-synthesized silver nanoparticles in shrimp aquaculture

Between 2010 and 2025, research on green-synthesized AgNPs in shrimp aquaculture has progressed in a non-linear and exploratory manner, characterized by a limited but conceptually rich body of *in vivo* study routes (see **Table 1**) supported by a broader base of *in vitro* antimicrobial evidence. Rather than representing a mature technological field, this literature reflects an emergent research trajectory in which proof-of-concept, biological plausibility, and biosafety considerations have evolved unevenly.

The first *in vivo* evidence supporting the application of green-synthesized AgNPs in shrimp culture was reported in 2010, when plant-mediated nanoparticles were shown to reduce *Vibrio harveyi*-associated mortality in juvenile shrimp *Fenneropenaeus indicus* under experimental challenge conditions [10]. These early studies established two foundational principles that continue to shape the field:

Year	Type of nanoparticles	Green synthesis source	Shrimp species	Exposure route	Main biological endpoints	Reference
2010	AgNPs	Plant extract (*Camellia sinensis*)	*Fenneropenaeus indicus*	Water exposure	Survival under bacterial challenge	[10]
2013	AgNPs	Plant extract (*Prosopis chilensis*)	*Penaeus monodon*	Water exposure	Survival and disease resistance	[11]
2016	AgNPs	Bacterial (*Bacillus subtilis*)	*Litopenaeus vannamei*	Dietary supplementation	Resistance to vibriosis	[12]
2017	AgNPs	Plant extracts (*Camellia sinensis* and *Azadirachta indica*)	*Litopenaeus vannamei*	Water exposure	Survival, following NHP-B infection	[5]
2018	AgNPs	Marine seagrass (*Cymodocea serrulate*)	*Penaeus monodon*	Water exposure	Survival and clinical signs of vibriosis	[13]
2019	AgNPs	Plant extract (*Rumex hymenosepalus*)	*Litopenaeus vannamei*	Water exposure	Survival under EMS/AHPND challenge	[14]
2020	Ag/AgCl NP	Algae extract (*Ulva clathrata*)	*Litopenaeus vannamei*	Dietary incorporation	Survival and silver depuration kinetics	[15]
2024	AgNPs	Plant extract (*Larrea tridentata*)	*Litopenaeus vannamei*	Dietary incorporation	Immune response and resistance to *Vibrio* spp.	[16]
2025	Ag/AgCl NP	Algae extract (*Ulva clathrata*)	*Litopenaeus vannamei*	Dietary incorporation	Growth performance, histopathology, microbiome, survival	[17]

Table 1.
In vivo *studies on green-synthesized AgNPs in farmed shrimp from 2010 to 2025.*

(i) biologically synthesized AgNPs retain antimicrobial activity in complex biological systems, and (ii) shrimp can tolerate short-term exposure to low doses without acute toxicity. Subsequent work in the early 2010s expanded this concept by testing different botanical reducing agents and shrimp species, confirming that the antimicrobial effect was not restricted to a single synthesis route or host species. However, these studies were primarily endpoint-driven, focusing on survival and pathogen inhibition, with limited mechanistic or toxicological resolution.

From 2016 onwards, the research emphasis shifted toward clinically relevant shrimp diseases, including vibriosis, necrotizing hepatopancreatitis (NHP-B), and early mortality syndrome (EMS), also called acute hepatopancreatic necrosis disease (AHPND). Several *in vivo* studies during this period demonstrated that green-synthesized AgNPs could reduce disease severity, improve survival, or enhance resistance when administered through water exposure or dietary routes. Notably, these studies began to acknowledge the dual role of AgNPs as antimicrobial and immunomodulatory agents. Nevertheless, most experimental designs still lacked integration across biological scales, as growth performance, immune markers, histopathology, and nanoparticle fate were rarely assessed simultaneously. This fragmentation limited the capacity to extrapolate laboratory findings to scaled production applications.

The most recent studies from 2020 to 2025 represent a qualitative shift toward systems-level evaluation. Dietary incorporation of green-synthesized AgNPs has been explored alongside endpoints such as growth, immune response, histological integrity, microbiome composition, and depuration kinetics routes. These works collectively highlight that biological effects are dose-dependent and context-specific, with beneficial outcomes observed within narrow exposure windows. Importantly, these studies also expose emerging concerns related to nanoparticle accumulation, sublethal tissue alterations, and microbiome modulation. As such, the narrative of AgNPs application has evolved from "antibiotic alternative" toward a more cautious framework of functional nanomaterials requiring rigorous risk–benefit assessment.

3. Green synthesis and characterization of silver nanoparticles: Principles and biological rationale

The green synthesis of AgNPs has emerged as a biologically inspired alternative to conventional chemical and physical methods, driven by the need to reduce environmental impact while enhancing biocompatibility for aquaculture applications. In the context of shrimp farming, green-synthesized AgNPs are particularly attractive due to their potential dual role as antimicrobial agents and immunomodulatory compounds, coupled with reduced reliance on toxic reagents. Green synthesis typically involves biological entities capable of acting simultaneously as reducing, stabilizing, and capping agents. These include plant extracts, marine macroalgae or seagrass, and bacterial systems, each conferring distinct physicochemical signatures to the resulting nanoparticles. Importantly, the biological origin of the reducing matrix has been shown to influence nanoparticle size, morphology, surface charge, and colloidal stability, parameters that ultimately condition biological performance (see **Figure 1**).

The presence of an organic corona on green-synthesized AgNPs plays a central role in modulating silver ion (Ag^+) release. Biomolecules adsorbed on the nanoparticle

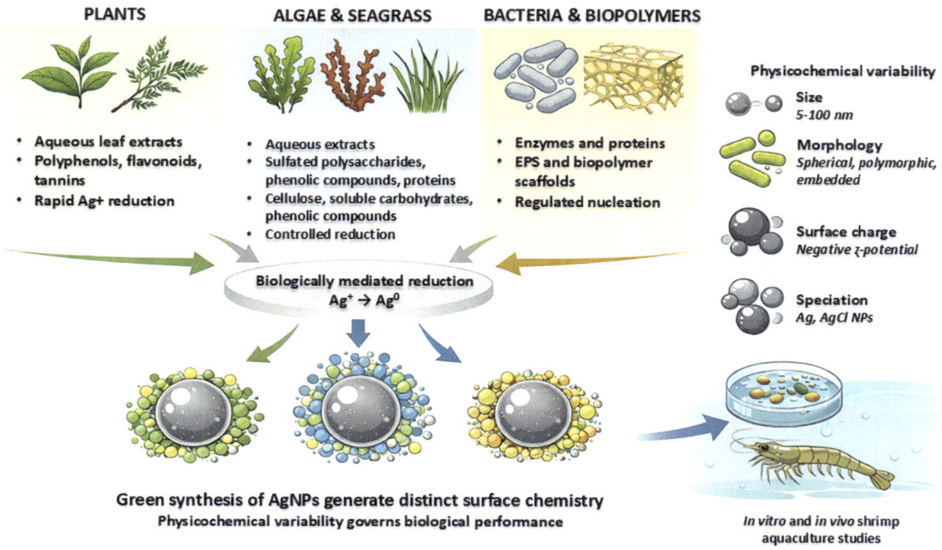

Figure 1.
Green synthesis routes of silver nanoparticles using plant, marine, and bacterial systems, where reducing and capping agents generate distinct organic coronas, leading to physicochemical variability in size, surface chemistry, and silver speciation. EPS: extracellular polymeric substances.

surface act as diffusion barriers, slowing Ag^+ dissolution and reducing acute cytotoxicity when compared with uncapped or chemically synthesized AgNPs [5, 10]. This moderated release has been associated with lower oxidative stress and improved tolerance *in vivo*, while still maintaining antimicrobial activity *in vitro* [15].

3.1 Green sources for AgNPs synthesis

3.1.1 Plant-mediated systems

Plant-derived systems represent the most extensively explored platforms for the green synthesis of AgNPs, owing to their chemical diversity, accessibility, and versatility. Across the literature, AgNPs have been synthesized using extracts obtained from a wide range of terrestrial and coastal plant species, including *Camellia sinensis* (tea leaves), *Prosopis chilensis, Avicennia marina*, *Larrea tridentata*, and *Rumex hymenosepalus*, among others [10, 11, 13, 14, 16]. Most plant-mediated syntheses employ aqueous extracts of leaves, although stems, roots, bark, and whole-plant materials have also been reported. Water is the predominant extraction solvent, favoring the recovery of polar phytochemicals such as polyphenols, flavonoids, tannins, reducing sugars, and organic acids, which act as primary electron donors during Ag^+ reduction. In some studies, hydroalcoholic or ethanolic extracts have been used to enhance the solubilization of less polar compounds, including terpenoids and certain alkaloids, thereby modifying reduction kinetics and nanoparticle morphology [18, 19]. These phytochemicals not only reduce silver ions but also adsorb onto the nanoparticle surface, forming a stabilizing organic corona that governs nucleation, particle growth, and aggregation behavior. In plant-mediated systems, the concentration and relative abundance of these biomolecules strongly influence reaction kinetics: rapid reduction typically favors the formation of smaller, quasi-spherical, and polymorphic AgNPs, whereas slower or heterogeneous reduction processes often yield broader size distributions and irregular morphologies. Variability in extract preparation, including solvent polarity, extraction temperature, pH, plant tissue type, and phenological stage, further contributes to inter-study heterogeneity, as reflected in the wide range of particle sizes and surface chemistries reported across both *in vitro* and *in vivo* investigations [10, 13, 16].

Plant-mediated synthesis resulted in AgNPs capped with polyphenols and flavonoids, which possess intrinsic antioxidant properties. These compounds can scavenge reactive oxygen species (ROS) generated during AgNPs–cell interactions, partially mitigating oxidative stress in host tissues *in vivo* while preserving antibacterial efficacy *in vitro* [11]. This dual antimicrobial–antioxidant functionality represents a distinctive advantage of plant-mediated AgNPs in aquaculture settings.

3.1.2 Algae and seagrass-mediated systems

Marine macroalgae and seagrass constitute chemically distinctive platforms for the green synthesis of AgNPs, particularly relevant for aquaculture-oriented research due to their natural adaptation to saline and ion-rich environments. These marine systems differ substantially in taxonomy and biochemical composition, differences that are directly reflected in the physicochemical properties of the resulting nanoparticles [13, 19, 20].

Algae-based AgNPs synthesis has been reported using brown, red, and green algal taxa, whose extracts are rich in structurally complex polysaccharides, phenolic compounds, proteins, and, in many cases, sulfated macromolecules. Aqueous extracts are predominantly employed, enabling the solubilization of alginates and fucoidans in brown algae, carrageenans, and agarans in red algae, and ulvans in green algae. These compounds act simultaneously as mild reducing agents and as effective capping materials, promoting relatively slow and controlled nucleation processes that generally yield spherical or polymorphic nanoparticles with comparatively narrow size distributions [20, 21]. The high density of sulfate and carboxyl functional groups associated with algal polysaccharides confers a strongly negative surface charge to algae-derived AgNPs, often reflected in highly negative zeta potential values. This electrostatic stabilization enhances colloidal stability under high-ionic-strength conditions and contributes to the frequent selection of algal systems in studies targeting marine or aquaculture applications. In some cases, interaction between silver ions and chloride-rich marine matrices also favors the partial formation of Ag/AgCl phases, further modifying surface chemistry and nanoparticle behavior [15, 17].

Seagrass-mediated synthesis, while sharing methodological similarities, relies on a chemically simpler matrix. Seagrass species contain cellulose, hemicellulose, soluble carbohydrates, and phenolic compounds, with a lower contribution of sulfated polysaccharides [19]. Aqueous leaf extracts are typically used, resulting in slower and less uniform reduction kinetics compared with macroalgal systems. Consequently, seagrass-derived AgNPs often display broader size distributions and more heterogeneous morphologies, ranging from spherical to irregular forms [13]. Surface functionalization in seagrass-derived nanoparticles is dominated by hydroxyl and carboxyl groups associated with cellulose derivatives and phenolic residues. While these groups provide steric stabilization, the lower sulfate content reduces electrostatic repulsion relative to algae-derived AgNPs, potentially increasing aggregation propensity under saline conditions. This contrast highlights how relatively subtle differences in biological composition translate into pronounced differences in nanoparticle surface chemistry, colloidal behavior, and physicochemical stability [13, 22].

Overall, algae- and seagrass-mediated green synthesis approaches illustrate the chemically rich yet inherently heterogeneous nature of marine-derived nanoparticle synthesis platforms. Variability in species, tissue type, extraction conditions, seasonal biochemical composition, and reaction parameters results in AgNPs spanning a wide size range, typically from below 10 nm to nearly 100 nm, with diverse surface chemistries. These features underscore the importance of detailed and standardized characterization when comparing nanoparticles synthesized from different marine sources and when interpreting downstream behavior in aquaculture-relevant environments [13, 15, 17, 22].

3.1.3 Bacteria and biopolymer-mediated systems

Bacterial and biopolymer-mediated systems constitute a comparatively more structurally controlled platform for the green synthesis of AgNPs, as microbial metabolism and natural polymer matrices provide defined biochemical environments for silver ion reduction and nanoparticle stabilization. In these systems, Ag^+ reduction is mediated by intracellular or extracellular enzymes (e.g., reductases), peptides, proteins, and other metabolites capable of donating electrons and coordinating silver nuclei during particle formation [12]. Bacterial synthesis is typically carried out using cell-free culture supernatants, biomass suspensions, or extracellular polymeric substances (EPS) in aqueous media. Compared with plant- or algal-derived extracts, bacterial systems tend to produce AgNPs with narrower size distributions and more uniform, predominantly spherical morphologies, reflecting the regulated nature of microbial redox processes. Reported particle sizes commonly fall within the 5–40 nm range, although strain-specific metabolism, growth phase, and culture conditions introduce variability [23]. Surface characterization studies consistently show that bacterial proteins, amino acids, and polysaccharides adsorb onto AgNP surfaces, forming an organic corona enriched in amide, carboxyl, and hydroxyl functional groups. FTIR analyses frequently confirm the presence of peptide and polysaccharide signatures, supporting their role as capping and stabilizing agents. This corona limits aggregation and modulates surface charge, often yielding moderately negative zeta potential values that enhance colloidal stability in aqueous systems [19, 23].

Biopolymer-assisted synthesis extends microbial approaches by incorporating natural polymers as chemical stabilizers and physical scaffolds. Bacterial cellulose has been explored due to its nanofibrillar architecture and high density of hydroxyl groups capable of coordinating silver ions. In these systems, AgNPs nucleation occurs within or on the polymer matrix, constraining particle growth and producing AgNPs–biopolymer composites rather than freely dispersed nanoparticles [24, 25]. While such composites exhibit enhanced structural stability and controlled silver release, their effective surface chemistry and apparent particle size are strongly influenced by the polymer matrix, complicating direct comparison with colloidal AgNPs synthesized via plant or algal routes [19]. Despite improved morphological control, bacterial and biopolymer-mediated systems remain sensitive to culture conditions, polymer purity, and post-synthesis processing. Residual biological macromolecules further contribute to surface heterogeneity, underscoring the need for comprehensive physicochemical characterization beyond particle size and morphology alone [24–26].

3.2 Characterization and physicochemical variability

Physicochemical characterization is essential for evaluating AgNPs, as their properties are intrinsically shaped by the biological matrices involved in synthesis. Unlike chemically synthesized nanoparticles, green AgNPs are formed in complex biochemical environments that introduce variability in size, morphology, surface chemistry, and silver speciation. As a result, particle diameter alone is insufficient to describe their behavior, and a multi-technique approach is required [10,14].

UV–visible spectroscopy is routinely used to confirm AgNPs formation through the detection of the surface plasmon resonance (SPR) band, typically between 400 and 450 nm. Shifts in peak position and bandwidth reflect differences in particle size distribution, aggregation state, and the surrounding organic corona [18, 27]. Transmission electron microscopy (TEM) remains the principal method for assessing core size and morphology, with most green-synthesized AgNPs exhibiting spherical or quasi-spherical shapes and reported diameters ranging from approximately 5 to 100 nm [22, 23, 26]. Dynamic light scattering (DLS) measurements commonly yield larger hydrodynamic diameters than TEM due to the contribution of capping biomolecules and particle agglomeration in suspension. Zeta potential analyses further indicate that green-synthesized AgNPs typically exhibit neutral to moderately negative surface charges, with algae-derived nanoparticles often showing more negative values owing to sulfated polysaccharides and carboxyl-rich macromolecules [15, 21]. Surface chemistry is most frequently characterized by Fourier-transform infrared spectroscopy (FTIR), which consistently reveals hydroxyl, carboxyl, amide, and sulfate functional groups associated with biological reducing and capping agents. These biomolecules form an organic corona around the metallic silver core, governing colloidal stability and aggregation behavior. Importantly, the composition of this corona varies markedly depending on whether plant metabolites, algal polysaccharides, or bacterial proteins dominate the synthesis process, contributing to substantial inter-study variability even among nanoparticles of similar nominal size [16, 22]. X-ray diffraction (XRD) analyses generally confirm the crystalline face-centered cubic (fcc) structure of metallic silver. Additional diffraction signals corresponding to AgCl phases have been reported, indicating partial transformation or co-existence of AgNPs and Ag/AgCl species, particularly in algal-mediated syntheses conducted under chloride-rich conditions [15, 17].

Overall, the studies summarized demonstrate that physicochemical variability is an inherent characteristic of green-synthesized AgNPs rather than an experimental limitation. Differences in biological source, extraction protocol, reaction conditions, and post-synthesis processing collectively shape nanoparticle properties, underscoring the need for comprehensive and standardized characterization to support meaningful comparison and translational interpretation.

3.3 Synthesis, structure, and functional performance outcomes

Recent advances in green, bio-assisted, and hybrid synthesis methods have moved beyond the initial objective of producing nanoparticles through highly sustainable processes, evolving toward the controlled modulation of nanoparticle structure and surface chemistry. These modifications directly influence their biological performance [19, 27, 28]. In plant-mediated systems, the abundance of polyphenols and flavonoids not only regulates reduction kinetics and particle morphology but also results in surface coatings with intrinsic antioxidant capacity. This dual functionality can mitigate oxidative stress in vivo while maintaining antimicrobial efficacy *in vitro* [16, 29]. Similarly, algae-mediated synthesis incorporates sulfated polysaccharides that enhance colloidal stability in saline environments and may contribute to immunomodulatory responses in shrimp, thereby favoring functional persistence under culture conditions [15, 30].

By contrast, bio-assisted and hybrid systems mediated by bacteria and biopolymers provide greater control over particle size distribution and morphology through enzymatically regulated reduction processes [24]. The resulting protein-rich coatings modulate silver ion release and reduce acute toxicity, allowing a more predictable balance between efficacy and safety [5, 13]. Hybrid systems, including the partial formation of Ag/AgCl phases in marine environments, further illustrate how compositional control modifies dissolution kinetics and long-term biological behavior [15, 17]. These examples demonstrate that improved performance in real aquaculture applications does not depend solely on antimicrobial potency; rather, the deliberate tuning of particle size, surface chemistry, colloidal stability, and silver speciation is crucial [10]. Therefore, translating advances in synthesis into truly effective functional performance requires understanding and optimizing structure–property–performance relationships under environmentally relevant conditions.

4. Application of silver nanoparticles in shrimp aquaculture

The application of AgNPs in shrimp aquaculture has emerged as a technological response to recurrent bacterial disease outbreaks and the increasing regulatory constraints on antibiotic use [7]. Green-synthesized AgNPs have gained particular attention due to their broad antimicrobial activity combined with a lower environmental footprint compared to conventionally synthesized nanoparticles. However, despite their growing experimental use, AgNPs-based interventions remain at an exploratory stage and require careful evaluation of both their potential benefits and inherent limitations. From a functional perspective, AgNPs exhibit broad-spectrum antibacterial activity and a relatively low propensity for inducing classical antibiotic resistance. Their nanoscale dimensions and surface chemistry enable multiple modes of action, while green synthesis routes often generate biologically derived surface coronas that enhance colloidal stability and may improve compatibility with aquatic organisms. Furthermore, AgNPs can be delivered through diverse application strategies, including waterborne exposure, surface coatings, and dietary supplementation, providing flexibility for experimental and applied use [8, 30, 31]. At the same time, significant uncertainties remain. Antimicrobial performance is strongly dose-dependent, and suboptimal or poorly controlled concentrations can lead to inconsistent outcomes. Moreover, the same physicochemical properties that underpin antimicrobial efficacy also raise concerns regarding silver accumulation, oxidative stress, and unintended effects on host tissues and associated microbiota. Variability in synthesis protocols and characterization further complicates cross-study comparison and standardization. These considerations highlight the need to clearly distinguish between antimicrobial potential demonstrated under simplified conditions and biological performance under realistic farming scenarios.

4.1 In vitro application of AgNPs against shrimp-associated pathogens

In vitro studies constitute the foundational body of evidence supporting the antimicrobial potential of AgNPs in shrimp aquaculture. Most investigations have focused on evaluating the activity of green-synthesized AgNPs against major bacterial pathogens, particularly species of the genus Vibrio. Target organisms commonly include *Vibrio harveyi, Vibrio parahaemolyticus, Vibrio alginolyticus*, and *Vibrio vulnificus*, which are closely associated with vibriosis and acute hepatopancreatic necrosis disease (AHPND). Using standard microbiological assays, such as minimum inhibitory concentration (MIC), minimum bactericidal concentration (MBC), agar diffusion, and biofilm inhibition tests, these studies consistently report effective bacterial growth suppression at relatively low nanoparticle concentrations [10, 11, 31, 32]. AgNPs have demonstrated activity against *Vibrio spp.,* frequently exceeding the antibacterial performance of the corresponding crude extracts. In several cases, enhanced efficacy has been attributed to the combined effects of the metallic core and the biologically derived capping layer [33]. While these findings establish antimicrobial promise and provide mechanistic insight, *in vitro* systems are inherently limited. They do not capture nanoparticle transformation in seawater, interactions with organic matter, host immune responses, or toxicity thresholds. As such, *in vitro* data alone are insufficient to predict safety or efficacy under farming-relevant conditions, reinforcing the need for controlled *in vivo* validation.

4.2 In vivo application of AgNPs in cultured shrimp

In contrast to the extensive *in vitro* literature, the *in vivo* applications of AgNPs in shrimp remain comparatively limited but are essential for assessing biological relevance, safety, and translational potential. Experimental studies have primarily focused on economically important cultured species, including *Litopenaeus vannamei*, *Penaeus monodon*, and *Fenneropenaeus indicus*.

Initial *in vivo* investigations demonstrated that green-synthesized AgNPs could reduce mortality in shrimp challenged with *Vibrio harveyi*, providing early proof-of-concept for their protective capacity [10]. Subsequent studies expanded this approach to include dietary supplementation and controlled exposure experiments. In *P. monodon* and *L. vannamei*, plant and algae-derived AgNPs were associated with improved survival under bacterial challenge and modulation of immune-related parameters, alongside reduced disease severity linked to vibriosis and AHPND [5, 13, 14]. More recent work has adopted a more integrative assessment framework, particularly in *L. vannamei*, evaluating not only survival but also growth performance, histopathological alterations, silver accumulation, and microbiome responses following exposure to algal-mediated AgNPs. These studies revealed that positive outcomes are tightly linked to nanoparticle dose, formulation, and exposure route, and that excessive or poorly controlled applications can induce sublethal effects, highlighting the narrow margin between efficacy and risk [15, 17]. *In vivo* evidence supports the potential of green-synthesized AgNPs as antimicrobial or immunomodulatory tools in shrimp aquaculture, while simultaneously underscoring the need for careful dose optimization, comprehensive biosafety assessment, and standardized experimental design. These observations provide the necessary framework for a more detailed discussion of toxicological thresholds, tissue-level alterations, and microbiome dynamics associated with AgNPs exposure.

5. Toxicological thresholds, histopathological alterations, and microbiome shifts

The increasing interest in AgNPs as functional tools in shrimp aquaculture has been accompanied by a growing recognition that antimicrobial efficacy must be evaluated alongside biosafety and long-term biological effects. While green-synthesized AgNPs are often perceived as inherently safer than chemically synthesized counterparts, experimental evidence indicates that their biological impact is strongly dependent on dose, exposure route, nanoparticle speciation, and physicochemical properties [19].

5.1 Toxicological thresholds and dose dependency

Across *in vivo* studies conducted in cultured shrimp, AgNPs exhibit a narrow window between beneficial and adverse effects. Low to moderate doses have been associated with improved survival under pathogenic challenge and limited physiological disruption, whereas higher concentrations frequently induce sublethal stress responses. Toxicological thresholds are influenced by particle size, surface chemistry, and the presence of organic capping agents, which modulate silver ion release and bioavailability [10, 17, 34, 35]. Importantly, dietary exposure studies suggest that chronic low-dose ingestion may result in silver accumulation in hepatopancreatic tissues, even in the absence of acute toxicity [17]. This accumulation raises concerns regarding physiological effects and potential carry-over into edible tissues, emphasizing the need for exposure scenarios that reflect realistic farming conditions rather than short- and long-term laboratory trials.

5.2 Histopathological alterations

Histopathological assessment has emerged as a critical endpoint for evaluating AgNPs biosafety in shrimp. Studies reporting beneficial outcomes at the organismal level have nevertheless identified dose-dependent structural alterations in the hepatopancreas, including epithelial degeneration, tubular atrophy, vacuolization, and hemocyte infiltration at elevated nanoparticle concentrations. These alterations are consistent with oxidative stress and impaired cellular homeostasis and may precede observable changes in survival or growth performance. The hepatopancreas, as the primary organ involved in digestion, detoxification, and immune function in shrimp, appears particularly sensitive to AgNPs exposure [5, 15, 17]. Consequently, histological integrity should be considered a primary criterion for defining safe application thresholds, rather than relying solely on mortality-based endpoints.

5.3 Microbiome modulation

Beyond direct host tissue effects, AgNPs exposure has been shown to alter the composition and structure of the shrimp-associated microbiome. While suppression of pathogenic *Vibrio* populations is often reported as a desirable outcome, recent evidence indicates that non-selective antimicrobial pressure may also reduce beneficial or commensal bacterial taxa, leading to shifts in microbial diversity and community balance [17, 36]. Such microbiome modulation may have cascading effects on nutrient assimilation, immune competence, and disease resilience. Importantly, microbiome responses appear to be dose- and formulation-dependent, suggesting that controlled delivery strategies may mitigate dysbiosis while preserving antimicrobial benefits [37–39]. These findings reinforce the need to integrate microbiome analysis into biosafety assessments of nanoparticle-based interventions.

6. Environmental implications of widespread AgNPs used in aquaculture

Based on the studies included in this chapter, the long-term environmental implications of AgNPs used in shrimp aquaculture are associated with three aspects: tissue accumulation, environmental transformation, and potential trophic transfer. Reported *in vivo* studies indicate that, under chronic low-dose dietary exposure, silver accumulation may occur in tissues such as gills and hepatopancreas, even in the absence of evident acute toxicity. This sublethal accumulation suggests that progressive Ag release and tissue retention constitute a dose- and time-dependent process [5, 13, 17]. Additionally, in saline environments, the cited studies show that AgNPs may undergo physicochemical transformations that modify solubility and release kinetics, reducing immediate bioavailability while potentially increasing environmental persistence. Interaction with organic matrices and suspended particles suggests that a fraction of nanoparticulate silver may associate with sediments or bioflocs in culture systems, potentially contributing to accumulation in benthic compartments.

Regarding trophic transfer, the analyzed studies do not report conclusive evidence of biomagnification; however, accumulation in digestive and hepatopancreatic tissues indicates a potential for mobilization toward higher trophic levels under continuous consumption scenarios [5, 17]. The absence of multigenerational or complete food-chain studies within the cited literature limits the possibility of drawing definitive conclusions but highlights the need for long-term evaluations under realistic farming conditions. Overall, the available evidence within the reviewed studies suggests that the main environmental concerns are not associated with large-scale acute toxicity events, but rather with sublethal accumulation, transformation in saline environments, and possible chronic effects derived from continuous exposure. The sustainability of AgNPs use in aquaculture will depend on controlled dosing strategies, formulations that minimize free release into surrounding waters, and assessment frameworks that integrate tissue accumulation, microbiome dynamics, and potential ecological transfer.

7. Emerging trends in feed formulation, biosafety, and regulatory considerations

Recent research trends indicate a progressive shift from direct waterborne exposure toward dietary incorporation of AgNPs in shrimp aquaculture. This transition is driven by the need to improve dose control, reduce environmental dispersion, and better align experimental exposure scenarios with practical farming conditions. Feed-based formulations, particularly those incorporating green-synthesized AgNPs nanoparticles, enable more precise delivery, lower effective doses, and reduced release of free nanoparticles into surrounding water bodies [17, 40, 41]. From a technological perspective, dietary inclusion opens new possibilities for functional feed design, including encapsulation strategies, surface-coated pellets, and slow-release matrices that modulate nanoparticle bioavailability. Such approaches may enhance antimicrobial efficacy while minimizing acute exposure peaks that are commonly associated with toxicity. However, feed formulation also introduces new challenges, including nanoparticle stability during feed processing, potential interactions with feed components, and the need to understand gastrointestinal transformation and absorption pathways in shrimp [34, 42]

In parallel, biosafety assessment frameworks for nanoparticle application in aquaculture are evolving toward integrative, multi-endpoint approaches. Beyond traditional mortality-based endpoints, recent studies emphasize the importance of incorporating histopathological evaluation, oxidative stress biomarkers, silver bioaccumulation profiles, and microbiome analyses to define safe operational windows [5, 41]. Such comprehensive assessments are increasingly recognized as essential for capturing sublethal and chronic effects that may not be apparent in short-term trials but are critical for long-term sustainability. Microbiome profiling has emerged as a particularly important component of biosafety evaluation. While suppression of pathogenic *Vibrio spp.* is often cited as a desirable outcome, non-selective antimicrobial pressure exerted by AgNPs may also disrupt beneficial microbial communities, leading to dysbiosis with potential consequences for nutrient assimilation and immune competence. Future biosafety frameworks must therefore balance pathogen control with microbiome resilience, moving beyond binary efficacy metrics toward ecosystem-informed evaluations [17, 38, 43].

From a regulatory standpoint, the application of nanoparticles in aquaculture remains poorly defined and largely unstandardized. Existing regulatory frameworks typically evaluate silver as a bulk chemical or ionic species, without accounting for the size-dependent behavior, surface chemistry, and transformation dynamics characteristic of nanomaterials [40, 44]. This regulatory mismatch creates uncertainty for both researchers and industry stakeholders and represents a major bottleneck for commercial translation. International bodies such as the OECD and EFSA have begun to propose guidance documents for the risk assessment of engineered nanomaterials in food and feed contexts; however, aquaculture-specific guidelines remain scarce, and few frameworks explicitly address nanoparticle fate, bioaccumulation, or trophic transfer in aquatic systems [40, 41, 45]. The absence of harmonized standards for nanoparticle characterization, exposure limits, and environmental risk assessment continues to hinder regulatory acceptance and cross-jurisdictional alignment.

8. Future perspectives: Toward the responsible integration of green-silver nanoparticles in shrimp aquaculture

The collective evidence presented in this chapter demonstrates that green-synthesized AgNPs represent a promising but complex technological tool for disease management in shrimp aquaculture. Over the last decade, research has progressed from initial proof-of-concept studies focused on antimicrobial efficacy toward more integrative evaluations encompassing host physiology, tissue integrity, nanoparticle fate, and microbiome dynamics. This evolution reflects a broader shift from viewing AgNPs solely as antibiotic alternatives to recognizing them as functional nanomaterials whose application requires careful balancing of efficacy, biosafety, and environmental responsibility.

The physicochemical diversity inherent to green-synthesized AgNPs also presents both a challenge and an opportunity. Variability in particle size, surface chemistry, organic corona composition, and silver speciation underscores the need for standardized synthesis and characterization protocols. Future research should aim to establish clear links between synthesis routes, physicochemical properties, and biological outcomes, enabling rational design rather than empirical trial-and-error approaches. Such standardization will be pivotal for improving reproducibility, facilitating cross-study comparisons, and generating data suitable for regulatory evaluation.

Additionally, a central future direction lies in the optimization of delivery strategies, particularly through dietary incorporation. Feed-based application has emerged as a more controllable and environmentally conservative alternative to waterborne exposure, enabling precise dose regulation and reduced nanoparticle dispersion into surrounding ecosystems. Advances in feed formulation, such as encapsulation, controlled-release matrices, and surface-coated pellets, offer opportunities to enhance bioavailability while minimizing acute exposure peaks and unintended toxicity. Future research should prioritize understanding nanoparticle stability during feed processing, gastrointestinal transformation, and tissue uptake under farming-relevant conditions, thereby aligning nanotechnology innovation with practical aquaculture operations.

Equally critical is the advancement of biosafety assessment frameworks. The chapter highlights that AgNPs' effects are highly dose-dependent and context-specific, with narrow margins separating beneficial outcomes from sublethal or adverse effects. Future studies must, therefore, move beyond short-term survival assays and incorporate multi-endpoint evaluations that include histopathological integrity, oxidative stress responses, silver accumulation, and microbiome structure. In particular, the hepatopancreas and the shrimp-associated microbiome emerge as sensitive indicators of nanoparticle impact and should be systematically integrated into risk assessment protocols. Long-term, low-dose exposure studies reflecting realistic production scenarios will be essential for defining safe operational windows and supporting translational credibility.

From a regulatory perspective, the responsible integration of AgNPs into shrimp aquaculture remains constrained by the absence of aquaculture-specific guidelines for nanomaterials. Current regulatory frameworks largely assess silver as a bulk chemical, overlooking the unique behavior of nanoscale materials in biological and aquatic systems. Future progress will require closer interaction between researchers, toxicologists, and regulatory agencies to develop harmonized standards addressing nanoparticle characterization, exposure limits, environmental fate, and food safety. Alignment with emerging international guidance on engineered nanomaterials provides a foundation, but sector-specific adaptation to aquaculture contexts is urgently needed.

Within this evolving landscape, the continued relevance of silver nanoparticles must also be considered in the context of competing nanomaterials (such as ZnO, TiO_2, and copper- and gold-based systems). Although alternative nanosystems may offer lower production costs or distinct antimicrobial mechanisms, the evidence reviewed in this chapter indicates that green-synthesized AgNPs retain a strategic advantage through their tunable surface chemistry, multifunctional biological effects, and demonstrated efficacy against key shrimp pathogens. However, sustaining this relevance requires addressing critical gaps, including standardization of synthesis and characterization protocols, improved definition of operational safety windows under farming-relevant conditions, environmentally informed formulation strategies, and integration of microbiome-aware management approaches. The future competitiveness of AgNPs will therefore depend not solely on antimicrobial potency but on their capacity to be rationally engineered, reproducibly characterized, and responsibly deployed within sustainable aquaculture systems.

Collectively, these future perspectives indicate that the key question facing the field is no longer whether AgNPs can exert antimicrobial effects, but under what conditions they can be applied safely, reproducibly, and sustainably. Advancing toward responsible integration will depend on coordinated progress in feed technology, biosafety science, physicochemical standardization, environmental stewardship, and regulatory development. When approached within this holistic framework, green-synthesized AgNPs may contribute meaningfully to resilient and sustainable shrimp aquaculture systems, rather than representing an isolated or short-lived technological solution.

9. Conclusion

This chapter reviews the current state of knowledge on green-synthesized silver nanoparticles (AgNPs) in shrimp aquaculture, outlining both their potential benefits and inherent limitations. Experimental evidence demonstrates that AgNPs can inhibit major shrimp pathogens and, under carefully controlled conditions, enhance survival and disease resistance, supporting their consideration as alternative tools within antibiotic-reduction strategies.

However, AgNPs' effects are strongly dependent on the synthesis route, physicochemical properties, dose, and exposure pathway. Beneficial outcomes are consistently confined to narrow operational ranges, beyond which sublethal toxicity, tissue alterations, and microbiome disruption may occur. These findings emphasize the need for integrative biosafety assessments rather than reliance on antimicrobial efficacy alone.

Recent trends highlight dietary delivery as a more controlled and environmentally responsible application strategy, alongside growing recognition of the importance of standardized characterization and multi-endpoint evaluation. Nevertheless, the absence of aquaculture-specific regulatory frameworks for nanomaterials remains a key obstacle to broader adoption.

Overall, green-synthesized AgNPs should be viewed not as direct antibiotic substitutes, but as functional nanomaterials whose responsible integration requires coordinated advances in formulation design, biosafety validation, and regulatory development. When applied within such a framework, AgNPs may contribute to more sustainable and resilient shrimp aquaculture systems.

Acknowledgments

The authors acknowledge that ChatGPT (OpenAI, San Francisco, CA, USA) was used to assist with language refinement, conceptual clarification, and partial generation of illustrative figures during manuscript preparation. Grammarly (Grammarly Inc., San Francisco, CA, USA) was additionally employed for grammar and stylistic editing. All AI-assisted text and figure drafts were critically reviewed, modified where necessary, and validated by the authors. All scientific content, data interpretation, critical analysis, and final conclusions were developed independently by the authors. The authors assume full responsibility for the accuracy, originality, and integrity of the work

Conflict of Interest

The authors declare no conflict of interest.

Author details

Maribel Maldonado-Muñiz[1*], Sonia A. Soto-Rodriguez[2], Martha Guadalupe Nieto Lopez[1] and Mireya Tapia Salazar[1]

1 Faculty of Biological Sciences (FCB), Autonomous University of Nuevo Leon (UANL), Nuevo Leon, Mexico

2 AC Mazatlán Unit for Aquaculture and Environmental Management, CIAD, Mazatlán, Mexico

*Address all correspondence to: maribel.maldonadomn@uanl.edu.mx

References

[1] FAO. El Estado Mundial de la Pesca Y la Acuicultura 2022. Roma, Italia: FAO; 2022. 1–288 p. DOI: 10.4060/cc0461es

[2] Soto-Rodriguez SA, Gomez-Gil B, Lozano-Olvera R, Betancourt-Lozano M, Morales-Covarrubias MS. Field and experimental evidence of Vibrio parahaemolyticus as the causative agent of acute hepatopancreatic necrosis disease of cultured shrimp (Litopenaeus vannamei) in Northwestern Mexico. Applied and Environmental Microbiology. 2015;**81**(5):1689–1699. DOI: 10.1128/AEM.03610-14

[3] Tran L, Nunan L, Redman R, Mohney L, Pantoja C, Fitzsimmons K, et al. Determination of the infectious nature of the agent of acute hepatopancreatic necrosis syndrome affecting penaeid shrimp. Diseases of Aquatic Organisms. 2013;**105**(1):45–55. DOI: 10.3354/dao02621

[4] Vincent A, Lotz J. Time course of necrotizing hepatopancreatitis (NHP) in experimentally infected Litopenaeus vannamei and quantification of NHP-bacterium using real-time PCR. Diseases of Aquatic Organisms. 2005;**67** (1–2):163–169. DOI: 10.3354/dao067163

[5] Acedo Valdez MR, Grijalva Chon JM, Larios Rodriguez E, Maldonado Arce AD, Mendoza Cano F, Sanchez Paz JA, et al. Antibacterial effect of biosynthesized silver nanoparticles in Pacific white shrimp Litopenaeus vannamei (Boone) infected with necrotizing hepatopancreatitis bacterium (NHP B). Latin American Journal of Aquatic Research. 2017;**45** (2):421–430. DOI: 10.3856/vol45-issue2-fulltext-17

[6] Vignesh R, Karthikeyan BS, Periyasamy N, Devanathan K. Antibiotics in Aquaculture: An Overview. South Asian Journal of Experimental Biology. 2011;**1**(3):114–120. DOI: 10.38150/sajeb.1(3).p114-120

[7] Luthman O, Robb DHF, Henriksson PJG, Jørgensen PS, Troell M. Global overview of national regulations for antibiotic use in aquaculture production. Aquaculture International. 2024;**32**(7):9253–9270. DOI: 10.1007/s10499-024-01614-0

[8] Khursheed S, Dutta J, Ahmad I, Rather MA, Badroo IA, Bhat TA, et al. Biogenic silver nanoparticles: Synthesis, applications and challenges in food sector with special emphasis on aquaculture. Food Chemistry X. 2023;**20**:101051. DOI: 10.1016/j.fochx.2023.101051

[9] Khan I, Saeed K, Khan I, Khan I, Saeed K, Khan I. Nanoparticles: Properties, applications and toxicities. Arabian Journal of Chemistry. 2019;**12** (7):908–931. DOI: 10.1016/j.arabjc.2017.05.011

[10] Vaseeharan B, Ramasamy P, Chen JC. Antibacterial activity of silver nanoparticles (AgNps) synthesized by tea leaf extracts against pathogenic Vibrio harveyi and its protective efficacy on juvenile Feneropenaeus indicus. Letters in Applied Microbiology. 2010;**50**(4):352–356. DOI: 10.1111/j.1472-765X.2010.02799.x

[11] Kandasamy K, Alikunhi NM, Manickaswami G, Nabikhan A, Ayyavu G. Synthesis of silver nanoparticles by coastal plant Prosopis chilensis (L.) and their efficacy in

controlling vibriosis in shrimp Penaeus monodon. Applied Nanoscience. 2013;**3**(1):65–73. DOI: 10.1007/s13204-012-0064-1

[12] Sivaramasamy E, Zhiwei W, Li F, Xiang J. Enhancement of vibriosis resistance in Litopenaeus vannamei by supplementation of biomastered silver nanoparticles by Bacillus subtilis. Journal of Nanomedicine and Nanotechnology. 2016;**7**(1):1–10. DOI: 10.4172/2157-7439.1000352

[13] RathnaKumari P, Kolanchinathan P, Siva D, Abirami B, Masilamani V, John G, et al. Antibacterial efficacy of seagrass Cymodocea serrulata-engineered silver nanoparticles against prawn pathogen Vibrio parahaemolyticus and its combative effect on the marine shrimp Penaeus monodon. Aquaculture. 2018;**493**:158–164. DOI: 10.1016/j.aquaculture.2018.04.061

[14] Alvarez-Cirerol FJ, López-Torres MA, Rodríguez-León E, Rodríguez-Beas C, Martínez-Higuera A, Lara HH, et al. Silver nanoparticles synthesized with rumex hymenosepalus: A strategy to combat early mortality syndrome (EMS) in a cultivated white shrimp. Journal of Nanomaterials. 2019;**2019**: 1–15. DOI: 10.1155/2019/8214675

[15] Maldonado-Muñiz M, Luna C, Mendoza-Reséndez R, Barriga-Castro ED, Soto-Rodriguez S, Ricque-Marie D, et al. Silver nanoparticles against acute hepatopancreatic necrosis disease (AHPND) in shrimp and their depuration kinetics. Journal of Applied Phycology. 2020;**32**(4):2431–2445. DOI: 10.1007/s10811-019-01948-w

[16] León-Valdez G, Valenzuela-Quiñonez W, Álvarez-Ruiz P, Soto-Robles CA, Nava-Perez E, López-Cervantes G, et al. The extract of larrea tridentata promotes the synthesis of silver nanoparticles and stimulates immune responses in penaeus vannamei against vibrio spp., causing acute hepatopancreatic necrosis disease. Microorganisms. 2024;**12**(11):2219. DOI: 10.3390/microorganisms12112219

[17] Maldonado-Muñiz M, Nieto-López M, Tapia-Salazar M, Gómez-Gil B, Guerrero A, Lozano-Olvera R, et al. Dietary silver nanoparticle effects on Penaeus vannamei growth, histopathology, faecal microbiome, and acute hepatopancreatic necrosis disease survival. Diseases of Aquatic Organisms. 2025;**162**:35–49. DOI: 10.3354/dao03848

[18] Eker F, Akdaşçi E, Duman H, Bechelany M, Karav S. Green synthesis of silver nanoparticles using plant extracts: A comprehensive review of physicochemical properties and multifunctional applications. International Journal of Molecular Sciences. 2025;**26**(13):6222. DOI: 10.3390/ijms26136222

[19] Mikhailova EO. Green silver nanoparticles: An antibacterial mechanism. Antibiotics. 2024;**14**(1):5. DOI: 10.3390/antibiotics14010005

[20] Dahoumane SA, Mechouet M, Wijesekera K, Filipe CDM, Sicard C, Bazylinski DA, et al. Algae-mediated biosynthesis of inorganic nanomaterials as a promising route in nanobiotechnology–A review. Green Chemistry. 2017;**19**(3):552–587. DOI: 10.1039/C6GC02346K

[21] Chan SS, Low SS, Chew KW, Ling TC, Rinklebe J, Juan JC, et al. Prospects and environmental sustainability of phyconanotechnology: A review on algae-mediated metal

nanoparticles synthesis and mechanism. Environmental Research. 2022;**212**:113140. DOI: 10.1016/j.envres.2022.113140

[22] Vemula P, Chatragadda R. Ecofriendly nanoparticles from marine flora: Green synthesis, applications, and future prospects. In: Mayra C, editor. SpringerBriefs in Applied Sciences and Technology [Internet]. Singapore: Springer; 2025. p. 1–135. DOI: 10.1007/978-981-96-5878-7_1

[23] Akdaşçi E, Eker F, Duman H, Bechelany M, Karav S. Microbial-based green synthesis of silver nanoparticles: A comparative review of bacteria- and fungi-mediated approaches. International Journal of Molecular Sciences. 2025;**26**(20):10163. DOI: 10.3390/ijms262010163

[24] Elayaraja S, Zagorsek K, Li F, Xiang J. In situ synthesis of silver nanoparticles into TEMPO-mediated oxidized bacterial cellulose and their antivibriocidal activity against shrimp pathogens. Carbohydrate Polymers. 2017;**166**:329–337. DOI: 10.1016/j.carbpol.2017.02.093

[25] Audtarat S, Hongsachart P, Dasri T, Chio-Srichan S, Soontaranon S, Wongsinlatam W, et al. Green synthesis of silver nanoparticles loaded into bacterial cellulose for antimicrobial application. Nanocomposites. 2022;**8**(1):34–46. DOI: 10.1080/20550324.2022.2055375

[26] Faried M, Shameli K, Ubaidillah, Miyake M, Hara H, Khairudin NBA. Green synthesis of silver nanoparticles in biopolymer stabilizer and their application as antibacterial efficacy. AIP Conference Proceedings. 2017;**1788**(1):30124. DOI: 10.1063/1.4968377

[27] Rajeshkumar S, Malarkodi C, Paulkumar K, Vanaja M, Gnanajobitha G, Annadurai G. Algae mediated green fabrication of silver nanoparticles and examination of its antifungal activity against clinical pathogens. International Journal of Metals. 2014;**2014**:1–8. DOI: 10.1155/2014/692643

[28] Iravani S. Green synthesis of metal nanoparticles using plants. Green Chemistry. 2011;**13**(10):2638. Available from: http://pubs.rsc.org/en/Content/ArticleHTML/2011/GC/C1GC15386B

[29] Gopinath V, Priyadarshini S, Meera Priyadharsshini N, Pandian K, Velusamy P. Biogenic synthesis of antibacterial silver chloride nanoparticles using leaf extracts of Cissus quadrangularis Linn. Materials Letters. 2013;**91**:224–227. Available from: http://linkinghub.elsevier.com/retrieve/pii/S0167577X12014073

[30] Camacho-Jiménez L, Álvarez-Sánchez AR, Mejía-Ruíz CH. Silver nanoparticles (AgNPs) as antimicrobials in marine shrimp farming: A review. Aquaculture Reports. 2020;**18**:100512. DOI: 10.1016/j.aqrep.2020.100512

[31] Palanisamy S, Anjali R, Rajasekar P, Kannapiran E, Vaseeharan B, Prabhu NM. Synthesis and distribution of bioinspired silver nanoparticles using spirulina extract for control of vibrio parahaemolyticus infection in aquaculture. Asian Journal of Chemistry. 2017;**29**(4):857–863. DOI: 10.14233/ajchem.2017.20335

[32] Roohi Fatima M, Dinesh S, Ram R, Perumal T, Ravi M, Chatterjii R, et al. Inhibitory effect of silver nanoparticles against pathogenic Vibrio anguillarum for the treatment of Vibriosis. South Indian Journal of Biological Sciences.

2016;**2**(4):451–459. DOI: 10.22205/sijbs/2016/v2/i4/103452

[33] Kumar P, Selvi SS, Prabha AL, Kumar KP, Ganeshkumar RS, Govindaraju M. Synthesis of silver nanoparticles from sargassum tenerrimum and screening phytochemicals for its antibacterial activity. Nano Biomedicine and Engineering. 2012;**4**(1):403–414. DOI: 10.5101/nbe.v4i1.p12-16

[34] Fondevila M. Potential use of silver nanoparticles as an additive in animal feeding. In: Perez DP, editor. Silver Nanoparticles. London: InTech; 2010. p. 325–335. DOI: 10.5772/8509

[35] Ribeiro F, Van Gestel CAM, Pavlaki MD, Azevedo S, Soares AMVM, Loureiro S. Bioaccumulation of silver in Daphnia magna: Waterborne and dietary exposure to nanoparticles and dissolved silver. Science of the Total Environment. 2017;**574**:1633–1639. DOI: 10.1016/j.scitotenv.2016.08.204

[36] Md Zoqratt MZH, Eng WWH, Thai BT, Austin CM, Gan HM. Microbiome analysis of Pacific white shrimp gut and rearing water from Malaysia and Vietnam: Implications for aquaculture research and management. PeerJ. 2018;**6**:e5826. DOI: 10.7717/peerj.5826

[37] See MS, Ching XL, Khoo SC, Abidin SZ, Sonne C, Ma NL. Aquatic microbiomes under stress: The role of gut microbiota in detoxification and adaptation to environmental exposures. Journal of Hazardous Materials Advances. 2025;**17**:100612. DOI: 10.1016/j.hazadv.2025.100612

[38] Adamovsky O, Buerger AN, Wormington AM, Ector N, Griffitt RJ, Bisesi JH, et al. The gut microbiome and aquatic toxicology: An emerging concept for environmental health. Environmental Toxicology and Chemistry. 2018;**37**(11):2758–2775. DOI: 10.1002/etc.4249

[39] Cornejo-Granados F, Lopez-Zavala AA, Gallardo-Becerra L, Mendoza-Vargas A, Sánchez F, Vichido R, et al. Microbiome of Pacific Whiteleg shrimp reveals differential bacterial community composition between Wild, Aquacultured and AHPND/EMS outbreak conditions. Scientific Reports. 2017;**7**(1):11783. DOI: 10.1038/s41598-017-11805-w

[40] Lowry GV, Gregory KB, Apte SC, Lead JR. Transformations of nanomaterials in the environment. Environmental Science & Technology. 2012;**46**(13):6893–6899. DOI: 10.1021/es300839e

[41] Petersen EJ, Diamond SA, Kennedy AJ, Goss GG, Ho K, Lead J, et al. Adapting OECD aquatic toxicity tests for use with manufactured nanomaterials: Key issues and consensus recommendations. Environmental Science & Technology. 2015;**49**(16):9532–9547. DOI: 10.1021/acs.est.5b00997

[42] Mackevica A, Skjolding LM, Gergs A, Palmqvist A, Baun A. Chronic toxicity of silver nanoparticles to Daphnia magna under different feeding conditions. Aquatic Toxicology. 2015;**161**:10–16. DOI: 10.1016/j.aquatox.2015.01.023

[43] Oetama VSP, Hennersdorf P, Abdul-Aziz MA, Mrotzek G, Haryanti H, Saluz HP. Microbiome analysis and detection of pathogenic bacteria of Penaeus monodon from Jakarta Bay and Bali. Marine Pollution

Bulletin. 2016;**110**(2):718–725. DOI: 10.1016/j.marpolbul.2016.03.043

[44] Kühnel D, Nickel C. The OECD expert meeting on ecotoxicology and environmental fate — Towards the development of improved OECD guidelines for the testing of nanomaterials. Science of the Total Environment. 2014;**472**:347–353. DOI: 10.1016/j.scitotenv.2013.11.055

[45] Peters R, Brandhoff P, Weigel S, Marvin H, Bouwmeester H, Aschberger K, et al. Inventory of Nanotechnology applications in the agricultural, feed and food sector. EFSA Supporting Publications. 2014;**11**(7):125. DOI: 10.2903/sp.efsa.2014.EN-621